T0250257

DRIVER BEHAVIOUR AND TRAINING

Human Factors in Road and Rail Transport

Series Editors

Dr Lisa Dorn
*Director of the Driving Research Group, Department of Human Factors,
Cranfield University*

Dr Gerald Matthews
Professor of Psychology at the University of Cincinnati

Dr Ian Glendon
*Associate Professor of Psychology at Griffith University, Queensland,
and President of the Division of Traffic and Transportation Psychology
of the International Association of Applied Psychology*

Today's society must confront major land transport problems. The human and financial costs of vehicle accidents are increasing, with road traffic accidents predicted to become the third largest cause of death and injury across the world by 2020. Several social trends pose threats to safety, including increasing car ownership and traffic congestion, the increased complexity of the human-vehicle interface, the ageing of populations in the developed world, and a possible influx of young vehicle operators in the developing world.

Ashgate's 'Human Factors in Road and Rail Transport' series aims to make a timely contribution to these issues by focusing on the driver as a contributing causal agent in road and rail accidents. The series seeks to reflect the increasing demand for safe, efficient and economical land-based transport by reporting on the state-of-the-art science that may be applied to reduce vehicle collisions, improve the usability of vehicles and enhance the operator's wellbeing and satisfaction. It will do so by disseminating new theoretical and empirical research from specialists in the behavioural and allied disciplines, including traffic psychology, human factors and ergonomics.

The series captures topics such as driver behaviour, driver training, in-vehicle technology, driver health and driver assessment. Specially commissioned works from internationally recognised experts in the field will provide authoritative accounts of the leading approaches to this significant real-world problem.

Driver Behaviour and Training
Volume IV

LISA DORN
Cranfield University, UK

Routledge
Taylor & Francis Group

LONDON AND NEW YORK

First published 2010 by Ashgate Publishing

Published 2016 by Routledge
2 Park Square, Milton Park, Abingdon, Oxon OX14 4RN
711 Third Avenue, New York, NY 10017, USA

Routledge is an imprint of the Taylor & Francis Group, an informa business

Copyright © 2010 Lisa Dorn

Lisa Dorn has asserted her right under the Copyright, Designs and Patents Act, 1988, to be identified as the editor of this work.

All rights reserved. No part of this book may be reprinted or reproduced or utilised in any form or by any electronic, mechanical, or other means, now known or hereafter invented, including photocopying and recording, or in any information storage or retrieval system, without permission in writing from the publishers.

Notice:

Product or corporate names may be trademarks or registered trademarks, and are used only for identification and explanation without intent to infringe.

British Library Cataloguing in Publication Data
  Driver behaviour and training.
  Vol. IV. -- (Human factors in road and rail transport)
  1. Motor vehicle drivers--Training of--Congresses.
  2. Motor vehicle drivers--Attitudes--Congresses. 3. Motor
  vehicle driving--Congresses. 4. Automobile driver
  education--Congresses. 5. Traffic safety--Congresses.
  6. Automobile drivers' tests--Evaluation--Congresses.
  I. Title II. Series III. Dorn, Lisa.
  363.1'25-dc22

  ISBN 9781409400844 (hbk)

Library of Congress Cataloging-in-Publication Data
Driver behaviour and training / edited by Lisa Dorn.
    p. cm.
  Includes bibliographical references and index.
  ISBN 978-1-4094-0084-4
  1. Traffic safety--Congresses. 2. Automobile drivers--Congresses. 3.
  Automobiles--Safety appliances--Congresses. 2. Automobile driver education--Congresses.
  1. Dorn, Lisa. II. Title.
  HE5614.I553 2003
  363.12'5--dc22

                                                                            2003058287

Contents

List of Figures   *ix*
List of Tables   *xiii*
Preface   *xvii*

PART 1   NEW THINKING IN DRIVER BEHAVIOUR AND EDUCATION

1   Driver Research Myths   3
        A. E. af Wåhlberg

2   Recent Findings on Risky Acts in Adolescence: Implications for
        Understanding European Drivers   7
        Divera Twisk and Willem Vlakveld

3   A Comparison of Inexperienced and Experienced Drivers' Cognitive
        and Physiological Response to Hazards   23
        Steve W. Kelly, Neale Kinnear, James Thomson and Steve Stradling

4   Development of the Driver Performance Assessment: Informing
        Learner Drivers of their Driving Progress   37
        Erik Roelofs, Marieke van Onna and Jan Visser

5   How Can Reflecting Teams Contribute to Enhanced Driving Teacher
        Learning?   51
        Hilde Kjelsrud

PART 2   DRIVER PERSONALITY AND DRIVER OFFENDING

6   Understanding the Unique Contribution of Aversion to Risk-Taking
        in Predicting Drivers' Self-Reported Speeding   61
        M. Anthony Machin and Janna E. Plint

7   Young Drivers: Investigating the Link Between Impulsivity and
        Problem Driver Status   75
        Fearghal O'Brien, Simon Dunne and Michael Gormley

8   Relationships Between Driving Style, Self-reported Driving
        Behaviour and Personality   89
        S.M. Skippon, N. Reed, T. Luke, R. Robbins, M. Chattington, and
        A.H. Harrison

9   Public Perception of Risk of Being Caught Committing Traffic
       Offences   105
       Isah Noradrenalina, M. Maslina and L. S. Kee

10   Rear Seatbelt Wearing in Malaysia: Public Awareness and Practice 119
       *Norlen Mohamed, Muhammad Fadhli Mohd Yusoff and Isah
       Noradrenalina*

11   The Continuous Evaluation of Driver Rehabilitation Programmes
       in Austria   131
       Julia Bardodej, Franz Nechtelberger and Martin Nechtelberger

12   Perceptions of the Spanish Penalty Point Law   139
       *Maria Eugènia Gras, Sílvia Font-Mayolas, Mark J.M. Sullman,
       Mònica Cunill and Montserrat Planes*

PART 3   POWERED TWO-WHEELER BEHAVIOUR

13   The Role of the Psychologist in a Moped Rider Training
       Programme   149
       Patrícia António and Manuel Matos

14   Interim Evaluation of the UK's National RIDE Scheme   161
       Cris Burgess, Paul Broughton, Fiona Fylan and Steve Stradling

15   Encouraging Rider Behaviour Change by Using Respected
       Communicators   179
       Paul S. Broughton, Sandy Allan and Linda Walker

16   The Motorcycle Rider Behaviour Questionnaire (MRBQ) and
       Commercial Motorcycle Riders in Nigeria   193
       Oluwadiya Kehinde Sunday and Ladoke Akintola

PART 4   AT WORK ROAD SAFETY

17   Contemporary Behavioural Influences in an Organisational Setting
       and Implications for Intervention Development   213
       Bevan Rowland, Jeremy Davey, James Freeman and Darren Wishart

18   A Review of the Effectiveness of Occupational Road Safety
       Initiatives   229
       Tamara Banks, Jeremy Davey, Herbert Biggs and Mark King

19   Developing Risk-assessment Tools for Fleet Settings: Where to
      From Here?   241
      James Freeman, Darren Wishart, Jeremy Davey and Bevan Rowland

20   From Research to Commercial Fuel Efficiency Training for Truck
      Drivers using TruckSim   257
      Nick Reed, Stephanie Cynk and Andrew M. Parkes

21   The Utility of Psychometric Testing for Predicting Bus Driver
      Behaviour   269
      Wendy Lord and Joerg Prieler

22   Identification of Barriers to and Facilitators for the Implementation
      of Occupational Road Safety Initiatives   275
      Tamara Banks and Jeremy Davey

PART 5   HUMAN FACTORS AND DRIVER ATTENTION

23   An Observational Survey of Driving Distractions in England   287
      Mark J.M. Sullman

24   Calibration of an Eye-tracking System for Variable Message Signs
      Validation   297
      *M. Claudia Guattari, Maria Rosaria De Blasiis, Alessandro Calvi
      and Andrea Benedetto*

25   Visual Behaviour of Car Drivers in Road Traffic   307
      Carmen Kettwich, Stefan Stockey and Uli Lemmer

26   Icons for Actions in a Driving Simulator   317
      Robert H Barbour

27   Contributory Factors for Incidents Involving Local and Non-local
      Road Users   327
      Linda Walker and Paul S. Broughton

28   Severity of Injury Outcomes for Older Drivers Involved in
      Intersection Crashes   339
      Peter Hillard

Index   *347*

List of Figures

Figure 3.1   Example pictures of (clockwise from top left) safe, developing hazard and hazard scenarios   27

Figure 3.2   Placement of skin conductance electrodes when measuring skin conductance   28

Figure 3.3   Comparison of driver groups' mean hazard ratings to safe, developing hazard and hazard pictures with standard error bars. Scale: extremely safe 1–7 extremely hazardous   29

Figure 3.4   Comparison of driver groups' SCR percentages to safe, developing hazard and hazard pictures with standard error bars   31

Figure 4.1   An eclectic model for the assessment of driving competence   38

Figure 4.2   Probabilities to pass the final exam based on the drivers' instructors' assigned DPA score (first pilot)   44

Figure 4.3   Probabilities to pass the final exam based on the drivers' examiners' assigned DPA score at moment 1 and 2 (second pilot)   45

Figure 4.4   Probabilities to pass the final exam based on the drivers' examiners' assigned DPA score at moment 3 (second pilot)   46

Figure 4.5   Individual growth of learner drivers in terms of pass probability   49

Figure 5.1   Kolb's model of learning   53

Figure 5.2   Reflecting team lessons   55

Figure 7.1   Reaction time to Go trials for both male and female problematic and non-problematic drivers   82

Figure 7.2   Number of omission errors committed by both male and female problematic and non-problematic drivers   83

Figure 8.1   Significant relationships between driving style, self-reported driving behaviours and domains and facets of personality   101

Figure 9.1   Perception of being caught at all locations surveyed   111

Figure 9.2   Percentage of perception of being caught on expressways   112

Figure 9.3   Percentage of perception of being caught on federal roads   113

Figure 9.4   Factors associated to the perception of being caught – expressways   115

Figure 9.5   Factors associated to the perception of being caught – federal roads   115

Figure 9.6   Effectiveness of enforcement approaches – expressways   116

Figure 9.7   Effectiveness of enforcement approaches – federal roads   116

Figure 10.1   New vehicle registrations in Malaysia, 1998–2007   120

Figure 10.2   A stepwise approach of the implementation of rear seatbelt
               policy   123
Figure 10.3   Rear seatbelt wearing rate by type of road and locality,
               May–August, 2008   125
Figure 10.4   Public awareness of impending enforcement of rear seatbelt
               law   125
Figure 10.5   Public awareness on benefit and attitude towards rear seatbelt
               wearing   126
Figure 10.6   Reasons for not wearing rear seatbelts   127
Figure 11.1   The four stages of effect measurement   132
Figure 11.2   Blood alcohol concentration level which participants of the
               first evaluation (in percentages) believed to be the adequate
               legal limit at three different times of measurement   133
Figure 12.1   Percentage of participants who supported the penalty point
               law, by gender and loss of points   141
Figure 14.1   Stage of change stage score (group by time-point)   170
Figure 14.2   Deviant beliefs (group by time-point)   172
Figure 20.1   Schematic of TruckSim   258
Figure 20.2   Interior and exterior of TruckSim   258
Figure 20.3   TruckSim road database   259
Figure 20.4   Screenshots from TruckSim   260
Figure 20.5   Mean percentage improvement in on-road fuel efficiency   263
Figure 24.1   Frame extracted with eye tracking software   300
Figure 24.2   Output of eye tracking software (frame by frame)   301
Figure 24.3   Points displayed during calibration   301
Figure 24.4   Example of not acceptable calibration   302
Figure 24.5   Example of acceptable calibration   303
Figure 25.1   Viewing directions   309
Figure 25.2   Average dwell time of each viewing direction in an urban
               environment   310
Figure 25.3   Dwell time percentage of the driving time for different viewing
               directions in an urban environment   310
Figure 25.4   Average dwell time of each viewing direction in a rural
               environment   311
Figure 25.5   Dwell-time percentage of the driving time for different
               viewing directions in a rural environment   312
Figure 25.6   Average dwell time of each viewing direction while driving
               on a motorway   313
Figure 25.7   Dwell-time percentage of the driving time for different
               viewing directions while driving on a motorway   314
Figure 26.1   CAS: Vehicle Movement Coding Sheet (reproduced with
               Permission of Ministry of Transport, New Zealand)   319

Figure 26.2   Vehicle icons: the icon elements, a saloon car, a two-seater,
               a sports utility vehicle, a van, a mini bus, a motorcycle, a
               truck, a truck and trailer (with a 60/100 kph tail, a passenger
               bus)   323
Figure 27.1   Crashes for locals and visitors by weekend and weekday   330

List of Tables

Table 2.1   Modified after the table presented by Steinberg at the Leiden
          Conference   11

Table 3.1   Mean hazard ratings for still images by experience group with
          standard deviations and Kruskal–Wallis analysis   30

Table 3.2   Mean percentage of responses showing an SCR to the stimulus
          item with standard deviations and Kruskal–Wallis analysis   31

Table 4.1   Mean and standard deviations on the consecutive assessments
          for learner drivers who failed and passed the final exam   43

Table 4.2   Classification table for DPA predictions (pilot 2) against the
          result of the final exam   47

Table 6.1   Mean, standard deviation, and Cronbach's alpha for all variables
          (N = 400)   68

Table 6.2   Intercorrelations among speeding, risk perception, personality
          characteristics and coping strategy variables (N = 400)   69

Table 6.3   Summary of hierarchical multiple regression analysis for
          predicting speeding (N = 400)   70

Table 7.1   Mean scores for all groups on the four Go/NoGo measures   81

Table 8.1   Significant correlations between DSI factors and driving style
          measures (SD = standard deviation)   93

Table 8.2   Significant correlations between DCQ factors and driving style
          measures   94

Table 8.3   Significant correlations between DBQ factors and driving style
          measures   95

Table 8.4   Significant correlations between driving style measures and
          scores on the domains of the NEO PI-R personality inventory 95

Table 8.5   Significant correlations between DSI factors and NEO PI-R
          domains and facets (facet names in this and subsequent tables
          are preceded by a letter indicating which domain they belong
          too: C for conscientiousness, A for agreeableness etc.)   97

Table 8.6   Significant correlations between DCQ factors and NEO PI-R
          domains and facets   98

Table 8.7   Significant correlations between the violations factor of the
          DBQ, and NEO PI-R domains and facets   99

Table 9.1   Ops Sikap Chinese New Year report year 2003–2007   106

Table 9.2   Respondents characteristics   110

Table 9.3   Traffic offences according to possibility of being detected
          among respondents   114

Table 10.1   Strategic Road Safety Interventions and Potential Fatality
              Reduction 2007–2010   121
Table 11.1   Results of the univariate analyses of the second evaluation   135
Table 12.1   Mean and standard deviations for the perceived effects of the
              penalty point law by gender, lost points (drivers) and support
              for the law   142
Table 14.1   Descriptive statistics, by group   164
Table 14.2   Scale reliabilities by time-point (combined sample)   166
Table 14.3   Factor solution for attitudinal/motivation items   167
Table 14.4   Comparison of speed preferences: group by time-point
              (ANCOVAs)   168
Table 14.5   Comparison of stage of change measures, group by time-point   169
Table 14.6   Scales scores, group and time-point   171
Table 15.1   Comparison of respondents' age profile against the Department
              for Transport data on PTW users   182
Table 15.2   Overall opinion of the website   185
Table 15.3   How often do you brake hard out of town (60 mph limit) on a
              typical journey?   186
Table 15.4   How often do you brake hard within town (30 mph limit) on a
              typical journey?   187
Table 15.5   'Around the Corner' has made me think about how I ride   188
Table 15.6   'Around the Corner' has made me a safer rider   189
Table 15.7   'Around the Corner' increased my knowledge on motorcycle
              safety   189
Table 16.1   Original MRBQ and the modified MRBQ used for the study   195
Table 16.2   Demographic and riding data of participants   199
Table 16.3   Factor structure of the MRBQ items   200
Table 16.4   Factor structure of the original MRBQ compared to the present
              study   202
Table 16.5   General linear modelling of the MRBQ and other factors on
              crash liability   204
Table 17.1   Factor structure of the MDBQ   218
Table 17.2   Mean and standard deviations for the MDBQ factors   219
Table 17.3   Logistic regression for crashes   221
Table 17.4   Logistic regression for offences   221
Table 20.1   Summary of full motion of truck simulator   259
Table 23.1   Driving only vs involvement in a secondary task   290
Table 23.2   Secondary activity while driving, by gender   291
Table 23.3   Driving only vs involvement in a secondary task, by age
              group   291
Table 23.4   Secondary activity while driving, by age group   292
Table 23.5   Driving only vs involvement in a secondary task, by time of
              day   292
Table 23.6   Secondary activity while driving, by time of day   293

Table 24.1   VMS tested   299
Table 24.2   VMS comprehension   304
Table 24.3   Percentages of time spent on VMS (picture and sentence)   305
Table 25.1   Distribution and frequency of traffic signs and traffic lights
             along the test track   308
Table 25.2   Dwell times and dwell-time percentage in an urban environment
             during the day and at night   311
Table 25.3   Dwell times and dwell-time percentage in a rural environment
             during the day and at night   312
Table 25.4   Dwell times and dwell-time percentage on a motorway during
             the day and at night   314
Table 27.1   Days of week crashes occur for car drivers   330
Table 28.1   Crash type by crash-involved driver age group (% of all injury
             crashes for cohort)   341
Table 28.2   Differential fatality rates by crash type and crash-involved
             driver age group (relative to cohort aged 30–49 years)   343
Table 28.3   Differential severe injury rates by crash type and crash-
             involved driver age group (relative to cohort aged 30–49
             years)   343

Preface

The World Health Organisation published a report in 2009 estimating that about 1.2 million people across the world die every year on our roads and between 20 and 50 million more are injured. Over 90 per cent of fatalities take place in low- to middle-income countries whereas death rates in higher-income countries have been declining for the past five years. However, there is still no room for complacency in high-income countries. A third of the people dying on UK roads are under the age of 25 and this equates to two young people being killed every day (Department for Transport, 2008).

Since the 1950s driver training and education has been implemented as a way of reducing accident risk. There have been several reviews of the evaluations of driver training and educational interventions that intuitively were expected to reveal a road safety benefit. The fact that they have concluded that there is little evidence that any form of education is beneficial is not surprising when these studies are closely inspected. The content of the driver training (often skills-based) neglects the behavioural issues that largely determine the decisions a driver makes moment by moment and their motives on a particular journey. Professor Ian Roberts of the Cochrane Injuries Group, London even went so far as to argue that driver educational interventions should not be implemented. This runs completely contrary to other health-related behaviours such as AIDS, smoking and obesity in which education is seen as a worthy endeavour. More often than not, the effectiveness of a driver educational intervention cannot be determined reliably on the basis of the currently available evidence, possibly because the majority of studies use sample sizes that are too small or fail to use a control group. It is imperative that researchers work with road safety professionals to ensure that evaluation studies are rigorously designed. What is also clear from these reviews is that road safety practitioners have yet to design appropriate education both in terms of their content and delivery that can demonstrate a road safety benefit. This is our challenge.

The International Conference on Driver Behaviour and Training unites researchers and road safety practitioners to debate some of the latest research on how to reduce the risk of road traffic collisions from an educational perspective. The conference was hosted in Amsterdam in November 2009 and brought together road safety professionals from around the world.

I would like to pay a special tribute to Professor Helen Muir who agreed that Cranfield University could host the first conference in 2003, the second in 2005 and the third in 2007. Twice named by *The Independent* as one of '10 Britons who shaped our world', Helen and her research team were the first to reproduce human behaviour in crash situations. In 2006, this work was recognised by Universities UK as one of the best 100 discoveries and developments in UK universities in the last 50 years that changed the world. Helen passed away on 20th March, 2010 and is sorely missed by her friends and family. Without Helen's encouragement and support, this conference would not have got off the ground.

I am grateful to my Head of Department, Professor Philip John, for agreeing to underwrite the event on behalf of Cranfield University and to a2om International Ltd for their sponsorship. Most of all, I am indebted to the contributors for sharing their ideas and research at the conference.

Lisa Dorn

PART 1
New Thinking in Driver Behaviour and Education

Chapter 1
Driver Research Myths

A. E. af Wåhlberg

Within all sciences, there are some beliefs that are accepted and beyond question. They are seen as basic facts, and are rarely tested. It is believed that previous researchers have established them, and we need not concern ourselves with the evidence that underpins these beliefs. This may be true for many branches of science, and many facts. For research into individual differences in traffic safety, however, it is a mistake to make this assumption. In fact, several of the most widespread explicit or implicit beliefs within this research area are myths. The conclusions from traffic safety research to date are often directly misleading, due to faulty methodology. At best they are unproven, but many are blatantly wrong. As these beliefs mainly concern research methodology, or have a bearing on it, this is a lamentable state of affairs.

This chapter will outline a number of these driver research myths. The arguments presented are all based upon the analyses, reviews and meta-analyses presented in af Wåhlberg (2009). The summary of this substantial piece of work shows that many of the current beliefs bandied about in traffic safety research are due to ignorance about the relevant literature. Instead of actually studying the fairly extensive number of studies that exist about some of the phenomena described here, or noticing the lack of research about other areas, traffic safety researchers tend to refer to just a few studies that support their claims about the method used, or at least this handful of studies are said to be supportive, or else they do not give any references at all. For some of the methods used there is often no research to refer to, but this lack of proof of validity does not seem to worry those involved in the kind of research that is criticised in af Wåhlberg (2009). That papers using techniques with unproven or disproved validity can actually be published would seem to be a further indicator of the widespread acceptance of various driver research myths.

The first myth to be described is that of the low stability over time of accident record, and the beliefs about the theory that predicts the opposite; accident proneness. Safety researchers have discarded accident proneness as a concept, without any real reasons for this. However, more importantly, traffic safety researchers have for decades claimed that a driver's future accident record cannot be predicted with any accuracy from previous crashes. This idea has apparently been based upon studies where low correlations between time periods have been found, while other studies have largely been ignored. However, what has not been recognised is that the size of the correlation between accidents in two different time periods for a group of drivers is very much dependent upon the variation in the sample. In fact, the low correlations reported in the literature are almost solely due to the use of low-risk populations and short time periods. Interpreting such data as proof of low stability of accident record is akin to saying that there is no accident proneness because a driver that had an accident yesterday did not have another one today. When a statistically more suitable time period is used, the resulting correlation between crashes is high. Therefore, accident proneness does exist, at least in terms of a stability of collision record over time.

The most widely spread and explicitly embraced myth in traffic safety would seem to be that self-reports can yield accurate and/or useful data for collision prediction, and that collision data in itself can be validly self-reported upon. Studies making these assumptions are very common, yet the authors seldom have any evidence for the criterion-related validity of their instrument (the predictors). Instead, they tend to focus on safe areas such as what factors can be extracted, homogeneity of these, and interrelationships between the (self-reported) variables of the study. The latter topic sometimes includes self-reported crashes, which is treated as a criterion.

It can easily be shown that accidents are forgotten and distorted, and that how people respond when asked about such events is strongly influenced by various cognitive mechanisms. Despite this, significant findings emerge when other self-reported data are used as predictors. How is this possible, when both independent and dependent variables have extremely low validities? Researchers using self-report measures refer to this fact as a basis for the claim that the real association must be much stronger than that found in their data.

However, as very few traffic accident prediction studies have controlled for common method variance, significant associations between various scales and crashes can always be suspected to be artefactual. As is well known in other research areas, cognitive mechanisms can create biases in responses to questionnaires that create associations that are not real. This is called common method variance, and the most well-known mechanism is social desirability. It can be suspected that this type of bias has affected the majority of findings using self-report data for crash prediction.

Turning to the next myth, it is often said that the association between individual traffic accident record and exposure (in reality mileage) is curvilinear, meaning that drivers who drive a lot (per year) have fewer crashes per kilometre. This has been interpreted as an effect of learning: high-mileage drivers have few crashes because they drive a lot. The fact that this idea has come to be accepted is profoundly strange, because not only are there very few studies that have reported on this aspect of the association between crashes and mileage (usually, only a correlation is reported), but these studies are also methodologically weak, to say the least. They have all used self-reported data for at least one of the variables (exposure), which create some of the curvilinearity by reporting bias. Furthermore, when the interpretation of the data is that more driving creates better drivers, the logic is faulty. Even if it were found in objectively gathered data that there is a certain nonproportionality in the association between accidents and exposure, this would not constitute evidence that drivers driving many miles have fewer crashes as a result of this. The opposite hypothesis is equally applicable to ecologically gathered data; good drivers drive more, because they are good drivers to begin with. To study this problem an experimental approach is needed, or a very peculiar naturally occurring situation, where drivers cannot influence how much they drive.

The most widespread implicit myth within research into individual differences in traffic safety is that proxy safety variables, like speed, can be used as criteria. As with the exposure myth, this one would seem to be based upon shallow knowledge of the available literature and faulty logic. The problem is that any possible (and some quite impossible) replacements for crashes as outcome variable are used, most often without any discussion of the validity of this variable, and the research community accepts this.

Regarding this use of unvalidated safety criteria, it can be noted that for the select few that have actually been tested against collisions, they have all been shown to be weak predictors. Furthermore, when results for several different criteria used in parallel are reported, it can be shown that the proxy variables yield different results from crashes. A driver who would be denoted as dangerous by one criterion would therefore not be so by another, and a variable that was found to be important for safety with a proxy criterion would not be so if crashes were used as outcome.

The reason for the use of proxy criteria would also seem to rest upon an erroneous interpretation of research results in neighbouring areas. It is often forwarded as a reason for using a proxy, for example speed, that this is an important factor in crash causation. This logical error is called base rate neglect, because it does not take into account how common the phenomenon is. Being present as a possible or real cause in a crash does not make this speed (for example) a strong predictor of individual differences in collision rate, because most drivers who drive fast do not crash.

A similar logical error is that of applying within-individual results to between-individuals situations. Although a driver who reduces his speed almost certainly becomes a safer driver, the same speed difference between two drivers does not have the same predictive power, because for the within-driver effect, all other variables are automatically held constant. Between individuals many things vary, some of which influences crash risk.

Actually, the use of safety criteria that have nothing whatsoever to do with traffic safety have become so widespread within driver research that the next step would be sheer randomness. Why even bother to try to measure safety in a valid way, when the journals can be flooded by industrially produced driver inventory studies, with scales constructed according to the cook book, and criteria that not even the authors seem to know whether they should be seen as dependent or independent variables? Today, what is predicted in most studies on individual differences in driver safety is simply what the drivers are prone to say in a questionnaire.

Finally, it can be noted that most researchers in individual differences driver research have no idea about how to construct an accident criterion. Here too, the methods used seem to be chosen largely at random, without anyone noticing that different methods might yield very different results. The length of the time period used, the type of incidents included, how exposure is controlled and a number of other factors, may influence what the results are from a given study. Yet no researchers have considered these critical factors. Again, a kind of implicit myth seems to have spread, where these factors are believed to be of no consequence.

How do myths within research start and spread? One reason would seem to be the distortion of facts by the use of secondary sources. If researchers do not take the time to actually read the sources they cite, but trust the interpretation of someone else that has cited it (and possibly read it), we have something that is similar to the whispering game: within a few transmissions, the facts have been distorted horrendously. One example of a simple but blatant error caused by the use of secondary sources, which is not from safety research but which is very instructive, is that of the bloody hands of Lady Macbeth.

Within clinical research and practice, obsessive-compulsive disorder is sometimes exemplified by the scene in Macbeth where the lady is (dry-) washing her hands, without being able to rid herself of the imaginary blood. However, although repeated hand washing is indeed one of the symptoms of obsessive-compulsive disorder, the example is erroneously interpreted, probably due to a shortcoming of those who use it; they have neither seen nor read the play. What they have at one time come across is just the scene where Lady Macbeth washes her hands. However, the washing is not for real; the lady is sleepwalking and dreaming of the murder she has coaxed her husband into committing, so there is no water. This behaviour is therefore not indicative of an obsession, but of post-traumatic stress syndrome, something which is likely to result if one has been involved in murder. The lesson learned from this example is important for traffic safety research – always read the sources you refer to.

In essence, the research into individual differences in traffic safety is of very poor quality. Sometimes, this can even be shown by using the researchers' own data, but for most problems, new research is needed, both regarding the biasing mechanisms and simply to re-do the previous studies. When this has been undertaken, we can finally see how much individual differences results can add to traffic safety. So far, the contributions have been pitiful.

Reference

af Wåhlberg, A.E. (2009). *Driver Behaviour and Accident Research Methodology. Unresolved Problems*. Aldershot: Ashgate.

Chapter 2

Recent Findings on Risky Acts in Adolescence: Implications for Understanding European Drivers

Divera Twisk and Willem Vlakveld

Introduction

This chapter reviews recent findings and theories in relation to our current understanding of the contributing factors to the crash risk of the 18–24-year-old novice driver and goes on to assess the implications of adolescent development and the effectiveness of countermeasures.

Most studies reviewing the literature on young drivers conclude that age and experience are contributing factors to their high crash risk (Engstrom et al., 2002; Gregersen and Bjurulf, 1996; OECD, 2006; Senserrick, 2006; Twisk, 1995, 2000; Vlakveld, 2005). These patterns are shown across the world, making car crashes the prime cause of death in highly motorised countries (Twisk and Stacey, 2007). Assessing the relative contribution of experience and age, several studies show that in adolescents aged over 18, inexperience influences crash risk to a greater extent than age (e.g. Maycock, 2002; Maycock and Forsyth, 1997; OECD, 2006; Vlakveld, 2005). These findings explain the effectiveness of measures such as graduated licensing (Shope, 2007) and supervised driving (Twisk and Stacey, 2007) that aim to facilitate the development of expertise in relatively safe traffic conditions. However, the progress in crash prevention is still too slow (OECD, 2006) and this is a particular problem for young male drivers (Lynam et al., 2002). This led the OECD (2006) to conclude that in order to advance the development of effective measures, our understanding of the actual underlying mechanisms of *how* factors such as age and experience affect risk, needs to be improved. Recently, relevant insights in these matters have come from fundamental research in the fields of neuro-psychology (Giedd, 2008), decision-making (Reyna and Farley, 2006) and risk taking (Gerrard et al., 2008). This has led to a renewed interest in deliberate risk-taking as a result of 'immaturity'.

This chapter reviews recent results from these fields and discusses the implications for our understanding of the role of experience and age. Keating (2007) has already discussed some of these findings in terms of the implications for licensing strategies for the very young (14–16) American novice driver. These analyses though are only partly relevant for Europe, as compared with the USA,

Australia and New Zealand, Europe licenses drivers relatively late at age 18 (OECD, 2006). Studies on the developmental trajectories within adolescence have concluded that 16-year-olds are significantly different from 18-year-olds (Arnett, 2000). Support for these conclusions also comes from recent findings in brain development (Giedd, 2008) and risk-taking (Steinberg et al., 2008).

As many different terms are used in the literature to refer to dangerous behaviours (Harré, 2000) we need to define these terms more precisely. In this chapter, 'risky act' refers to the statistical *risk of an activity*, 'risk-taking' to intentionally *engaging in an activity while being aware of the high statistical risk* and 'error' to *an unintended inadequate action*.

Neuropsychological Findings

Previously, studies on brain development showed that at the age of five the brain was fully developed and did not structurally change after that. Reviews on adolescent crash risk therefore concluded that adolescent performance was not limited by biological immaturity (e.g. Eby and Molnar, 1999; Lynam and Twisk, 1995). However, recent findings from fundamental research into adolescent brain development (e.g. Giedd, 2008), decision-making strategies (e.g. Reyna and Farley, 2006) and risk-taking (e.g. Gerrard et al., 2008; Steinberg, 2008) have challenged existing understanding of the nature of risky activities in adolescence. Within the scope of this chapter we will only be able to provide a broad overview of these findings. More detail can be found in recent special issues of *Developmental Review*, and the *Handbook of Adolescent Psychology* (Lerner, 2004).

Before discussing the relevant findings we need to define 'adolescence'. Unlike puberty, which is completely defined by the biological processes, adolescence is not a well-defined period. The term is used to indicate the period of transition that starts with the hormonal changes of puberty and ends when the psychosocial status associated with adulthood is reached. It roughly covers the period of age 10 to 24 (Susman et al., 2004), and as the puberty-related hormonal changes start later in boys than in girls, adolescence starts later in boys. This stage of life is frequently described as a troublesome period of deviancy, mood swings, and high risk-taking (e.g. Donovan and Jessor, 1985; Vollebergh et al., 2006). Such descriptions date as far back in history as Rousseau's *Emile* (Koops and Zuckerman, 2003).

Among the many physical changes in adolescence, the recent findings on structural changes in the adolescent brain are probably the most relevant for risky behaviour. In puberty two major developments influence the brain: changes in *brain structure* and changes in the brain's *emotional and motivational systems* that are hormone-induced (Keating, 2004). The third influence comes from the person's interaction with his environment: the so-called 'experienced induced changes'.

Around age ten, the change in structure starts with a *growth in grey matter,* which is followed by *an increase in the density and organization of the white matter.* The grey matter, also often referred to as the cortex, consists of the brain cells and short axons. This part of the brain is responsible for information processing. The white matter consists of the long axons and dendrites and is responsible for the communication between the brain areas. The described changes in the grey and white matter in adolescence probably increase the efficiency of information processing and many aspects of cognitive control such as response inhibition. This period of growth is followed by a period of '*synaptic pruning*', during which the brain discards grey matter at a rapid rate (Giedd, 2008). This pruning raises the probability that experience-based brain development also continues during adolescence (Keating, 2004). In particular the prefrontal cortex (PFC) undergoes far more changes during adolescence than at any other stage of life. Recent studies show that its development continues into a person's early twenties, as a linear process. So the older you are the better! The PFC controls the brain's most advanced functions, such as prioritising thoughts, imaging, thinking in the abstract and anticipation of consequences, planning and impulse control. The second change results from puberty-related *hormonal* processes. These alter many bodily functions, but in particular the *emotional and motivational* systems in the brain, leading to poor impulse control, high reward sensitivity and an urge for immediate gratification of needs. As a result, adolescents strive for short-term gains, often at the expensive of more valuable future gains.

These brain developments and associated behaviour are not unique to humans and have also been observed in primates. It has been hypothesised that it facilitates the acquisition of skills necessary for independence and survival away from parental care-takers. A second hypothesis is that increased interactions with peers and exploration of novel areas may help the dispersal of adolescents away from the natal family unit, and subsequently reducing the chance of inbreeding (Spear, 2000). A third hypothesis proposed by Keating (2004) stresses the possibility that these brain developments also increase the opportunity for behavioural and physiological recovery from negative developments in childhood: a second chance to benefit from experiential learning.

Dual Processes Underlying Behaviour

A second line of research that has generated new insights relevant for risky behaviour is the shift in focus away from the idea that behaviour is a result of conscious and deliberate processes (e.g. Ajzen, 2001; Parker et al., 1992) to paradigms accounting for two processes underlying behaviour: an 'automatic/ heuristic/intuitive' route (System 1) and a 'reasoned route' (System 2). Each process has its own characteristics. See Evans (2008) for a review of these so-called dual process theories. Although dual process theory has been successfully used in 1970–1980 in the study of acquisition of skills and skilled performance (Rasmussen, 1985; Shiffrin and Schneider, 1977), it has recently been applied to decision-making (Kahneman, 2002) and more specifically to decision making in adolescence (Reyna and Farley, 2006) and risk-taking (Gerrard et al., 2008).

Findings resulting from these studies challenge the adequacy of the most frequently used model in research of adolescent of determinants of adolescent driver risk taking: the theory of planned behaviour (TPB) (e.g. Fischhoff, 2008; Gerrard et al., 2008; Klaczynski and Robert, 2004; Reyna, 2004; Séguin et al., 2007; Steinberg et al., 2008). To study the 'popularity' of TPB as a model to the study adolescent unsafe behaviour, Twisk (in preparation) categorised the peer reviewed articles used in Strecher and colleagues systematic review (2007) and concluded that TPB was used in more than half of the studies. This hegemony may be unjustified, and after a more detailed description of the dual process theories in the following section some relevant findings on TPB will be discussed in more detail.

Evans (2003) describes the dual processes as follows. System 1 is old in evolutionary terms and shared with other animals. It comprises of a set of autonomous subsystems that include both innate input modules and domain-specific knowledge acquired by learning mechanism. System 1 is fast, often error-free, automatic and often not consciously monitored. In contrast, System 2 is evolutionarily recent and distinctively human. It permits abstract reasoning and hypothetical thinking and is constrained by working memory capacity and correlated with measures of general intelligence. System 2 is slow and uses cognitive processing based on *effort-consuming systematic reasoning*. This system *requires* attention and controlled information processing.

Understanding Risky Activities in Adolescence

Adolescent development and dual processes

To understand adolescent risk-taking, Steinberg (2008) has integrated findings from brain research and dual process theory. He postulates that adolescent risk-taking is a function of two systems: the cognitive control system (System 1) and the socio-emotional system (System 2). In adolescence the biological brain structures responsible for these two systems change dramatically, but do so at a different pace. The socio-emotional system is the first to develop and includes those regions of the brain that are associated with judgments of attractiveness, recognition of relevant stimuli (such as faces), and other forms of 'social processing'. These changes lead to an increase in sensation-seeking. Rewards and social information become highly attractive. This system peaks between ages 13 and 16. As adolescents grow into adults, the cognitive control system increases steadily in strength.

This system is responsible for self-regulation and deliberative thinking – weighing costs and benefits, thinking ahead, regulating impulses and inhibition. It develops gradually and linearly from preadolescence onwards well into the late teens. It is hypothesised that adolescent risky activities result from an early and sudden increase in sensation-seeking and reward-seeking, alongside a slower increase in the person's ability to engage in self-regulation (Sunstein, 2008).

Table 2.1   Modified after the table presented by Steinberg at the Leiden Conference

Curvilinear development Follows pattern suggestive of maturation of socio-emotional system (system I)	Linear development Follows pattern suggestive of maturation of cognitive control system (system II) and connections between systems
• Risk preference: highest at 16 • Risk perception: lowest at 16 • Sensation-seeking: highest at 13 • Attention to rewards: highest at 16	• Impulse control • Delay discounting • Future orientation • Resistance to peer influence • Avoidance of costs • Planning ahead

According to Steinberg the socio-emotional network is not continuously (over) activated. In the low activation conditions, it is easily controlled by the cognitive control system. Only in high activation conditions like the presence of peers or stress factors, is the system aroused. Experiments where the *emotional* context of the decision-making has been manipulated showed that adolescent and adult decisions do not differ under low emotional conditions (cold decision-making), but strongly differ in high emotional conditions (hot decision making) (Séguin et al., 2007). Peers may affect these emotional conditions, and the magnitude of their influence is dependent on the age of the adolescent. Using self-report techniques Steinberg and Monahan (2007) assessed the resistance to peer influence, on the willingness to engage in antisocial activities. Results showed that resistance to peer influences increased linearly between ages 14 and 18. In contrast, between ages 10 and 14 and between ages 18 and 30 resistance to peer pressure did not change. The authors concluded: 'middle adolescence is an especially significant period for the development of the capacity to stand up for what one believes and resist the pressures of one's peers to do otherwise'. The influence of peer presence has also been studied in a sort of simulated driving task (Gardner and Steinberg, 2005), in which a person had to drive as quickly as possible through a series of traffic light-regulated intersections. Compared to a solo trip, trips in the presence of peers had a higher incidence of risky acts and more crashes.

These findings demonstrate the role of the emotional system in decision-making and the importance of cold versus hot decisions. In the next section the implications of the dual systems approach for the prediction of risky actions are discussed.

Behaviour antecedents and risky actions

In traffic safety and in particular in studies among drivers studies have primarily focused on the controlled processing system (System 1), using theoretical models such as the Theory of Reasoned Action (TRA)/Planned Behaviour (TPB).

According to the TRA/TPB people act in accordance with their intentions and perceptions of control over the behaviour, while intentions in turn are influenced by attitudes toward the behaviour, subjective norms and perceptions of behavioural control (Ajzen, 2001). In line with the theory, reviews showed relatively strong relationships between intention and behaviour (e.g. Armitage and Conner, 2001). If intention predicts behaviour sufficiently then the dual process theories might not add to the understanding of adolescent behaviour, as such a result indicates that behaviour is controlled by behavioural intentions and therefore a result of a 'rational' decision-making process. However, the high predictive power of intentions has mainly been demonstrated in correlational studies, and the weakness of such studies is that the causal *relationship* of two phenomena is not known. Theoretically it is possible instead of *the intention* is controlling the behaviour, *the behaviour* is controlling the intention in that the behaviour is rationalised after the act. To study the *causal relationship* between intention and behaviour and to test the magnitude of this effect, Webb and Sheeran (2006) reviewed intervention studies that all used controlled randomized trials. Such study designs allow causal inferences. Their study showed that although the hypothesis was confirmed that intention significantly changed behaviour, the effect size was considerably smaller than previous reviews based on correlational designs had suggested. A medium-to-large change in intention (d < 0.66) resulted in a small-to-medium change in behaviour (d < 0.36). Apparently, behaviour is not only controlled by intentions. Although this review did not include intervention programmes in traffic safety there is no reason to assume that this would be different in traffic safety. Moreover taking the recent findings in adolescent maturation into account and in particular the important role of System 2, most likely the relationship between intention and behaviour will be even weaker in adolescence.

Some studies have explicitly looked at the relationship between intention and behaviour in adolescents. Gibbons and colleagues (2006) discussed in Gerrard et al. (2008) showed in a prospective study on driving under the influence (DUI) that 40 per cent of adolescents who engaged in DUI in a period of one year, a year before had expressed a strong intention not to drink and drive. Similarly, Twisk (1989) studied intention to DUI in adolescent military servicemen. Intention not to drink and drive was high, yet many of them admitted to having DUI in the previous week. To explore this relationship further, Gerrard et al. (2008) reviewed the literature on the relationship between intention and age in the health domain, and concluded that the relationship is weaker in younger samples than in older samples, and that this relationship is also weaker in adolescents without actual experience with the behaviour.

In conclusion, the most frequently used model has strong limitations in explaining adolescent decision-making processes. Two models that use dual process theories aim to overcome the limitation of models like TPB, namely the *prototype willingness model* (Gerrard et al., 2008) and *fuzzy trace theory* (Reyna and Farley, 2006).

Gerrard's 'prototype willingness model of adolescent risk behaviour' (2008) postulates two processes. A reasoned path similar to that described in TRA and a social reaction path, that is image-based involving more heuristic processing and intuition. This image-based system primarily operates outside of one's own awareness of salient cues and heuristics (System 1), and incorporates two elements: 'risk prototypes' and 'behavioural willingness'. *Risk prototypes* are images of people who engage in risky behaviour versus people who do not. The other element, *behavioural willingness,* refers to a person's openness/readiness to engage in risky behaviour. Both are subconsciously influencing the likelihood that a person might engage in such risk behaviour. According to Gerrard, overall favourability of images (e.g. of substance users) predicts risky behaviour better than do specific attributes described by subsets of adjectives. For example, when asked to describe a typical teenage smoker, only the overall positivity or negativity of the description matters; the details do not predict risk-taking behaviour. Although 'perceived risks' and 'expected benefits' predict behavioural intentions and risk-taking behaviour, in adolescence 'behavioural willingness' is an even better predictor of susceptibility to risk-taking, as in real life adolescents actually do riskier things than they either intended or expected to do.

Fuzzy-trace theory assumes the operation of two systems in decision-making processes. According to the theory, people process and memorize risk related information in two ways: the actual *detailed verbatim and visual information* and *the fuzzy gist* representations. The first is a more cognitive rational assessing of information, while the second is a more or less automatic and immediate response to the risky conditions. In theory, for the drink-driving decision these two systems operate as follows. The verbatim system processes and retrieves detailed factual information, like blood alcohol concentration (BAC) limits, gender differences in alcohol metabolism, enforcement and actual risk levels. Processing of this type of information is slow, and requires attention. In contrast, the fuzzy gist representation only contains the essential meaning: 'it is unwise to drive after drinking'. The retrieval and application of such gists is fast and almost automatic. Over time, verbatim representations rapidly fade and become hard to retrieve, and simultaneously judgment and decision-making become more and more based on these fuzzy gists. As demonstrated in studies with children, adolescents and adults, with increasing age experience and expertise, decisions are increasingly based on fuzzy gists (and automatic). These studies also show that the tendency to take risks decreases when decisions are based on gists. When we generalise this finding to adolescent drivers, we can conclude that because of being inexperienced, these drivers will not benefit from gist-like decision making yet (Reyna and Farley, 2006) and probably still rely on the slow verbatim processes.

The presented results demonstrate the relevance of 'the emotional/intuitive' component in decision-making. Both the prototype willingness model and the fuzzy trace theory incorporate this element. Although both theories are applied to risky decision-making, the question is to what extent this pattern is unique to decisions about risk, and not to any other type of decision-making. Nor have we excluded the notion that the differences between adolescents and adults may not be a result of differences in the appreciation of risk but results from the greater experience of adults with a wide range of activities and outcomes.

Implications for Research and Countermeasures

The presented findings are of utmost relevance in the licensing of novice drivers in Europe. They show that several characteristics of adolescent behaviour are at least partly a result of their biological development.

Adolescent drivers 'maladapted'?

However, unlike perhaps earlier accounts of the 'stupidity' of adolescent behaviour, we cannot regard adolescents are being 'miswired' (Males, 2009). It is a period that is characterised by accelerated development, learning from experiences and possibly also a period that creates a second chance and new opportunities. In this respect, adolescent behaviour is highly functional and adaptive, and for driver licensing creates a great opportunity to accommodate those typical adolescent needs, such as the search for novel experiences, exploring new domains, meeting new people. The downside is that the current traffic system is inherently unsafe, putting novices and their potential crash partners at risk of injury and death.

Age makes a large difference

Interpretation from these studies suggests that the age of the subject is of critical importance. As adolescence covers an age period of more than ten years and the individual developmental processes peak at different ages, a youngster of 18 may therefore fundamentally differ from a 16-year-old. The implication of this observation is well demonstrated with reference to the phenomenon of peer pressure. Although laboratory studies have shown that adolescent drivers in the presence of their peers are more risky, recent analyses of crash data do not confirm this for 18–24-year-old drivers (Engström et al., 2008). The analyses of Swedish crash data in combination of studies from other European countries led to the conclusion that for drivers in the ages 18–24 passengers had a protective effect (Rueda-Domingo et al., 2004; Vollrath et al., 2002). Increased risks were most often reported in studies in the United States, Australia and New Zealand (Doherty et al., 1998; Chen et al., 2000; Williams, 2001) involving younger drivers than in the European studies. In line with such findings of a questionnaire study on peer pressure (Steinberg and Monahan, 2007) possibly these apparently contradictory results can be explained by actual age differences of the driver population. However, alternative explanations are possible like differences in passenger ages (Engström et al., 2008). Despite this, the studies demonstrate the relevance of driver age, and the world of difference it makes whether the driver is 16 or 18.

As a result countermeasures effective for the 16-year-old US driver may not be effective for the 18-year-old European driver.

Exposure as a double-edged sword

The relationship between biological immaturity and experiential learning is also relevant for traffic safety. One may argue the biological limitations and the adolescent preferences for exploration, rewards and novelties which put them at risk sometimes, means they should be protected from these risks by tighter (parental) control. However, exposure to new tasks and environments is also a necessary condition for the brain to develop. This process is often referred to as *experiential* development. Limiting these experiences may be counterproductive and because of the phenomenon 'use it or lose it', limiting experiences may waste the opportunity for learning new advanced skills. Currently we see that schoolchildren are losing their independent mobility because parents find it too dangerous for them to go places on their own. Some have suggested that because of these restrictions on their mobility modern children are less 'streetwise' than earlier generations. To find the balance between protection on the one hand and stimulating growth by allowing exposure to challenging conditions on the other is hard to achieve. In these decisions age is a crucial factor. Measures that are effective for the 16-year-old novice driver in the USA, may be totally ineffective and even detrimental for an 18-year-old European driver. Therefore, with respect to the development of expertise such as learning to drive, two conclusions in opposite directions can be drawn. First, as the brain is still developing and as a result of this is highly adaptive and able to learn, adolescence appears to be the best period to learn new skills like car driving. This is also confirmed by the observation that the older a person is, the greater the number of driving lessons required to pass the driving test. An opposite conclusion is reached when the late development of the frontal brain lobe is also taken into account, as this area is responsible for the integration of the information and controlling the execution of actions. This may imply that not only the control of emotions and interferences is poor, but also the selection and integration of relevant information. This might mean that learning a psychomotor skill like driving becomes easier with age.

Understanding the acquisition of driving skills

With regard to the insights gained from the dual process theory, we can conclude that from the early 1970s onwards, dual process theory has been extensively applied in studies of the development of expertise (e.g. Rasmussen, 1985; Schneider and Chein, 2003). In learning new skills, two qualitatively different systems are in operation: a system that is analytic, requiring controlled processing (later coined by Evans as System 2) and a system which is automatic and more or less subconscious (System 1), characterized by quick and easy processing of information. Novices become experts as a result of practice on the task. This leads to a gradual transition from attention and energy-consuming error-prone controlled processes (System 2) to automatic processes that do not require attention and are relatively error-free (System 1). This explains the high incidence of unintended errors and the high mental load in the first month of training, the detrimental effects of dual tasks and reduced fitness (e.g. fatigue (Groeger, 2006; Hutchens et al., 2008), and effects of alcohol (Hooisma et al., 1988; Mathijssen and Twisk, 2001). On the basis of the new findings, there is a need to reconsider our understanding of the strengths and weaknesses of the current practices in driver training and the licensing processes. The need to address these so-called higher-order skills is reinforced by these new findings: e.g. the role of self-regulation (Craen et al., 2008; Kuiken and Twisk, 2001), the ability to reflect on performance, the ability to set goals and to monitor, regulate, and control cognition, motivation and behaviour (Westenberg, 2008). As in learning to drive, the feedback about the quality of the performance is poor, errors may be overlooked and become routine. In the case of deliberate risk-taking the absence of negative outcomes may give rise to the belief that this behaviour is not dangerous after all. Experience in the absence of negative consequences may increase feelings of invulnerability and thus explain the decrease in risk perceptions from early to late adolescence, as exploration increases (Reyna and Farley, 2006). If this also applies to learning to drive, we would expect to see that risk levels increase in the first month after licensing. The opposite appears to be the case: crash risk in Europe drops in the first month after licensing (Lynam et al., 2002). Moreover, a detailed analysis of the influence of age versus that of experience (read the automation of driving skills) indicates that age has some impact on crash likelihood, but that driving experience has a large protective effect (Vlakveld, 2005).

Although more studies are needed, these patterns of results indicate that driving skill acquisition may only be partly limited because of immature brain structures, and that at least partly these limitations can be overcome by driving experience. This means that the challenge is to facilitate the development of experience in relatively safe conditions. Measures such as accompanied driving and graduated driving license systems are examples of potentially effective measures aiming to stimulate experience development in safe driving conditions.

Adolescents as a problem group

Popular discussions about the new insights may create an image of adolescence as group of youngsters only doing crazy things, and putting themselves and others at risk. Such an image is reinforced by the common belief that adolescence is a period of *Sturm und Drang* (storm and stress). Such descriptions date as far back in history as Rousseau's *Emile* as quoted in Koops and Zuckerman (2003) when he describes:

> A change in humour, frequent anger, a mind in constant agitation, makes the child almost unmanageable. He becomes deaf to the voice that makes him docile. His feverishness turns him into a lion. He disregards his guide; he no longer wishes to be governed.

The question is to what extent these observations are supported by data about the present-day adolescent. A Dutch review about problem behaviour (Junger et al., 2003) Westenberg (2008) concludes that only 15 per cent of all youngsters show problematic behaviour. Koops and Zuckerman (2003) analysing responses of modern adolescents, reach similar conclusions showing that adolescents feel happy and do not report problematic behaviours. So, more information is needed about the incidence of adolescents acting out of control while driving. Naturalistic driving studies like the Forty-Teen-Study (Simons-Morton personal communication) may shed more light on that.

Conclusion

New insights from studies in adolescent development have great relevance for the understanding of risky driving. Because of the speed and the magnitude of the changes in brain and behaviour, the exact age range to which the findings apply needs to be taken into account. It is not clear-cut yet how to deal with exposure. On the one hand, exposure to challenges is a necessity for learning. On the other hand conditions might be too hard to handle for an adolescent driver. Our current understanding of the process by which novices become experts is not challenged. This is in contrast to the application of the theory of planned behaviour to understand, predict and change adolescent risk-taking. More research is needed to assess whether dual theories will improve our understanding of antecedents of risk-taking. An issue not discussed in this chapter is the high risk of male novice drivers. Neither dual theories nor brain research has yet improved our understanding of why in particular young male drivers have a far higher risk than female drivers. Although research discussed here has focused on adolescents as a risk group, recent studies show that adolescence is also a period of great opportunities, and in contrast to popular belief, the majority of adolescents are happy and do not show problem behaviours.

References

Ajzen, I. (2001). Nature and operations of attitudes. *Annual Review of Psychology*, 52(1), 27–58.

Armitage, C.J. and Conner, K.R. (2001). Efficacy of the theory of planned behaviour. *British Journal of Social Psychology*, 40(4), 471–499.

Arnett, J.J. (2000). Emerging adulthood: a theory of development from the late teens through the twenties. *American Psychologist*, 55(5), 469–480.

Chen, L.-H., Baker, S.P., Braver, E.R. and Li, G. (2000). Carrying passengers as a risk factor for crashes fatal to 16- and 17-year-old drivers. *Journal of the American Medical Association*, 283, 1578–1582.

Craen, S.D., Twisk, D.A.M., Hagenzieker M.P., Elffers, H. and Brookhuis, K.A. (2008). The development of a method to measure speed adaptation to traffic complexity: identifying novice, unsafe, and overconfident drivers. *Accident Analysis and Prevention*, 40(4), 1524–1530.

Doherty, S.T., Andrey, J.C. and MacGregor, C. (1998). The situational risks of young drivers: the influence of passengers, time of day and day of week on accident. *Accident Analysis and Prevention*, 30(1), 45–52.

Donovan, J.E. and Jessor, R. (1985). Structure of problem behaviour in adolescence and young adulthood. *Journal of Consulting and Clinical Psychology*, 53, 890–904.

Eby, D.W. and Molnar, L.J. (1999). *Matching Traffic Safety Strategies to Youth Characteristics: A Literature Review of Cognitive Development*. The University of Michigan Transport Research Institute

Engström, I., Gregersen, N.P., Granström, K. and Nyberg, A. (2008). Young drivers – reduced crash risk with passengers in the vehicle. *Accident Analysis and Prevention*, 40(1), 341–348.

Engstrom, K., Diderichsen, F. and Laflamme, L. (2002). Socioeconomic differences in injury risks in childhood and adolescence: A nation-wide study of intentional and unintentional injuries in Sweden. *Injury Prevention*, 8(2), 137–142.

Evans, J. (2003). In two minds: dual-process accounts of reasoning. *Trends in Cognitive Sciences*, 7(10), 454–459.

Evans, J. (2008). Dual-processing accounts of reasoning, judgment, and social cognition. *Annual Review of Psychology*, 59(1), 255–278.

Fischhoff, B. (2008). Assessing adolescent decision-making competence. *Developmental Review*, 28(1), 12–28.

Gardner, M. and Steinberg, L. (2005). Peer influence on risk taking, risk preference, and risky decision making in adolescence and adulthood: an experimental study. *Developmental Psychology*, 41(4), 625–635.

Gerrard, M., Gibbons, F.X., Houlihan, A.E., Stock, M.L. and Pomery, E.A. (2008). A dual-process approach to health risk decision making: the prototype willingness model. *Developmental Review*, 28(1), 29–61.

Giedd, J.N. (2008). The teen brain: insights from neuroimaging. *Journal of Adolescent Health*, 42(4), 335–343.

Gregersen, N.P. and Bjurulf, P. (1996). Young novice drivers: towards a model of their accident involvement. *Accident Analysis and Prevention*, 28(2), 229–241.

Groeger, J.A. (2006). Youthfulness, inexperience, and sleep loss: the problems young drivers face and those they pose for us. *Injury Prevention*, 12(suppl 1), 19–24.

Harré, N. (2000). Risk evaluation, driving, and adolescents: a typology. *Developmental Review*, 20(2), 206–226.

Hooisma, J., Twisk, D.A.M., Platälla, S., Muijser, H. and Kulig, B.M. (1988). Experimental exposure to alcohol as a model for the evaluation of neurobehavioural tests. *Toxicology*, 49(2–3), 459–467.

Hutchens, L., Senserrick, T.M., Jamieson, P.E., Romer, D. and Winston, F.K. (2008). Teen driver crash risk and associations with smoking and drowsy driving. *Accident Analysis and Prevention*, 40(3), 869–876.

Junger, M., Mesman, J. and Meeus, W. (2003). *Psychosociale problemen bij adolescenten: Prevalentie, risicofactoren en preventie*. Assen: Van Gorcum.

Kahneman, D. (2003). Maps of bounded rationality: A perspective on intuitive judgment and choice. In T. Frangsmyr, ed., *Les Prix Nobel: The Nobel Prizes 2002*. Stockholm: The Nobel Foundation, pp. 449–489.

Keating, D.P. (2004). Cognitive and brain development. In R.M. Lerner and L.D. Steinberg, eds., Handbook of Adolescent Psychology, 2nd edn. Hoboken, NJ, John Wiley and Sons, pp. 45–85.

Keating, D.P. (2007). Understanding adolescent development: implications for driving safety. *Journal of Safety Research*, 38(2), 147–157.

Klaczynski, P.A. and Robert, V.K. (2004). A dual-process model of adolescent development: Implications for decision making, reasoning, and identity. *Advances in Child Development and Behavior*, 32, 73–123.

Koops, W. and Zuckerman, M. (2003). Introduction: a historical developmental approach to adolescence. *The History of the Family*, 8(3), 345–354.

Kuiken, M.J. and Twisk, D.A.M. (2001). *Safe Driving and the Training of Calibration: Literature review*. Leidschendam, SWOV Institute for Road Safety Research, R-2001–29.

Lerner, R.M. and Steinberg, L.D., eds (2004). *Handbook of Adolescent Psychology*. Hoboken, NJ, John Wiley and Sons.

Lynam, D., Nilsson, G., Morsink, P., Sexton, B., Twisk, D.A.M., Goldenbeld, C., et al. (2002). *Sunflower: A Comparative Study of the Development of Road Safety in Sweden, the United Kingdom, and The Netherlands*. Leidschendam, The Netherlands: SWOV Institute for Road Safety Research/Crowthorne, Berkshire, Transport Research Laboratory TRL/Linköping, Swedish National Road and Transport Research Institute VTI.

Lynam, D. and Twisk, D.A.M. (1995). *Car Driver Training and Licensing Systems in Europe: Report Prepared by Members of Forum of European Road Safety Research Institutes (FERSI) and Supported by European Commission Transport Directorate, DG VII*. Crowthorne, Berkshire, Transport Research Laboratory TRL, TRL Report No. 147.

Males, M. (2009). Skeptical appraisal: does the adolescent brain make risk taking inevitable? *Journal of Adolescent Research*, 24(3), 3–20.

Mathijssen, M.P.M. and Twisk, D.A.M. (2001). *Opname en afbraak van alcohol in het menselijk lichaam: Verslag van een demonstratie naar aanleiding van een 'experiment' in het tv-programma 'Blik op de weg'. In opdracht van het ministerie van justitie, directoraat-generaal rechtshandhaving, directie handhaving*: Leidschendam, Stichting Wetenschappelijk Onderzoek Verkeersveiligheid (SWOV) R-2001–19.

Maycock, G. (2002). *Estimating the Effects of Age and Experience on Accident Liability Using Stats19 Data*. Behavioural Research in Road Safety XII, London, Department of Transport.

Maycock, G. and Forsyth, E. (1997). *Cohort Study of Learner and Novice Drivers*. TRL Research.

OECD. (2006). *Young Drivers: The Road to Safety*. Crowthorne, Transport Research Laboratory.

Parker, D., Manstead, A.S.R., Stradling, S.G. and Reason, J.T. (1992). Determinants of intention to commit driving violations. *Accident Analysis and Prevention*, 24(2), 117–131.

Rasmussen, J. (1985). Trends in human reliability analysis. *Ergonomics*, 28(8), 1185–1191.

Reyna, V.F. (2004). How people make decisions that involve risk a dual-processes approach. *Current Directions in Psychological Science*, 13(2), 60–66.

Reyna, V.F. and Farley, F. (2006). Risk and rationality in adolescent decision making: implications for theory, practice and public policy. *Psychological Science in the Public Interest*, 7(1), 1–44.

Rueda-Domingo, T., Lardelli-Claret, P., Luna-del-Castillo, J.D., Jiménez-Moleón, J.J., García-Martín, M. and Bueno-Cavanillas, A. (2004). The influence of passengers on the risk of the driver causing a car collision in Spain: Analysis of collisions from 1990 to 1999. *Accident Analysis and Prevention*, 36(3), 481–489.

Schneider, W. and Chein, J.M. (2003). Controlled and automatic processing: behavior, theory and biological mechanisms. *Cognitive Science*, 27(3), 525–559.

Séguin, J.R., Arseneault, L. and Tremblay, R.E. (2007). The contribution of 'cool' and 'hot' components of decision-making in adolescence: implications for developmental psychopathology. *Cognitive Development*, 22(4), 530–543.

Senserrick, T.M. (2006). Reducing young driver road trauma: guidance and optimism for the future. *Injury Prevention*, 12(suppl 1), i56–60.

Shiffrin, R.M. and Schneider, W. (1977). Controlled and automatic information processing ii. Perceptual learning, automatic attending and a general theory. *Psychological Review*, 84(127), 127–190.

Shope, J.T. (2007). Graduated driver licensing: review of evaluation results since 2002. *Journal of Safety Research*, 38(2), 165–175.

Spear, L.P. (2000). The adolescent brain and age-related behavioral manifestations. *Neuroscience and Biobehavioral Reviews*, 24(4), 417–463.

Steinberg, L. (2008). A social neuroscience perspective on adolescent risk-taking. *Developmental Review*, 28(1), 78–106.

Steinberg, L., Albert, D., Cauffman, E., Banich, M., Graham, S. and Woolard, J. (2008). Age differences in sensation seeking and impulsivity as indexed by behavior and self-report: evidence for a dual systems model. *Developmental Psychology*, 44(6), 1764–1778.

Steinberg, L. and Monahan, K.C. (2007). Age differences in resistance to peer influence. *Developmental Psychology*, 43(6), 1531–1543.

Strecher, V.J., Shope, J., Bauermeister, J.A., Chang, C., McHale, N.-B., Boonin, A. et al. (2007). *Review of Judgement and Decision-making Literature Pertinent to the Development of Traffic Offender Training Improvement Programmes (s301q) Technical Report*. London, Department for Transport.

Sunstein, C.R. (2008). Adolescent risk-taking and social meaning: a commentary. *Developmental Review*, 28(1), 145–152.

Susman, E., J. and Rogol, A. (2004). Puberty and psychological development. In R.M. Lerner and L.D. Steinberg, eds, *Handbook of Adolescent Psychology*, 2nd edn. Hoboken, NJ, John Wiley and Sons, pp. 15–45.

Twisk, D.A.M. (1989). *Verkeersveiligheidsproblematiek van rijden onder invloed: Cursus en examen in cbr-stijl : Een toepassing in de militaire rijopleiding.* Leidschendam, Stichting Wetenschappelijk Onderzoek Verkeersveiligheid (SWOV) R-89-42.

Twisk, D.A.M. (1995). *Factors contributing to the high accident liability of novice drivers and the role of driver training.* Paper presented at the seminar Behavioural Research in Road Safety vi. Royal Holloway, University of London, 4–5 september 1995. Leidschendam, SWOV Institute for Road Safety Research, D-95-9.

Twisk, D.A.M. (2000). Why did the accident involvement of young (male) drivers drop about 50%? In *Behavioural Research in Road Safety X: Proceedings of the 10th Seminar on Behavioural Research in Road Safety.* Esher, Surrey, 3–5 April, pp. 109–117.

Twisk, D.A.M. and Stacey, C. (2007). Trends in young driver risk and countermeasures in European countries. *Journal of Safety Research*, 38(2), 245–257.

Vlakveld, W.P. (2005). *Jonge beginnende automobilisten, hun ongevalsrisico en maatregelen om dit terug te dringen: Een literatuurstudie.* Leidschendam, Stichting Wetenschappelijk Onderzoek Verkeersveiligheid SWOV.

Vollebergh, W., van Dorsselaer, S., Monshouwer, K., Verdurmen, J., van der Ende, J. and ter Bogt, T. (2006). Mental health problems in early adolescents in the Netherlands. *Social Psychiatry and Psychiatric Epidemiology*, 41(2), 156–163.

Vollrath, M., Meilinger, T. and Krüger, H.-P. (2002). How the presence of passengers influences the risk of a collision with another vehicle. *Accident Analysis and Prevention*, 34(5), 649–654.

Webb, T.L. and Sheeran, P. (2006). Does changing behavioral intentions engender behavior change? A meta-analysis of the experimental evidence. *Psychological Bulletin*, 132(2), 249–268.

Westenberg, P.M. (2008). De jeugd van tegenwoordig *De psycholoog, oktober*, 10(43), 546–552.

Williams, A.F. (2001). *Teenage Passengers in Motorvehicle Crashes: A Summary of Current Research.* Arlington, VA, Insurance Institute for Highway Safety.

... ...
... ...

Chapter 3

A Comparison of Inexperienced and Experienced Drivers' Cognitive and Physiological Response to Hazards

Steve W. Kelly, Neale Kinnear, James Thomson and Steve Stradling

Introduction

Learning to drive has been conceptualised as a series of stages which take the learner from mastery of the basic mechanics of driving, through anticipation of other road users' behaviour, to the development of a consistent driving style which incorporates all the acquired skills (Parker and Stradling, 2001). Hazard perception is one of the main skills to be acquired in the development of event anticipation (Deery, 1999), and several studies have indicated that lack of hazard perception skills is related to the likelihood of being involved in a road traffic accident (Pelz and Krupat, 1974; Quimby et al. 1986). In addition, novice drivers have been found to show poor hazard perception skills (Deery, 1999). With several studies reporting a link between driving experience and awareness of hazards (e.g. Ahopalo, 1987; Pelz and Krupat, 1974; Quimby and Watts, 1981), early under-development of this skill may be an important contributing factor to the young novice group being over-represented in accident statistics (Cooper et al., 1995; Maycock et al., 1991).

Driver behaviour modelling is one way in which the psychological processes of hazard perception and learning to drive can be conceptualised and studied. Fuller (2008) summarised recent developments in driver behaviour modelling while promoting the *risk allostasis theory*. Risk allostasis theory has developed from the previous *task difficulty homeostasis* (see Fuller, 2005) and endorses the role of feelings in driver behaviour and decision-making. In an experimental study of task difficulty homeostasis, Fuller (2005) found that drivers' self-reported feelings of risk correlated highly with their perceived demand of the driving task, when viewing video clips of driving at different speeds. Surprisingly, drivers' objective risk ratings of the risk of collision or loss of control were not related to task demand in the way that feelings of risk were. These results were replicated by Kinnear and colleagues (2008), where it was further reported that novice drivers may rely on objective risk estimates rather than their feelings of risk. Fuller (2008) reports that risk allostasis theory is in line with other work, notably Summala (2007) and Vaa (2007) in considering the role of feelings in risk appraisal and driver behaviour, and concludes 'Above all else, there is a current convergence in recognizing the primacy of the role of feeling in driver decision-making and this recognition opens up a whole new set of exciting and promising research questions' (p. 27).

Slovic and Peters (2006) suggest that risk is processed in two fundamental ways: risk as analysis and risk as feelings. The 'analytic system' uses logic and normative rules, such as the probability calculus and risk assessment. It is relatively slow, requiring effort and conscious control (Slovic et al., 2004). Meanwhile,

> The 'experiential system' is intuitive, fast, mostly automatic, and not very accessible to conscious awareness. The experiential system enabled human beings to survive during their long period of evolution and remains today the most natural and most common way to respond to risk. It relies on images and associations, linked by experience to emotion and affect (a feeling that something is good or bad). This system represents risk as a feeling that tells us whether it is safe to walk down a dark street. (Slovic et al. 2004, p. 311)

In relation to driving, it would not be out of place to reword the last sentence as, 'This system represents risk as a feeling that tells us whether it is safe to continue driving at a certain speed'. These two definitions of risk assessment can be applied to the findings from Fuller (2005) and Kinnear et al. (2008) whereby risk measured by feelings was found to be distinct from risk measured as probability. It could further explain how novice drivers may rely more on the analytic system whereas experienced drivers are able, by way of their experience, to determine risk through the experiential system.

Epstein (1994) believes these two systems work in parallel and refers to 'affect' as being subtle feelings which are intimately associated with the experiential system. It is further stated that affect is a major motivating factor in behavioural response – a position supported by other authors including Zajonc (1980) and Damasio (1994, 2003). Zajonc (1980) was one of the earliest proponents of emotion in decision-making and argued that affective reactions to stimuli in our environment are often the very first, occurring automatically and subsequently guiding information processing and behaviour.

Damasio's *somatic marker hypothesis* (SMH) has interestingly already been discussed in relation to driving behaviour by Fuller (2008), Summala (2007) and Vaa (2007). Damasio (1994, 2003) argues that the basis of the SMH is that unconscious processes occur before reasoning and a cost–benefit analysis takes place. If, for example, a situation appears to be developing that could advance into something threatening or dangerous, a feeling of unpleasantness will be produced in the body (i.e. a gut feeling). Damasio (1994, 2003) labels this a 'somatic marker'; *soma* being Greek for 'body'. It is a marker because this bodily feeling will be marked against the developing scenario so that the organism will learn that should this scenario begin to be built up again, the body can respond earlier (Damasio, 1994, 2003). The process put forward by Damasio is one of an evolved socio-emotional learning system that is utilised by humans to avoid danger within their environments. It would therefore appear likely that if such a socio-emotional process does exist it could easily be applied to the driving scenario.

In testing the SMH, Damasio and colleagues cite experiments involving the Iowa Gambling Task (IGT) as demonstrating support for the process (see Bechara and Damasio, 2005). One important source of support involved participants playing the IGT while their skin conductance response was measured. Skin conductance response (SCR) is commonly used as a measure of minute physiological changes that can demonstrate an emotional or psychological response through the sympathetic component of the autonomic nervous system (Dawson et al., 2000). It is reported that as healthy participants became experienced with the task, they began to generate SCRs in anticipation of selecting a card (patients with damage to the ventromedial prefrontal cortex and the amygdala failed to generate SCRs) (Bechara et al., 1997, 1999). In addition, these SCRs were greater before picking a card from the risky decks. Bechara and Damasio (2005) consider the SCRs to be demonstrations of learned somatic markers to actions with anticipated negative consequences.

Development of substantial fear and anxiety in real life, which may have dramatic consequences for the individual, would seem intuitively to be the type of situation where emotional learning would have an enormously influential role in guiding behaviour. Interestingly, some historical studies have already examined emotional responses during driving. Taylor (1964) examined SCRs of volunteers driving predetermined routes with variation of road, traffic and lighting conditions. Driving produced the highest SCRs of any other voluntary activity examined, suggesting a large emotional component being involved in driving behaviour. Helander (1978), using four different routes and participants with a wide range of ages and experience in driving, found a rank correlation of 0.95 between SCR and brake pressure. Analysis of electromyographic (EMG) responses from the leg muscles involved in braking demonstrated that the SCR was not a consequence of or concomitant with braking but an antecedent.

In a more recent driving related study where SCR was used as a measure, Crundall and colleagues (2003) compared police pursuit drivers, experienced drivers and novice drivers. Participants' SCR and hazard ratings were recorded while viewing police pursuit and emergency response video clips. The study reports that mean hazard ratings showed no difference across driver groups. However, frequency of SCR was found to be greater for the police drivers than for the other two groups. This suggests that there is a distinction between hazard perception as a cognitive ability and hazard perception as a function of emotional response, at least in highly trained drivers. The distinctions between cognitive ratings and emotional appraisal of hazardous situations demonstrated in this study would further support Slovic et al's (2004) definition of the analytic and experiential forms of risk appraisal. It would seem, therefore, that further investigation to determine the existence of an emotional component of risk appraisal when driving would be useful.

This Study

The current study examined SCRs of experienced and inexperienced drivers to three types of still picture presented on a computer screen. These still pictures were of 'safe', 'hazardous' and 'developing hazard' situations. Cognitive appraisals of how hazardous the situation appeared to be were also collected.

If development of emotional responsiveness to potentially hazardous situations is a critical component of becoming an 'experienced' driver then two things could be expected. First, similar to the results of Crundall et al. (2003), it would be expected that *there will be no difference between inexperienced and experienced drivers' hazard ratings of safe, developing hazard or hazard pictures.* Second, it would also be expected that *there will be a difference between inexperienced and experienced drivers' emotional response to either developing hazard or hazard scenarios, or both.*

Method

Design

A 2 × 3 mixed design was used. The between-groups factor was driving experience (inexperienced vs experienced) and the within-groups factor was type of situation (safe, hazard and developing hazard). Hazard ratings were taken in response to images of the three scenarios, while physiological measures included participants' skin conductance response (SCR) and respiration amplitude.

Fifteen images (5 × Safe; 5 × Developing Hazard; 5 × Hazard) were presented at timed intervals in random order.

Participants

Twenty-one inexperienced drivers (9 male; 12 female) and 18 experienced (10 male; 8 female) drivers of a similar age range took part in the experiment. Inexperienced drivers were defined as having held a UK driving licence for less than three years and experienced drivers as having held their licence for three years or more.

The mean age for inexperienced drivers was 21.7 years (SD = 3.6, range = 17.8–33.8); whilst the mean age for experienced drivers was 25.4 years (SD = 2.9, range = 20.2–31.0). Inexperienced drivers reported to have held their UK driving licence for a mean of 13.3 months (SD = 8.7; range = 1–29); while experienced drivers reported to have held their licence for a mean of 86.2 months (SD = 43.3; range = 36–168).

Materials

Fifteen still images were taken, with permission, from a commercially available CD-ROM (Focus Multimedia Driving Test Success: Hazard Perception). The images were chosen to portray examples of safe, hazardous and potentially hazardous situations (five images per category). An example of an image from each category can be seen in Figure 3.1. A pilot study was conducted to ensure that these pictures depicted situations which were safe, dangerous or ambiguous with respect to inherent risk. Twenty random volunteer participants, who all held a current UK driving licence, were asked to rate each image as either 'safe' or 'hazardous'. The mean number of participants rating the images within each category as 'hazardous' was: Safe 1.8 (SD = 1.3); Developing hazard 12.2 (SD = 2.8); and Hazardous 18 (SD = 1.0). The results of the pilot study suggested that the images in the three categories reasonably reflected safe, developing hazard and hazard scenarios.

Figure 3.1   Example pictures of (clockwise from top left) safe, developing hazard and hazard scenarios

The images were randomly presented full screen on a 19″ computer monitor using Superlab 4 experiment generator software. A Cedrus RB-730 button box was used to record participants' hazard ratings data. Participants' SCR and respiration were measured by the Biopac MP35 system using electrodermal pre-settings with

Biopac EL507 EDRS isotonic gel disposable electrodes and a respiratory belt and transducer. The SCR and respiration traces were recorded and analysed using Biopac BSL Pro software.

Participants also completed a questionnaire about themselves and their driving history.

Procedure

Participants were asked to read an experiment information sheet and sign the consent form if they were happy to proceed with the experiment. Participants were seated approximately 60cm from the computer monitor with the button box at a comfortable distance on the desk. Electrodes were attached on the palmer surface of the medial phalanx of the middle and index fingers of the non-preferred hand (see Figure 3.2). Participants were also asked to position a belt attached to a respiratory transducer around their chest and take several large breaths in order to check the recording equipment was operational and to provide a comparison respiration trace.

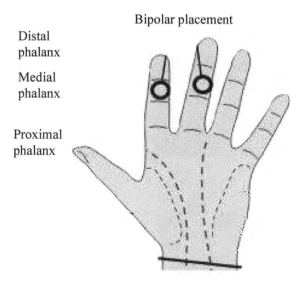

Bipolar placement

Distal
phalanx

Medial
phalanx

Proximal
phalanx

Figure 3.2   Placement of skin conductance electrodes when measuring skin conductance

Participants were told that they would see 15 images of various road scenarios. They were asked to imagine that they were the driver of the vehicle encountering these scenarios and that when the image disappeared from the screen they would be required to make a rating from 1 (safe) to 7 (extremely hazardous) for how hazardous that situation appeared to them. Each image remained on screen for five seconds and was replaced by a screen which prompted participants to provide a rating. This screen was displayed for ten seconds after which the next road scene was shown. Images were presented randomly via the Superlab 4 experiment generator package.

Once all images had been shown, the electrodes and respiration belt were removed and participants were asked to complete the questionnaire.

Ethical approval for the study was granted by the Psychology Ethics Board at Strathclyde University, where the experiment was performed.

Results

1. There will be no difference between the driver groups' hazard ratings of safe, developing hazard or hazard pictures.

Analysis of hazard ratings

Figure 3.3 shows the mean hazard ratings for the driving scenarios. The increase in ratings across hazard type is statistically significant for both experienced and inexperienced drivers (Page's L-trend test, L = 230 and 252, respectively, $P <$ 0.01 for both). As can be seen, both the experienced and inexperienced driver groups gave similar ratings to all categories of pictures, although specifically the developing hazard category.

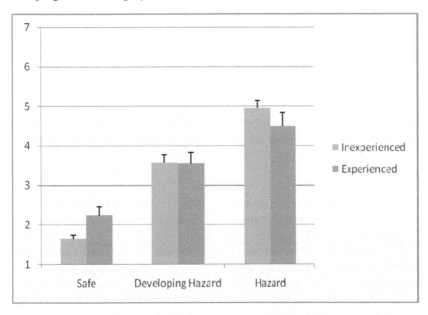

Figure 3.3   **Comparison of driver groups' mean hazard ratings to safe, developing hazard and hazard pictures with standard error bars. Scale: extremely safe 1–7 extremely hazardous**

A Kruskal–Wallis test (see Table 3.1) shows that only the ratings for the safe images were significantly different with experienced drivers judging the 'safe' situations as being more hazardous than the inexperienced drivers were judging them. The hypothesis that there would be no difference between the groups' hazard ratings is therefore only partially supported. However, the crucial comparison is the developing hazard condition where the mean ratings for the inexperienced and experienced driver groups are numerically almost identical and do not show a statistical difference.

Table 3.1   Mean hazard ratings for still images by experience group with standard deviations and Kruskal–Wallis analysis

	Safe		Developing hazard		Hazard	
	Mean	SD	Mean	SD	Mean	SD
Inexperienced	1.65	0.45	3.57	0.90	4.95	0.87
Experienced	2.23	0.97	3.54	1.23	4.48	1.49
χ^2	4.78*		0.11		0.4	

* p <0.05.

Hazard ratings and gender

T-tests of gender and hazard ratings were performed and found no significant difference between males and females across any picture category for their hazard ratings (safe: t(38) = 0.406; developing hazard: t(37) = –0.979; hazard: t(37) = -1.58; P = ns for all).

2. There will be a difference between inexperienced and experienced drivers' emotional response to either developing hazard or hazard scenarios, or both.

Analysis of SCRs

Before SCR data was analysed, participants' initial deep breath respiration trace was compared with their overall respiration. Any SCR that was preceded by respiration which approximated the amplitude of the initial deep breath was excluded from analysis. There were very few instances (less than 1 per cent) where this occurred and SCR data is reported proportionately to take account of the missing data points. A SCR to a particular image was taken as any rise in trace amplitude over 0.05 microSeimens (μS) beginning between one and three seconds after stimulus presentation (Levinson and Edelberg, 1985; Barry, 1990). A latency of one to five seconds has often been used to measure SCRs, however, Levinson and Edelberg (1985) report that using a narrower gap of one to three seconds improves the reliability of measuring only SCRs to the stimulus. The use of a one to three second latency period is supported by

Barry (1990) and Boucsein (1992). Given the investigative nature of the experiment, it was considered important to ensure that any SCRs recorded as responses to the stimulus were as reliable as possible.

Figure 3.4 shows the mean number of SCR responses, calculated as percentages, per stimulus item condition for experienced and inexperienced drivers. Experienced drivers numerically show more SCRs across all stimulus conditions. However, this difference is only statistically significant for developing hazard items, as shown by a Kruskal–Wallis analysis (see Table 3.2). This therefore supports the hypothesis that there will be a difference between inexperienced and experienced drivers' emotional response to either developing hazard or hazard scenarios. Although the results support the hypothesis, the low percentage response across all conditions, specifically the hazard pictures, is a concern to the validity of the current stimulus method.

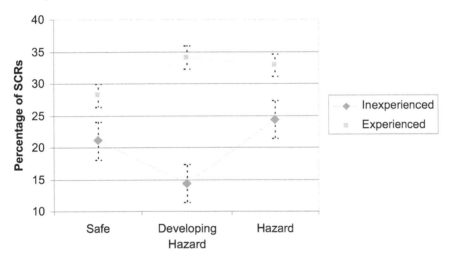

Figure 3.4   **Comparison of driver groups' SCR percentages to safe, developing hazard and hazard pictures with standard error bars**

Table 3.2   **Mean percentage of responses showing an SCR to the stimulus item with standard deviations and Kruskal–Wallis analysis**

	Safe		Developing hazard		Hazard	
	Mean	**SD**	**Mean**	**SD**	**Mean**	**SD**
Inexperienced	21.1	20	14.4	20.4	24.4	22.3
Experienced	28.2	23.5	34.1	27.2	32.9	30.8
χ^2	0.68		5.74*		0.47	

* $p<0.05$

Peak to peak analysis

In addition to the presence or absence of SCRs, the size of the psycho-physiological response, where generated, could also provide useful information about driver differences. Where an SCR was evoked, peak to peak measurements were compared across groups. One outlier (defined as more than two standard deviations from the mean) was replaced with the recalculated group mean for the inexperienced-hazard condition. Numerically, experienced and inexperienced drivers show similar peak to peak increases in SCR to safe and developing hazard clips, with inexperienced drivers showing a larger increase to hazardous images than experienced drivers. However, t-tests demonstrated no significant difference between experienced and inexperienced drivers for safe, developing hazard or hazard scenes (t(22) = 0.059; t(20) = –0.199; t(21) = 0.811, respectively, *P* >0.10).

SCR and gender

T-test comparisons of gender SCR scores were performed and results showed no significant differences between males and females across any picture category (safe: t(37) = 1.62; developing hazard: t(36) = 1.26; hazard: t(36) = 0.17; *P* = ns for all).

Results summary

In summary, a difference in hazard ratings was found in safe images between experienced and inexperienced drivers, yet no difference was found on psycho-physiological measures for these types of stimuli. A more striking pattern emerged for images depicting a developing hazard which showed almost identical ratings for cognitive estimations of risk but displayed a significant difference in number of SCRs between experienced and inexperienced drivers. Hypothesis 1 was therefore supported for the developing hazard and hazard pictures although not for the safe pictures. Meanwhile, hypothesis 2 was supported as there was a significant difference between the groups whereby experienced drivers produced significantly more SCRs to developing hazards than inexperienced drivers. There was no difference in the proportion of SCRs to safe or hazardous pictures by experience level.

Discussion

Results demonstrated that inexperienced drivers show a marked decrement in psycho-physiological response to developing hazard scenarios. This is in spite of the fact that their cognitive assessment of risk for that particular type of road scene is statistically no different to that of the experienced driver group. Similar to the results of Crundall and colleagues (2003), and theory proposed by Slovic et al. (2004) this suggests that emotional and cognitive components of hazard perception are dissociable and vary with driver experience. Further evidence for this is shown in the responses to the safe images where both groups did not differ in terms of emotional response but the experienced drivers gave higher ratings of risk. These results rule out an interpretation of skin conductance response simply being a product of participants' estimation of risk and lend support for greater consideration of the potential role of feelings in driving behaviour (e.g. Vaa, 2004, 2007; Fuller, 2005; 2008) and in particular, the role of feelings in anticipating hazards.

Developing hazards and SCRs

Though previous studies (e.g. Crundall et al., 2003) have examined SCRs to driving scenarios and hazards using experienced and inexperienced drivers, no previous study has specifically examined psycho-physiological responses to situations where a hazard is not immediately apparent but where the scenario would indicate an increased likelihood of a hazard developing within a short time frame. It is this type of scenario where, logically, an experienced driver would benefit from an emotional signal alerting them to potential danger and, theoretically, where Damasio's (1994) somatic marker hypothesis (SMH) would produce the maximum effect of emotion on behaviour. The anticipatory component of the SMH is possibly the most interesting aspect of the theory.

Though much of the data in support of the SMH have been collected using laboratory-based gambling tasks, Damasio's (1994) basis for the SMH is that the basic motive of all organisms is survival and therefore their primary task is risk monitoring (Vaa, 2004). Consequently, it might be predicted that a system such as SMH or similar should show greater behavioural effect for a potentially life-threatening behaviour such as driving and to be functionally useful should therefore be anticipatory in nature. This study was not designed to determine whether emotional responses were anticipatory in nature. However, with psycho-physiological difference being found only for the developing hazard stimuli, the results are suggestive of that interpretation.

Peak to peak SCRs

The current study did not find a difference in peak to peak measures of skin conductance response between groups. It is possible that real-life development of emotional markers will vary in nature to those elicited by a single session laboratory task, especially to still images. The presence or absence of a response to developing hazards is potentially more indicative of simply whether or not hazardous scenarios have been emotionally connected through discrete experiences over a significant driving timescale. Crundall et al. (2003) also reported an effect of frequency of SCRs but not amplitude, although they did not report a significant difference between inexperienced and experienced drivers. A crucial difference between the current study and Crundall et al (2003) may be that the current study separated hazards and developing hazards whereas Crundall and colleagues' (2003) study does not differentiate between hazards and developing hazards. In the current study, no difference in emotional response was found between experienced and inexperienced drivers to clearly hazardous situations; it was specifically the developing hazards that demonstrated this difference. Amalgamating responses to both types of scenario may have masked differences between novice and experienced drivers in Crundall et al.'s (2003) study. However, the result that police drivers show a difference in skin conductance response suggests that experience, and/or specialised training, may be the critical factor in developing appropriate emotional responses to hazards.

Limitations of the current study

Whether emotion is epiphenomenal to cognitive decision-making or has a causal role in guiding behaviour, as Damasio (1994, 2003) suggests, is yet to be examined in the context of driving. The current study's use of still images has the advantage of tying the psycho-physiological response to a particular type of visual scene. However, its major weakness is that it lacks ecological validity, as possibly demonstrated by the low percentage SCR responses in the hazard category. Future studies could look to use either more stimuli or to use more dynamic stimuli. Cohen (1981) suggests drivers' visual fixation patterns to still and dynamic driving scenarios vary considerably, with drivers fixating on many more factors when using dynamic stimuli. More naturalistic, dynamic stimuli may provide a beneficial avenue for investigating an emotional component to hazard awareness in future.

Despite the limitations of the current study, the findings, when placed alongside recent relevant theory, paint an intriguing picture for future research.

References

Ahopalo, P. (1987). *Experience and Response Latencies in Hazard Perception.* Helsinki, Finland: Traffic Research Unit, University of Helsinki.

Barry, R.J. (1990). Scoring criteria for response latency and habituation in electrodermal research: a study in the context of the orienting response. *Psychophysiology*, 27(1), 94–100.

Bechara, A. et al. (1997). Deciding advantageously before knowing the advantageous strategy. *Science*, 275(5304), 1293–1294.

Bechara, A. et al. (1999). Different contributions of the human amygdala and ventromedial prefrontal cortex to decision-making, *Journal of Neuroscience*, 19(13), 5473–5481.

Bechara, A. and Damasio, A.R. (2005). The somatic marker hypothesis: A neural theory of economic decision. *Games and Economic Behavior. Special Issue on Neuroeconomics*, 52(2), 336–372.

Boucsein, W. (1992). *Electrodermal Activity*. New York, Plenum Press.

Cohen, A.S. (1981). Car drivers' pattern of eye fixations on the road and in the laboratory. *Perceptual and Motor Skills*, 52(2), 515–522.

Cooper, P.J. et al. (1995). An examination of the crash involvement rates of novice drivers aged 16 to 55. *Accident, Analysis and Prevention*, 27(1), 89–104.

Crundall, D. et al. (2003). Eye movements and hazard perception in police pursuit and emergency response driving. *Journal of Experimental Psychology: Applied*, 9(3), 163–174.

Damasio, A.R. (1994). *Descartes' Error: Emotion, Reason and the Human Brain*. New York, Putnam.

Damasio, A.R. (2003). *Looking for Spinoza: Joy, Sorrow and the Feeling Brain*. London, Heinemann.

Dawson, M.E. et al. (2000). *The Electrodermal System*. New York, Cambridge University Press.

Deery, H.A. (1999). Hazard and risk perception among young novice drivers. *Journal of Safety Research*, 30(4), 225–236.

Epstein, S. (1994). Integration of the cognitive and the psychodynamic unconscious. *American Psychologist*, 49, 709–724.

Fuller, R. (2005). Driving by the seat of your pants: a new agenda for research. In *Behavioural Research in Road Safety 2005*. London, Department for Transport.

Fuller, R. (2008). Recent developments in driver control theory: from task difficulty homeostasis to risk allostasis. Paper presented to International Conference on Traffic and Transport Psychology, Washington, DC, 31 August–4 September 2008.

Helander, M. (1978). Applicability of drivers' electrodermal response to the design of the traffic environment. *Journal of Applied Psychology*, 63(4), 481–488.

Kinnear, N. et al. (2008). Do we really drive by the seat of our pants? In L. Dorn, ed., *Driver Behaviour and Training Volume III*. Aldershot: Ashgate, pp. 349–365.

Levinson, D.F. and Edelberg, R. (1985). Scoring criteria for response latency and habituation in electrodermal research: a critique. *Psychophysiology*, 22(4), 417–426.

Maycock, G. et al. (1991). *The Accident Liability of Car Drivers*. No. 315. Crowthorne: Transport Research Laboratory.

Parker, D. and Stradling, S.G. (2001). *Influencing Driver Attitudes and Behaviour, Road Safety Research Report No. 17*, March. London, Department of the Environment, Transport and the Regions, HMSO.

Pelz, D.C. and Krupat, E. (1974). Caution profile and driving record of undergraduate males. *Accident Analysis and Prevention*, 6, 45–58.

Quimby, A.R. et al. (1986). *Perceptual abilities of accident involved drivers, Transportation and Road Research Laboratory Report RR27*. Crowthorne, Transportation and Road Research Laboratory.

Quimby, A.R. and Watts, G.R. (1981). *Human Factors and Driving Performance.* Crowthorne, Transport and Road Research Laboratory

Slovic, P. et al. (2004). Risk as analysis and risk as feelings: some thoughts about affect, reason, risk, and rationality. *Risk Analysis*, 24(2), 311–322.

Slovic, P. and Peters, E. (2006). Risk perception and affect. *Current Directions in Psychological Science*, 15(6), 322–325.

Summala, H. (2007). Towards understanding motivational and emotional factors in driver behaviour: comfort through satisficing. In P. Cacciabue, ed., *Modelling Driver Behaviour in Automotive Environments: Critical Issues in Driver Interactions with Intelligent Transport Systems*. London, Springer-Verlag, pp. 189–207.

Taylor, D.H. (1964). Drivers' galvanic skin response and the risk of accident. *Ergonomics*, 7(4), 439–451.

Vaa, T. (2004). *Survival or Deviance? A Model for Driver Behaviour.* No. TOI 666/2003 (In Norwegian with summary in English). Oslo, Norway, Institute of Transport Economics.

Vaa, T. (2007). Modelling driver behaviour on basis of emotions and feelings: intelligent transport systems and behavioural adaptations. In P. Cacciabue, ed., *Modelling Driver Behaviour in Automotive Environments: Critical Issues in Driver Interactions with Intelligent Transport Systems*. London, Springer-Verlag, pp. 208–232.

Zajonc, R.B. (1980). Feeling and thinking: preferences need no inferences. *American Psychologist*, 35(2), 151–175.

Chapter 4

Development of the Driver Performance Assessment: Informing Learner Drivers of their Driving Progress

Erik Roelofs, Marieke van Onna and Jan Visser

Introduction

In response to the sad statistics regarding the over-representation of young drivers in traffic accidents scholars have argued for a more focused look at driver training. Until recently the drivers' task was perceived as a set of elementary driving tasks pertaining to vehicle control and applying traffic rules. More recently, driving is increasingly being considered as a broad domain of competence. An important step for defining driving as a domain of competence was made by the appearance of the Goals for Driver Education (GDE) matrix.

The matrix stresses the overriding significance of the higher levels of driver behaviour with regard to accidents, and the need for drivers to possess not only knowledge and skills, but also risk awareness and self-evaluation skills at multiple levels (Hatakka et al., 2002).

Nowadays the GDE matrix is increasingly being used for developing curricula for driver training at both pre- and post-licence stages. In some European countries a two-phased driver training programme, including a pre-licence and post-licence phase, is now under consideration. In addition, initial driver training programmes have been further improved by using lesson designs led by professional driving instructors aimed at systematic teaching of meaningful driving tasks. In addition, initiatives for permanent (ongoing) road safety education have emerged. Driver education is considered to be a long-lasting or even a life-long process (Vissers et al., 2007).

This shift towards competence-oriented driver training asks for new forms of driving assessment which inform and support the learner driver and the instructor about the acquisition of driving competence (Dierick and Dochy, 2001). Until recently, and in line with traditional views on driver training, assessment instruments took the form of rather isolated testing of knowledge about traffic rules in theory tests and technical driving skills in predominantly examiner-led road tests. Following the competence based view on driving as initiated by the GDE matrix it is increasingly advocated to assess higher order aspects of driving: risk tolerance, reflection on one's own driving behaviour and hazard perception (Vissers, 2004).

In 2007, the Dutch national institute for educational measurement (Cito) formed a consortium with organisations active within the field of driver training. The aim of this consortium is to develop assessment instruments which monitor driving competence for educational purposes throughout peoples' traffic career.

As a first step a general model for driving competence was developed, which could serve as a basis for the construction of future assessments. The general model was developed within the context of (student) teacher evaluation (Roelofs and Sanders, 2007), and adapted to the field of driving, by fitting in psychological notions of driving (Roelofs et al., 2008). By applying the model, different measures of driving performance were described and classified. The full description of the development of model and accompanying view on driving assessment is to be published in the near future.

In essence, competent driving performance is seen as contributing by means of decisions and driving actions to valued effects such as aiding own safety and those of others, aiding traffic flow, preventing harmful emissions and noise pollution, efficient use of fuel, comfort, and prevention of premature wear. The effects can be considered as measures of quality of the driving performance. These effects are brought about by applying a repertoire of (strategic, tactical and vehicle manoeuvring) actions which are to be applied within specific traffic situations. The actions are carried out based on a conscious or automated decision-making process, in which the driving task at hand is appraised. This process involves perceptual processes, the prediction of future events and the decision to act in a certain way. The process of decision-making and acting is based on a state of knowledge, skills and attitude as present in the mind of the driver, and is constantly shaped by changing traffic contexts. All components are summarised in Figure 4.1.

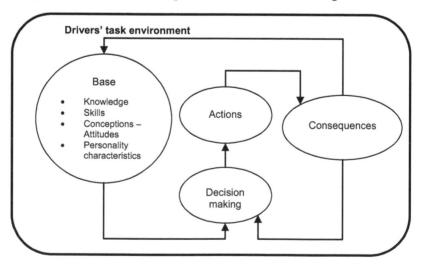

Figure 4.1   An eclectic model for the assessment of driving competence (based on Roelofs and Sanders, 2007)

This chapter describes the development and application of an on-road driver performance assessment (DPA), reflecting the competence-based view of driving.

By applying the general (heuristic) model of driving competence, a domain of driving competence, accompanying performance criteria were elaborated. In addition an assessment and scoring procedure for driving instructors and driver examiners was constructed. The assessment procedure was tested within the context of initial driver training. Because it is meant to inform drivers about their progress towards initial driver proficiency, the DPA is regarded as a monitoring instrument, to be used on multiple occasions throughout the training period. To evaluate the adequacy and appropriateness of inferences based on DPA scores, an initial validation study was set up.

Although the entire validation study addressed a wider range of aspects of validity (Kane, 2006) this chapter is restricted to the following research questions:

1. How reliable are judgments of learner drivers' driving performance based on the DPA?
2. To what extent do DPA scores yield correct predictions about the final driving exam result?

Method

A framework for driver performance assessment

Based on a framework for competent driving a scoring procedure was developed, in which driving tasks are judged against strict criteria (Cito, 2008). Five performance criteria were used, which were derived from desirable effects of driving mentioned above:

1. *Safe driving* refers to one's ability to drive in such a way that the risk of injury or damage to the driver or other road users is kept to a bare minimum. The driver is expected to constantly drive at an appropriate speed: adapted to the circumstances and to the tasks at hand, and to his/her own ability. The driver maintains sufficient safety margins in relation to other road users, correctly assesses risk, recognises danger in time and then chooses to act accordingly.
2. *Consideration for other road users* means that the driver pays attention to his/her own safety and progress of other road users. This means that the driver does not stick dogmatically to his/her own rights and is able to disregard other people's mistakes. The driver avoids surprising others with his/her actions by making his/her intentions clear well in advance. The driver gives others space to correct their mistakes and considers the

position of other road users and judges whether his/her own behaviour causes irritation or nervousness to others.

3. *Facilitating traffic flow* implies the ability to drive in traffic without impeding the progress of other road users. The driver's actions are not only safe but also vigorous and smooth. A driver chooses where to stop or where to turn in a way that causes the least inconvenience to other traffic. The same applies to performing manoeuvres, such as reversing and parking. These tasks are carried out in such a way that they do not hinder other road users.

4. *Environmentally responsible driving* involves driving in such a way that emission of harmful gases and noise levels are kept to a minimum and that optimal use of fuel is achieved. Such driving involves avoiding sudden changes in speed: calm accelerating and decelerating, changing up to a higher gear in good time and making use of the car's rolling momentum, gentle use of the accelerator and clutch when carrying out manoeuvres. The engine should be switched off during long stops.

5. *Controlled driving* means smooth technical vehicle handling and control. This means operating navigational systems skilfully, carrying out actions smoothly: is the car being steered and controlled smoothly, is the driver turning, stopping and driving off without stutters and jerks? On an operational level this means flexible and skilful execution of manoeuvres: steering, accelerating, using the clutch and braking are all done automatically and without fault.

A scoring procedure

Experienced driving instructors or examiners are responsible for the performance assessment. As part of a practical training session, the learner driver drives along a route on public roads to enable a judgement of their driving performance. Part of the session consists of driving without intervention from the driver instructor, whose task is to observe the learner driver's driving skills. To collect evidence of driving competence the driving instructor instructs the participant to drive along a representative route through five different areas, requiring a full range of typical driving actions: residential access roads (1) inside and (2) outside built-up areas, roads connecting towns (3) inside and (4) outside built-up areas, and (5) highways.

In order to make a systematic and comparable judgement of driving proficiency, scoring forms were developed in which the various driving tasks were judged against the five criteria described above. Two versions were used. First, a fine-grained version in which each performance criterion is elaborated further into specific performance indicators and in which 13 different specific driving tasks are discerned. These tasks can be categorised under five main tasks: preparing for driving (e.g. prepare vehicle and driver), making progress (e.g. driving off and stopping), crossing intersections, moving laterally (e.g. changing lanes), carrying out special manoeuvres (e.g. turning). The resulting scoring form consisted of 126 cells to be scored, after a combination of 13 driving tasks and 12 performance criteria, and excluding 30 cells which do not yield unique information about environmentally responsible driving. The assessor is expected to score each cell on a rating scale ranging from 1 (very unsatisfactory) to 4 (optimal).

The second version, the 'global task area version', was more general in nature. In this version, driving through a specific type of area is considered as a task. The task performances were scored directly according to the five performance criteria. The resulting scoring form consisted of 30 cells to be scored, after a combination of six driving tasks (five areas plus 'carrying out special manoeuvres') and five performance criteria. The assessor assigned scores for each cell on a five point rating scale ranging from 4 (unsatisfactory) up to 8 (optimal). The reason for using this scoring range is that the assessors' work within an educational context, described below, where scoring levels 1 up to 3, refers to preliminary driving proficiency levels. The lowest scoring level was set at level 4, which is fully equivalent with score level 1, as used in the fine-grained version.

The driving instructors acting in the role of assessors were trained to carry out the performance assessments in a series of three three-hour workshops. A detailed scoring manual was developed to support the scoring procedure. Inter-rater reliability was tested by using a set of 12 video-clips showing critical parts of the task performance of four drivers, to be scored individually by each of the assessors. Inter-rater reliability coefficients were calculated to indicate assessors' mastery of the assessment procedure. For both versions of the form the inter rater reliability reached an acceptable level (mean Gower coefficients for similarity above 0.70).

Participants

The performance assessment procedure was carried out within two separate pilot studies. The first pilot took place within a Dutch driving school, delivering short and condensed training programmes which culminated, approximately 15 days after the start, with a final exam consisting of a 55-minute practical driving test. A total of 11 driving instructors participated. Their instructional experience was on average 15 years (SD = 7.5). Their mean age was 43.2 (SD = 8.1). The instructors used the DPA within a two-month period during their regular training programme which was administered to each learner driver on two occasions. The first occasion was a driving assessment administered during the training programme, which yielded a partial dispensation on the final exam (the so-called 'learner interim test'). On this occasion a professional driver examiner from the Dutch National Driving Examination Institute (CBR) administered an official driving assessment lasting approximately 55 minutes. In the meantime the driver instructor independently judged the performance of the learner driver using the DPA. The second occasion on which the DPA was used by the driver instructor was during the final exam, again administered by a CBR examiner. In sum, 41 female (mean age 20.6 years) and 50 male learner drivers (mean age 19.8 years) participated in the first study. The sample was representative in terms of their prior school education.

The second pilot involved 26 Dutch driving instructors who work according to the method of Driver Training Stepwise (DTS) (Nägele and Vissers, 2003). DTS is a modular driver training programme which differs from the traditional (driving school-based) training in two fundamental respects: (1) the pupil learns how to drive in a series of highly structured steps (driving scripts) and (2) the pupil is only allowed to enter the next stage of learning if they show complete mastery of the previous stage. At the end of each learning stage there is a test to assess whether the pupil has obtained the required level. In DTS-programmes learner drivers are taught an average of 38 practical driving lessons of 60 minutes each.

The driving instructors administered the DPA to each learner driver after the first two learning stages or modules of DTS had been completed (score 1). In addition to this the DPA was also used by the driver examiner at the learner interim test (score 2) and at the official driving test (score 3). In the second pilot 35 female learner drivers (mean age 20.3) and 26 male learner drivers (mean age 19.6) participated. The sample was representative in terms of their prior school education.

Analysis

Overall scores on the DPA formed the basis for further analyses. To arrive at an overall score for driving performance on the fine-grained version the individual cell scores were aggregated across the tasks towards indicator scores. Five criteria scores and an overall score was then calculated. For the global task area version, the first step in the aggregation procedure was the aggregation towards criteria scores, which in turn were aggregated towards overall scores.

In order to estimate a pass/fail prediction boundary for the DPA score, logistic analyses were carried out. In this analysis the odds ratio between passing and failing on the final examinations is predicted by the DPA score.

$$\ln \frac{p_{pass}}{p_{fail}} = a + bx$$

where:
p_{pass} = the chance to pass
p_{fail} = the chance to fail
a = a constant
b = a regression coefficient belonging to x, referring to the DPA score.

The cut-off score for the DPA is determined by choosing a cut-off point where the probability to pass rises rapidly. In the present analyses the cut-off was set at $P = 0.50$. The cut-off score for the DPA can be determined in such a way that the number of misclassifications is minimised. Misclassified learner drivers can either be those who fail the exam with a 'pass' prediction, or those who pass the exam with a 'fail' prediction. Using a relatively high cut-off score for the prediction based on the DPA, such as one corresponding to an 80 per cent pass probability, leaves less failed candidates with a pass prediction. However, more learner drivers will find themselves passed with a fail prediction.

Results

Reliability and discriminative power

To estimate the test–retest reliability of the DPA measurements the scores on the two last measurement moments were correlated. In pilot 1 the correlation between the first and last moment was 0.80. In the second pilot the correlation between the second and the third measurement score was 0.70. Note that due to differential growth of driving proficiency during the training period, this correlation is an underestimation of the test–retest reliability.

Comparisons of mean DPA scores for learner drivers who passed and failed on the final exam show large and significant differences compared to the advantage of those who passed. Most of the effect sizes (Cohen's d) are above 1, depicting large effects (see Table 4.1). In pilot 2 the differences between failing and passing learner drivers amount to a 1.5 standard deviation on the third DPA assessment, which is closest to the exam. In the first pilot, DPA score 1 is assigned a few days before the final exam. Its discriminative power is also large ($ES = -1.07$).

Table 4.1   Mean and standard deviations on the consecutive assessments for learner drivers who failed and passed the final exam

	Failed			Passed					
	N	Mean	SD	N	Mean	SD	t	df	ES
Pilot 1									
DPA score 1	23	2.45	0.42	23	2.85	0.53	−2.85*	44	−1.07
DPA score during final exam	45	2.56	0.50	47	3.13	0.56	−5.05†	90	−0.84
Pilot 2									
DPA score 1	6	5.49	0.42	50	6.13	0.48	−3.13†	54	−1.20
DPA score 2	7	5.63	0.34	52	6.15	0.47	−2.83*	57	−0.98
DPA score 3	7	5.82	0.28	52	6.57	0.45	−6.12†	57	−1.55

* p<0.05; † p<0.1.

Predictions of results on the final exams

Figures 4.2 up to 4.4 show scatter plots in which the DPA score is related to the predicted probability to pass, using the results of logistic regression analyses. Note that in the first pilot and second pilot different scales were used for the DPA. Cut-off scores corresponding a pass probability $P = 0.50$ were calculated as a basis for a fail-pass dichotomisation prediction.

Above the cut-off probability to pass (50 per cent representing a specific DPA score) it is assumed that candidates receive the prediction that they will pass the final exam. Below the cut-off point, they will be expected to fail the exam. The actual results of each candidate are represented by a circle-shaped symbol for failed candidates and a square-shaped symbol for passed candidates.

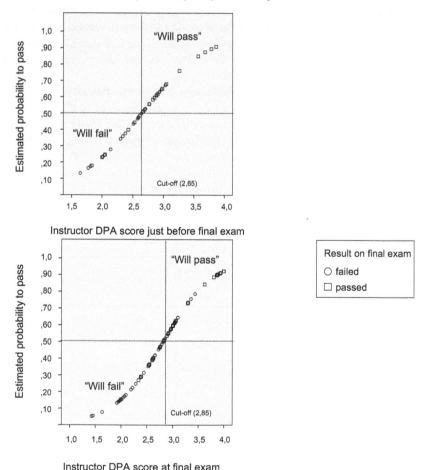

Figure 4.2   Probabilities to pass the final exam based on the drivers' instructors' assigned DPA score (first pilot)

Figure 4.2 shows that during the first pilot, the compact driver training, driving instructors assigned a wide range of DPA scores to their learner drivers, ranging from 1.4 up to 3.9. In the second pilot, in which the driver training stepwise method was employed, the driver examiners stayed within a relatively narrow range of scores (see Figures 4.3 and 4.4). In all cases only a few learner drivers received a score low on the DPA scale. It seems that the learner drivers in the second pilot as a group performed at higher levels compared to the learner drivers in the first pilot.

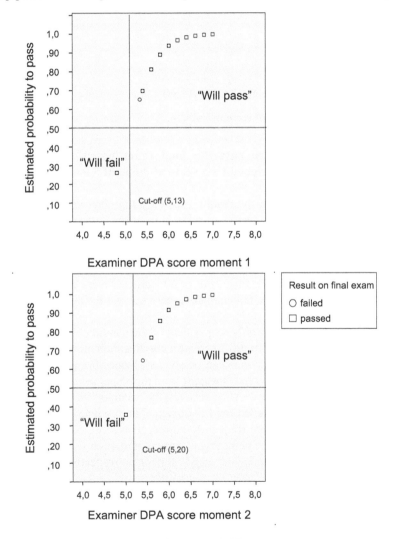

Figure 4.3   **Probabilities to pass the final exam based on the drivers' examiners' assigned DPA score at moment 1 and 2 (second pilot)**

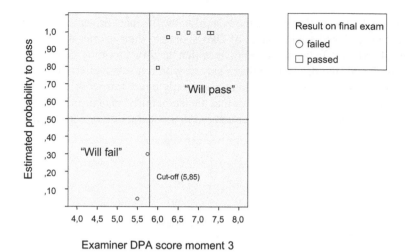

Figure 4.4   Probabilities to pass the final exam based on the drivers' examiners' assigned DPA score at moment 3 (second pilot)

This result is reflected in the results regarding the correctness of pass–fail predictions. Starting with the results of pilot 2, from the data it appears that the percentage of correct predictions (91, 90, 88 per cent respectively) is higher than was the case for pilot 1 (65 and 64 per cent respectively). Most of this result is attributable to the large number of actual passing learner drivers in pilot 2, leaving not many candidates to misclassify: 7 out of 60 learner drivers failed on the final exam. The correctness of their 'fail' prediction varies across the three measurement moments.

On the contrary, in pilot 1 only 43 per cent of the candidates passed the final exam. This gives a good opportunity to create a classification table from which the correctness of pass–fail predictions can be studied. Inspecting this classification table (Table 4.2) it can be observed that apart from the 65 per cent overall accuracy of predictions there is some difference between the accuracy for those candidates who passed and those who failed the final exam. The predicted results of those that actually passed were more frequently correct than those of candidates who actually failed. For Time 1, the percentages correct prediction were 69.6 and 60.9 per cent respectively, for Time 2, 70.2 and 57.8 per cent respectively.

Discussion

For educational purposes a formative driver performance assessment (DPA) was developed to support training decisions and to enable predictions about pass probabilities on the final driving exam. To evaluate the adequacy and appropriateness of inferences based on DPA scores, a first validation study was set up. In the

Table 4.2   Classification table for DPA predictions (pilot 2) against the result of the final exam

		Predicted result		Percentage correct
Actual result on final exam		**Will fail**	**Will pass**	
	Failed	14	9	60.9
Moment 1	Passed	7	16	69.6
	Total			65.2
	Failed	26	19	57.8
Moment 2	Passed	14	33	70.2
	Total			64.1

present study only part of a full validity argument could be evaluated. Building on Kane (2006) a validity argument can be set up in which the chain of inferences when interpreting the outcomes of a performance assessment is evaluated. More specifically, three inferences form the core of the validity argument: (1) reliable scoring of performance by assessors, (2) generalisation from the observed score on a specific assessment task to a full range of assessment tasks, (3) extrapolation of assessment results to practice. In the subsequent discussion these inferences will be addressed.

A first condition for performance assessments to yield valid score interpretations is score reliability (research question 1). Results of inter-rater reliability analyses during assessor training indicated satisfying levels of rater agreement.

A limitation is that no direct inter-rater reliability data could be collected for a large number of assessors doing their job within a real driving situation. Some instructors in the assessor role indicated that using video episodes of driving limited their ability to observe all aspects of driving, for instance in determining the adequacy of speed choice, and vehicle control.

Another issue pertains to the nature of the performance scoring. Different versions of the DPA were employed, involving rating scales of different lengths and a different degree of decomposition into subtasks. The fine-grained version involves a highly analytical judgement, whereas the global task area form requires a more holistic judgment (Clauser, 2000). The question can be raised whether both versions are equally decisive and informative about the state of the drivers' driving proficiency.

The second inference in the validity argument was an unaddressed issue, the generalisability of performance scores. We did not estimate the effects of differences between the assessors as a potential source of measurement error, nor did we estimate the effects of differences in the driving tasks that learner drivers were exposed to. The question is whether the tasks which learner drivers carried out were representative for the full range of possible tasks. It is a well-known finding that task variation accounts for much of the variance in performance (Brennan, 2000).

The third inference, extrapolation, addressed the question whether the assigned DPA scores are predictive of success on the final driving exam (research question 2). The results show meaningful differences in DPA scores between passed and failed learner drivers. High effect sizes were found, indicating high differential capacity. This relationship is probably underestimated, considering that the examiners' judgements on the final exam will not be perfectly reliable.

Another way to look at the predictive validity of the DPA is to consider misclassifications regarding the 'fail' and 'pass' predictions on the (future) final exam, based on the DPA scores. The results show different pictures for the two applications of the DPA. In the first pilot a fine-grained version of the DPA was used in the context of condensed driver training. Results show that 65 per cent of the learner drivers were correctly classified. In the second pilot, the context of Driver Training Stepwise (DTS) a more holistic version of the DPA was employed. Using DPA scores 90 per cent of the learner drivers could be classified correctly in terms of the predicted outcome of the final exam. The higher percentage of correct decisions compared to the first DPA pilot is mainly attributable to high pass rate and consequently the high correctness of pass predictions. This finding could be explained by the nature of the DTS training programme. Within this programme the driver instructor collects a great deal of information about the learner driver regarding the progress on script mastery, by means of specially designed progress cards. They do so over a relatively long period which lasts nearly five months on average. The instructor will only send their learner driver to the final exam once they have mastered all driving scripts and passed the formative assessments. These drivers will receive DPA scores above the cut-off which instructors (implicitly) use in their judgements.

Looking back at the extrapolation inference, the quality of the external validation criterion used can be questioned. For purposes of initial driver training the pass–fail prediction is a sufficient criterion. For the instructor and the learner the growth towards exam preparedness is an important issue.

However, DPA was also meant to inform drivers about their driving proficiency beyond the final exam, in the context of lifelong learning. This would require an external criterion that relates to actual driving outcomes during the subsequent driving career, such as involvement in collisions and damage claims for one's own and other vehicles. For now, it seems at least that the DPA is an appropriate tool to inform learner drivers about their progress towards obtaining their driving licence.

Replication of this study on a larger sample of learner drivers at various stages during the training programme enables the estimation of growth curves in terms of the increasing probabilities to pass an exam. Figure 4.5 shows how typical growth curves could look, given cut-off scores at various times for different pass probabilities. Individual learner drivers can be plotted, as is done for the

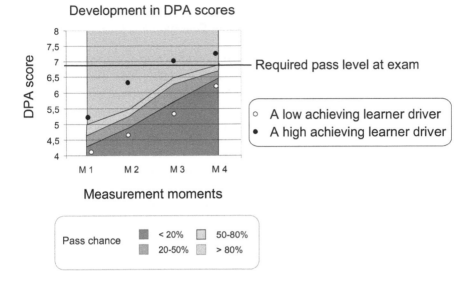

Development in DPA scores

Figure 4.5 **Individual growth of learner drivers in terms of pass probability**

(hypothetical) cases of a learner driver who is learning relatively quickly and relatively slowly.

If driving instructors are well trained as assessors, the DPA can be used as a monitoring instrument enabling instructors to adapt their instruction to the growth of the individual learner driver. Future research will have to show whether the use of this assessment use can be validated in real-world settings.

References

Brennan, R.L. (2000). Performance assessments from the perspective of generalizability theory. *Applied Psychological Measurement*, 24(4), 339–353.

Cito (2008). *Driver Performance Assessment. Scoring Manual*. Arnhem, Cito.

Clauser, B.E. (2000). Recurrent issues and recent advances in scoring performance assessments. *Applied Psychological Measurement*, 24, 310–323.

Dierick, S. and Dochy, F. (2001). New lines in edumetrics: new forms of assessment lead to new assessment criteria. *Studies in Educational Evaluation*, 27(4), 307–329.

Gregersen, N.P. and Bjurulf, P. (1996). Young novice drivers: towards a model of their accident involvement. *Accident Analysis and Prevention*, 28, 229–241.

Hatakka, M. et al. (2002). From control of the vehicle to personal self-control; broadening the perspectives to driver education. *Transportation Research Part F*, 5, 201–215.

Kane, M.T. (2006). Validation. In R.L. Brennen, ed., *Education Measurement*, 4th edn. Westport, CT, American Council on Education and Praeger Publishers, pp. 17–64.

Nägele, R.C and Vissers, J.A.M.M. (2003). *Rijopleiding in Stappen (RIS). Evaluatie van de vervolgproef in de provincie Gelderland 2002–2003. Driver Training Stepwise (DTS). [Evaluation of the follow-up in the province of Gelderland].* Veenendaal, Traffic Test.

Roelofs, E.C. et al. (2008). Development of multimedia tests for responsive driving. In L. Dorn, ed., *Driver Behaviour and Training*, Volume III. Farnham, Ashgate Publishing Limited, pp. 251–264.

Roelofs, E. and Sanders, P. (2007). Towards a framework for assessing teacher competence. *European Journal for Vocational Training*, 40(1), 123–139.

Vissers, J.A.M.M. (2004). Testing and teaching of the higher order skills of the GDE-matrix. CIECA Congress, Warsaw.

Vissers, J.A.M.M. et al. (2007). *Learning Goals for Lifelong Road Safety Education.* Amersfoort: DHV.

Chapter 5

How Can Reflecting Teams Contribute to Enhanced Driving Teacher Learning?

Hilde Kjelsrud

Introduction

Working to improve road safety in Norway involves working with all elements in the transportation system; the vehicle, roads, environment and drivers. Our focus is on the driver and how driving teachers should best be educated to deal with the human factor in driving. At Nord-Trøndelag University College in Norway, the education of driving teachers was extended from one to two years and increased from an upper secondary level to two years at University College level in 2004. The aim of extending this education was to develop wider range of competences including professional educational, social, professional ethics and change and development competence. This chapter proposes that critical self-evaluation and good self-awareness is also a key competence that driving teachers need.

This chapter reports on testing a method to help driving teacher students to be more aware of how they can use reflection in teaching drivers. The rationale for the method is that it may assist learner drivers to be more aware of their own limitations and possibilities. The main question was how can a driving teacher teach a learner driver to reflect if they are uncertain about how to do it themselves? Therefore, our focus has been on reflection concerning how and what students are thinking, how they are feeling and their actions. This chapter reports on a method of helping driving teachers to improve their ability to reflect on their own teaching skills by sharing knowledge and experiences, and to work on self-awareness in a process called 'reflecting teams'.

Usually in the practical in-car part of educating driving teachers, there is only one real learner driver and one student (the driving teacher), one fellow student in the back seat and one teacher for guidance. Students are not allowed out on a driving lesson alone without a fellow student in the back seat for safety reasons. The back seat student and the student driving teacher evaluate the lesson and use the reflecting team method, which is defined as a group of persons communicating with each other about something they have seen or experienced in action during the lesson. The one or those who have been involved in the action are present in the room but they are not allowed to participate in the communication, only listen to the team.

Theoretical Framework

Our main focus is on knowledge and learning, different learning models, reflection, reflecting teams, care, mutual trust, active empathy, access to help, go-ahead spirit and no condemnation of approach. We think these are important criteria to make reflecting teams to work effectively.

Knowledge and experience are important ways in which we change the way we look at things. Definition of knowledge is mostly based on Polanyi's work (1962, in Newell et al., 2002), where Plato's original definition: 'Justified true belief' is also referred to. Anette Baches' definition (Nordhaug, 2006, p. 249) is that knowledge is a mixture of experiences, values, context information and expert insight, which makes it possible to assess and incorporate new experiences and information.

In the academic literature knowledge is often divided into tacit and explicit knowledge (Newell et al., 2002). Tacit knowledge resides within the individual, and is not easy to articulate or communicate. This is often referred to as 'know-how'. It is knowledge in our heads, in our practical skills and in our actions. Explicit knowledge is easier to grasp, because it can be codified and communicated to others. In our reflecting teams we wanted to focus on both of these aspects of knowledge, but it seems easier to reflect on explicit knowledge than tacit knowledge – it is safer and possibly easier to comprehend. Some of our aims are to make more of the tacit knowledge explicit.

Gottschalk (2004) explains the difference between data, information, knowledge and wisdom. Information is interpreted data and when this information combines with experience, context, understanding and reflection it transforms into knowledge. Knowledge starts action or no action; this is what Gottschalk calls wisdom. We also want to look at the difference between different strategies of action built upon exposed theory and theory in use (Argyris et al., 1985, in Rennemo 2006). This is important for the reflecting team because it shows the possible difference between what the driving teacher says they plan on doing and what they actually do in the driving lesson.

As a student you can accept new knowledge in different ways. Two ways that explain the acquisition of knowledge are: 'single loop' learning by Argyris and Schøn and Jean Piaget's 'assimillasjon' (Rennemo, 2006). These two ways of handling new knowledge do not change our knowledge, they simply confirm and add knowledge to old existing knowledge, perhaps to change our basic assumptions. This was called 'double-loop learning' by Argyris and Schøn or 'accommodation' by Piaget (Rennemo, 2006). In this way new knowledge changes our existing knowledge.

In focusing on the learning process we also want to bring in Kolb's process of learning, which illustrates various dynamics in the learning process (Rennemo, 2006).

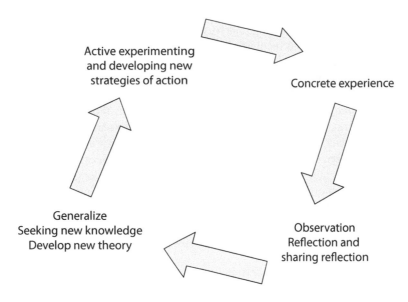

Active experimenting and developing new strategies of action

Concrete experience

Generalize Seeking new knowledge Develop new theory

Observation Reflection and sharing reflection

Figure 5.1   Kolb's model of learning (Rennemo, 2006)

This model provides an overview of the processes involved as part of a reflecting student in the process of learning. According to Coghlan and Brannick (2006, p. 35) *'Reflection is the critical link between the concrete experience, the interpretation and taking new action.'*

Another way of reflection has been represented by Tiller (1999) in 'the stairs of learning':

Experiences connected to theory | Connecting experiences | Getting experiences in order | Idle talk about experiences

The Stairs of Learning

According to Tiller there are various steps the student must take to learn from experience: if the student stops at the first, second or third step, they will not connect theory to experiences. What about a fifth step containing development of new theory and/or change of practice?

Reflection means throwing something back or to mirror something. We can divide reflection into in-action and on-action. Reflection in-action is defined as what you do during action, and reflection on-action is defined as what you do after action. To reflect during or after action the student needs time, space and motivation. We also need a 'toolbox' to help us reflect. Possible 'tools' are a reflection book or a reflecting team (Rennemo, 2006). In our project we were using reflecting notes after each session for the driving teacher teachers, the driving teacher student and reflecting team (students learning to drive). According to Rennemo (2006) the aim of the reflecting team is to feedback information to someone, and for the present study it is to the driving teacher student. The reflecting team is different because it gives the student time to listen and think about how others have experienced their action without spending time defending their actions.

We will split reflection into three different elements: thoughts, actions and feelings (Sewerin, 1996). Reflection often concerns action and thoughts, but we also want to focus on feelings and emotions. A safe and caring environment is essential. It is often safer to reflect on action and thoughts, because telling others about our feelings may make us vulnerable. Developing knowledge requires good relations within an organisation (Krogh et al., 2005) and a caring environment is important for the present study. Students needed to know their fellow students well as we worked on introducing the new method into the group, focusing on mutual trust, active empathy, access to help, go-ahead spirit and no condemnation (Von Krogh and Ichijo, 2005).

Trust is about handling uncertainty and accepting vulnerability (Newell et al., 2002). A high degree of trust is necessary to get to a level of communication that makes it possible to share tacit knowledge. But trust is not easy to gain, and there are different sources of trust and processes of establishing trust. Different reasons for developing trust are: a contractual agreement that binds the parties in the relationship, belief in competence among participants and a belief in the 'goodwill' of those involved (Sako, 1992, in Newell et al., 2002). Taking students from different parts of Norway together in small groups requires work to develop real trust. There are different types of trust, but some are fragile – they are easily built and may be easily broken down. Others are resilient, take time to build and are not that easy to break down.

Another important issue to consider is empathy within the group. What is empathy? Again there is an ambiguous conception, but it is about being able to enter into somebody's feelings and ideas in situations they are exposed to, in other words to understand the situation through the other person. In Krogh et al. (2005) we find active empathy as one of five dimensions of care and it has been used proactively to understand the other person. Active empathy is to try to understand someone else through observation and communication. Communication as an 'active listener' is essential. The main focus is on the other person, not on the person doing the reflecting. Using a reflecting team method may help both the reflecting team members and the driving teacher student towards being aware of their own empathetic characteristics and help develop this personal quality.

Methodology

Students work in groups consisting of four and five students during the two-year driving teacher programme at Nord-Trøndelag University College. We asked two of these small groups to take part in the study as they were already established groups that had been worked together on their studies for a year, they trusted each other, and felt safe accompanying each other in driving lessons. Therefore 9 out of 84 possible participants voluntarily took part in the project. Participants were halfway towards being a qualified driving teacher. There were seven male students and two female students.

Our method used live driving lessons with driving teacher students and learner drivers involving nine learning to drive students in seven driving lessons, including a pilot driving lesson at the beginning. Two of the students were the reflecting team, asking the driving teacher student in advance about their plans and reflecting together during the driving lesson. After the lesson, the reflecting team discussed the lesson while the driving teacher student only listened to their reflections. Our presumption is that the reflecting team approach can be an additional pedagogical help for students to improve their own self-awareness, and also develop better self-awareness within their students.

For the pilot, the six driving lessons were structured as shown in Figure 5.2.

Time schedule	What	Who	Where
15 minutes	Pre-guidance before driving lesson	Reflecting team (2 students) and driving teacher (1 student)	Classroom
50 minutes	Driving lesson, 45 minutes, in the middle of the lesson the reflecting team gets 5 minutes outside the car to talk	Learner driver and driving teacher stay in the car. Reflecting team talks about if there is anything they need to clarify between themselves	In car, 7 seats
20 minutes	Reflection talk after driving lesson	Between reflecting team. Driving teacher is not allowed to talk, only listen	Classroom
5 minutes	Driving teacher is allowed to briefly tell about their driving lesson and how they felt about the situation	Driving teacher in focus, but reflecting team is allowed to answer and ask questions	Classroom
15 minutes	Writing a reflection note (some wrote a page, some wrote a couple of sentences)	All three students: driving teacher and reflecting team	Classroom (no master students in the room)

Figure 5.2   Reflecting team lessons

During each session two master students were observing and making notes. After each session, a reflection note was written for ourselves, since we chose not to do a video recording.

Results and Conclusions

The results from our research show that using reflecting teams is an additional method which helps students in their learning process. The results suggest that they improve in their assessment of others and in their self-assessment. This is an important quality in being a driving teacher, since our students teach learner drivers to assess themselves, according to the GDE matrix.

In the reflection note we asked both the reflecting team and the driving teacher two questions:

1. What was the difference between using reflecting team and an ordinary lesson? (In an ordinary lesson the driving teacher student gets direct feedback/guidance from a teacher/fellow student)
2. What did you learn during the process?

In the first question they responded that they needed more time, that they felt obliged to do well themselves, and had to focus a lot more on the task they had agreed on during the pre-guidance. Participants also felt there was a need for pre-guidance. They reported that they took more notes during the driving lesson. The reflecting team led to different kinds of reflection after the lesson; they had deeper thoughts, more thorough understanding and greater awareness of their technical skills.

For the second question, almost every student reported that they learned to have more focus, an increased awareness about their own part in helping others to get better, reflections concerning how things we do and say affect others, self-awareness when it comes to their own work as a teacher and that thoroughness is important when educating drivers. Almost all students commented on the importance of trust. They reported that using a reflecting team would be more difficult and more frightening if they did not know each other so well.

Conclusions

Based on this study, we consider that the reflecting team method could be additional to traditional instruction and coaching and will be implemented in the theoretical part of the education of driving teachers in Norway. However, our research shows that trust must be established in the group for the method to work and the driving teacher student should give the reflecting team a written plan for the lesson, before or during the pre-guidance.

References

Coghlan, D. and Brannick, T. (2006). *Doing Action Research in your Own Organization.* London, Sage.

Gottschalk, P. (2004). *Informasjonsteknologi i kunnskapsledelse.* Oslo: Universitetsforlaget.

Newell, S. et al. (2002). *Managing Knowledge Work.* Basingstoke, Palgrave Macmillan.

Nordhaug, O. (2006). *Kunnskapsledelse, trender og utfordringer.* Oslo, Universitetsforlaget.

Nord-Trøndelag University College (2007). *Faculty of Education for Driving Instructors, Course Curriculum; Two-year Basic Training for Driving Instructors.*

Rennemo, Ø. (2006). *Lever og lær.* Oslo, Universitetsforlaget.

Sewerin, T. (1996). *En plass i stolen, en arbeidsbok for grupper om grupper.* Dalby, MiL Publishers.

Tiller, T. (1999). *Aksjonslæring. Forskende partnerskap i skolen.* Kristiansand, Høyskoleforlaget.

Von Krogh, G. and Ichijo, K.I. (2005). *slik skapes kunnskap.* Oslo, N.W. Damm og Sønn.

PART 2
Driver Personality and Driver Offending

Chapter 6

Understanding the Unique Contribution of Aversion to Risk-Taking in Predicting Drivers' Self-Reported Speeding

M. Anthony Machin and Janna E. Plint

Introduction

There is considerable interest in the role of risk perceptions in determining driving behaviour. Speeding is an example of a risky driving behaviour that has been studied by many researchers (Aarts and van Schagen, 2006; Jonah, 1997; Lam, 2003). Risk perceptions in relation to driving are 'the subjective experience of risk in potential traffic hazards' (Deery, 1999, p. 226) and have been identified as one of the strongest predictors of speeding (Machin and Sankey, 2008). Models of the predictors of driving behaviour have included dispositional characteristics and coping strategies reflecting the different factors that combine to influence the appraisal of risk when driving. In particular, the transactional model of driver stress and coping developed by Matthews (2001) indicates that risk perceptions are probably a function of the driver's appraisal of their environmental demands and their choice of coping strategies, both of which are influenced by the driver's personality characteristics. Therefore, the unique role of risk perceptions in predicting risky driving behaviour depends on the type of model that is proposed, which could contribute to differing conclusions about the importance of risk perceptions. This study focuses on the unique contribution of one measure of risk perceptions (aversion to risk-taking) in the prediction of speeding whilst controlling for a range of other predictors of speeding. It also examines whether this outcome will change depending on the age, gender and the frequency of driving.

Assessing Drivers' Risk Perceptions

There have been a number of approaches to assessing drivers' risk perceptions, mainly reflecting a cognitive-based assessment process. However, when measuring perceived risk, Rundmo and Iversen (2004) considered it was important to distinguish between cognitively-based and affective-based subjective assessments. Rundmo and Iverson discovered that drivers' probability judgements pertaining to negative outcomes and level of concern about traffic risks were not related to risky driving behaviour (including speeding). However, being worried about negative outcomes, feeling unsafe and other emotional reactions were predictors of risky driving behaviour, leading the authors to conclude that the affective component of risk perception is more important than the cognitive component when predicting risky driving behaviour.

One approach to assessing risk perceptions involves assessing how dangerous various activities are perceived to be. Based on a scale developed by Dalziel and Job (1997), Machin and Sankey (2008) compared the predictive strength of aversion to risk-taking with three other risk perception variables and five measures of personality. The combined worry and concern items used by Rundmo and Iversen (2004) were used to measure the affective aspect of risk perception but did not contribute to the prediction of speeding. Likelihood of an accident, driving efficacy and aversion to risk-taking were significant unique predictors of speeding accounting for 6, 3 and 15 per cent of the variance respectively. Two personality variables were also significant predictors of speeding, with excitement-seeking and altruism accounting for an additional 2 and 3 per cent of the variance respectively. Further analysis using structural equation modelling demonstrated that the impact of two personality variables was equal to the participant's aversion to risk-taking in influencing speeding behaviour, given that the effects of the personality variables on speeding was partially mediated by aversion to risk-taking.

One of the difficulties in drawing conclusions from the previous study relates to the variables that were *not* included in the model that was being tested in that study. While there is clearly a strong relationship between aversion to risk-taking and speeding, we recognise that drivers' risk perceptions may also be related to their choice of coping strategies. A second issue concerns the possible moderating effects of demographic characteristics such as age, gender and frequency of driving. These two issues are discussed and then the proposed analyses for the current study outlined.

Impact of Drivers' Coping Strategies

Matthew's (2001) transactional model proposed that drivers' appraisal of their environment and their assessment of their capacity to cope influences their perceptions of risk and subsequent coping strategies. Some drivers will adopt more maladaptive coping mechanisms, which may contribute to greater speeding.

Matthews and colleagues (1996) identified five coping styles applicable to driving: confrontive coping, task-focused coping, emotion-focused coping, reappraisal and avoidance. Confrontive coping strategies involve antagonising other drivers or risk-taking and are therefore potentially dangerous. Task-focused strategies are safety-enhancing because they involve coping efforts related to driving safely. Emotion-focused coping represents strategies of self-criticism and worry, which may cause cognitive interference and distract the driver. Avoidance may also be associated with reduced attention to task, whilst reappraisal may be more adaptive because it is associated with positive cognitions of the driving experience.

Matthews et al.'s (1996) research suggested that confrontive and emotion-focused coping were maladaptive coping styles associated with more negative outcomes. For example, they found that confrontive coping is linked to violations, errors and loss of safety. They also found that emotion-focused strategies such as self-criticism have the potential to distract the driver and confirmed that confrontive coping is correlated with greater speeding. Therefore, the conceptual model must include these coping strategies in order to understand the contribution of risk perceptions to risky driving behaviour.

Another group of influences on driving behaviour are drivers' personality characteristics which can contribute in two ways: as direct contributors to risky driving behaviour, or as indirect effects. Machin and Sankey (2008) included the same personality variables used by Ulleberg and Rundmo (2003), assessing anxiety, anger, excitement-seeking, altruism and normlessness. As described above, the impact of excitement-seeking and altruism on speeding was partially mediated by aversion to risk taking. Matthews et al. (1997) developed the Driver Stress Inventory (DSI) specifically to measure differences in drivers' personality. The DSI includes scales assessing aggression, dislike of driving, hazard monitoring, thrill seeking and fatigue proneness. They found that thrill seeking and aggression are associated with more risky driving, in particular, speeding. These results suggest the conceptual model should also incorporate measures of personality in addition to coping strategies and risk perceptions when predicting risky driving behaviour.

Other Factors Influencing Speeding

The conceptual model should also incorporate the demographic characteristics that are related to risky driving behaviour. There is considerable support for the link between being male and being younger with an increased level of risky driving behaviour. Yagil (1998) found that younger male drivers expressed lower motivation to obey traffic or road laws, compared with older and female drivers. Mast and colleagues (2008) linked 'masculinity' with increased speeding. In their study of 83 males, participants were randomly primed by actively listening to either feminine, masculine, or neutral words from a radio whilst driving a car simulator. Results from the study demonstrate that once the selected participants began listening to the masculine words, their speed dramatically increased from start to end of the driver simulator.

The frequency of driving might also influence drivers' risk perceptions and driving behaviour. In particular, it is likely that more frequent drivers will evaluate the demands of driving differently, may assess their capacity to cope differently and therefore develop different risk perceptions.

In order to determine whether these demographic characteristics impact on the conceptual model of the predictors of risky driving behaviour, a series of moderator analyses will be conducted in which the overall fit of the conceptual model is evaluated separately for males and females, for younger and older drivers and for more and less frequent drivers. This process can be conducted using a multiple group analysis within a structural equation model. However, in order to specify the structural equation model, we would need to refine the set of predictor variables as described in Machin and Sankey (2008). Therefore, for this chapter, the whole set of predictors will be used in a standard multiple regression analysis with all predictors entered simultaneously. Subsequent standard multiple regression analyses will be conducted for each of the subgroups so that the unique contribution of aversion to risk-taking in the prediction of speeding can be assessed. More elaborate analysis based on structural equation modelling will be conducted and reported in another chapter.

Method

Participants

The 402 participants, who completed the online (web-based) Driving Attitudes Survey, consisted of a sample of first- to third-year psychology students from the University of Southern Queensland (USQ). The data were collected between 2007 and 2008.

There was a high proportion of female respondents who completed the survey (80.3 per cent). Approximately 80.3 per cent of the total participants fell between the ages of 17 to 40. The remaining 19.7 per cent of participants fell between the ages of 41 and 75. In total, approximately 20.8 per cent of the participants were young drivers, aged between 17 and 19. Of the participants, 71.8 per cent held open drivers licences, and 20.8 per cent held provisional drivers licences. Of the remaining participants, 7 per cent held learner licences, whilst less than 1 per cent held disqualified licences. A high majority of the participants had held their respective licences for more than three years (62.8 per cent). The remaining participants had held their licences for less than three years (37.3 per cent), with an even spread across each six-month period in between. Most respondents drove often, with 73.8 per cent driving every day, and 15.8 per cent driving more than three times a week: 4.8 per cent of respondents drove once a week, whilst 5.8 per cent drove less than once a week.

Measures

An online survey questionnaire, titled the Driving Attitudes Survey, consisting of 126 items and 5 sections, was used for this study. The questionnaire was used to examine Australian driver's self-reported risk perceptions, personality characteristics and coping strategies as predictors of speeding.

Demographics

The first section of the survey consisted of eight items intended to collect basic demographic information. The first question was designed to determine whether the participant was a member of the Australian Drivers Training Association (ADTA) (e.g., Yes/no). If the participant responded 'yes' to the question, they were required to answer an additional question intended to collate years of membership. This was performed by the participant typing the appropriate number into the box provided. If the participant responded 'no', they were required to answer an additional question indicating whether they were a USQ student (e.g., Yes/no). Remaining items gathered basic demographic information including age, gender, type of licence held (e.g., learner, open, provisional or disqualified), how long the driver had held their licence (e.g., 0–6 months, 6–12 months, 1–2 years, 2–3 years, > 3 years), and how often they drove (e.g., every day, once a week, more than three times a week and less than once a week).

Driver coping scales

The Driver Coping Questionnaire (DCQ) (Matthews et al., 1997), consisting of 35 items and 5 scales, was used to examine participants' cognitive responses to driving when it is difficult, stressful, or upsetting. Each scale consisted of seven items designed to measure a particular coping strategy. The coping strategies measured included confrontive coping (e.g., relieving feelings by taking risks or driving fast), task-focused coping (e.g., avoiding reckless or impulsive actions), emotion-focused coping (e.g., wishing that one was a more confident and forceful driver), reappraisal (e.g., trying to gain something worthwhile from the drive) and avoidance coping (e.g., staying detached or distanced from the situation). All items were all positively scored, with a scaling factor used to give an overall score from 0–100. From a UK sample, Cronbach's alpha coefficients for the scales were found to fall within 0.72 to 0.84 (Matthews et al., 1997). As the levels fell below the recommended acceptability ($\alpha \geq 0.70$), the internal consistencies of the scales were deemed acceptable (Steiner, 2003).

The Driver Stress Inventory

The Driver Stress Inventory (DSI) (Matthews et al., 1997), consisting of 47 items and 5 scales, was used to assess participant's typical feelings experienced whilst driving. The scales measured the following characteristics: aggression, hazard monitoring, thrill seeking, dislike of driving and fatigue proneness.

Using an 11-point visual-analogue scale (VAS), participants were asked to respond by stating their agreement with each question, which ranged from 0 (*not at all*) to 10 (*very much*). Some of these items were reverse-scored to help prevent random responding, from 0 (*very much*) to 10 (*not at all*). Total scores were calculated using a scaling factor, which could theoretically range from 0–100. Cronbach's alpha coefficients for the scales were between 0.73 to 0.87 in a UK sample, and 0.73 to 0.85 in a US sample (Matthews et al., 1997). These scales were therefore deemed acceptable (Steiner, 2003).

Risk perceptions

The measures of risk perceptions included in this survey included an affective-based scale, and three cognitively-based scales (Machin and Sankey, 2008). The affective-based Worry and Concern scale included six items designed to measure the participant's perception of traffic injury and risks (e.g., to what extent are you feeling unsafe that you yourself could be injured in a traffic accident?). Scores on each item were summed to obtain a total score ranging from 6 to 30. The Cronbach's alpha coefficient was found to be 0.88 by Machin and Sankey (2008) which was acceptable.

The three cognition-based scales consisting of Likelihood of Accident, Efficacy, and Aversion to Risk-taking were also taken from Machin and Sankey (2008). The Likelihood of Accident scale consisted of two items, in which the driver was required to rate his/her chance, as well as other driver's chances of an accident in the next 12 months. The scale items were both positively keyed, and scored on a ten-point rating scale. Increments of 10 per cent were used for the scale and ranged from 1 (*0–10 per cent, no chance*) to 10 (*90–100 per cent, extremely likely*). Combined overall Likelihood of Accident score range was 2 to 20. Machin and Sankey (2008) only reported the results for the single item relating to likelihood of the drivers themselves having an accident and therefore there was no Cronbach's alpha coefficient for that one item.

The Efficacy scale consisted of five items designed to measure the participant's confidence whilst driving in certain conditions. Participants were required to respond to each question by stating their agreement on a five-point Likert type rating scale from 1 (*not at all*) to 5 (*extremely*), (e.g., How confident are you on unfamiliar roads). All items were positively keyed, with scores for the five items summed together to provide a total score ranging from 5 to 25. The Cronbach alpha coefficient for Efficacy reported by Machin and Sankey (2008) was 0.88 which was acceptable.

The Aversion to Risk-taking scale consisted of eight items designed to assess how dangerous participants thought specific actions are whilst driving. Participants were required to answer by stating their agreement to each question on a five-point Likert type rating scale from 1 (*not at all dangerous*) to 5 (*extremely dangerous*), (e.g., how dangerous is running a red light). All items were positively keyed, with scores for the eight items summed together to provide a total score ranging from 8 to 40. The Cronbach's alpha coefficient for the Aversion to Risk-taking scale reported by Machin and Sankey (2008) was 0.79 which was acceptable.

Risky driving behaviour

The same scales that Ulleberg and Rundmo (2003) used to measure self-reported risky driving behaviours were included in this study. These measures assessed self-assertiveness, rule violations and speeding. However, for the purposes of this study, only speeding was considered. The Speeding scale consisted of six items designed to measure the rate participants engaged in speeding related behaviour (e.g., I overtake cars in front when they are driving at the speed limit). All items were positively keyed, with participants required to answer by stating their agreement to each item on a five-point Likert type rating scale from 1 (*never*) to 5 (*very often*). Scores for the six items were summed together to provide a total possible score for speeding, ranging from 6 to 30. The Cronbach's alpha coefficient for speeding reported by Machin and Sankey (2008) was 0.82 which was acceptable.

Procedure

A link to the Driving Attitudes Survey was posted onto the USQ Psychology Online Survey System (OLS). This permitted the first- to third-year psychology students to start participation in the study. The students, who were enrolled in specific psychology courses, were initially informed of the study by information presented in their introductory materials. The study was given the Ethics Approval from the USQ Psychology Department Ethics Committee (EP200733) prior to commencement. The standard procedure for gaining informed consent was performed, with a title page at the beginning of the web survey notifying participants of their rights. Participants were informed their results would be kept confidential, and were notified that they were free to withdraw from the study at any time.

Results

The initial sample size consisted of 402 cases. An alpha level of 0.05 was used for all statistical analyses conducted. Before any data screening or analyses were performed, reliabilities were calculated by computing Cronbach's alpha, to measure the internal consistencies for all scale items. For the present study, all of the scales obtained reasonable internal consistency reliabilities ($\alpha > 0.70$) apart from Likelihood of Accident ($\alpha > 0.65$). As in Machin and Sankey (2008), only the results for the single item relating to likelihood of the drivers themselves having an accident will be reported. Table 6.1 includes the mean, standard deviation, and coefficient alpha values for all 15 scales used in the analysis. Initial data screening revealed no data were missing. However, two cases were deleted as those responses contained an identical answer for each question which is indicative of a response set. The final sample size was 400. The intercorrelations among the variables are presented in Table 6.2.

Table 6.1   Mean, standard deviation, and Cronbach's alpha for all variables (N = 400)

Variable	Number of items	Mean	SD	α
Speeding	6	11.38	4.44	0.84
Worry and concern	6	15.57	5.47	0.92
Likelihood of accident	1	2.44	1.67	–
Efficacy	5	17.29	4.05	0.88
Aversion to risk-taking	8	30.54	4.68	0.78
Aggression	12	48.78	15.24	0.85
Dislike of driving	12	42.10	16.11	0.85
Hazard monitoring	8	67.44	13.54	0.78
Fatigue proneness	7	43.86	18.17	0.80
Thrill-seeking	8	26.73	21.79	0.89
Confrontive coping	7	29.33	18.04	0.84
Task-focused coping	7	76.70	15.47	0.83
Emotion-focused coping	7	40.19	17.49	0.79
Reappraisal coping	7	52.13	16.30	0.79
Avoidance coping	7	42.62	14.39	0.70

Number of items = final number of items in each measure.

Standard multiple regression analysis was used to predict speeding from the risk perception, personality characteristics and driver coping strategy variables. All variables were entered simultaneously and the unique contribution of each predictor was assessed by examining the significance of the Beta weight and the magnitude of the squared semi-partial correlation coefficient (sr^2). These results are reported in Table 6.3. The overall model explained 50 per cent of the variance in Speeding ($R^2 = 0.50$), which was significant with $F(14, 385) = 27.25, P < 0.001$. There are five variables that uniquely add to the prediction of speeding with the greatest unique contribution from aversion to risk-taking ($sr^2 = 0.07$) with a beta weight of -0.31 ($t = -7.06, P < 0.001$).

Table 6.2 Intercorrelations among speeding, risk perception, personality characteristics and coping strategy variables (N = 400)

Variable	1	2	3	4	5	6	7	8	9	10	11	12	13	14
1. Speeding	1.00													
2. Worry and concern	0.04	1.00												
3. Likelihood of accident	0.15	0.27	1.00											
4. Efficacy	0.14	-0.34	-0.22	1.00										
5. Aversion to risk-taking	-0.50	0.15	0.01	-0.12	1.00									
6. Aggression	0.46	0.24	0.17	-0.04	-0.27	1.00								
7. Dislike of driving	-0.11	0.50	0.32	-0.72	0.09	0.14	1.00							
8. Hazard monitoring	-0.30	0.11	-0.14	0.11	0.40	-0.20	-0.08	1.00						
9. Fatigue proneness	0.03	0.15	0.09	-0.32	-0.09	0.14	0.36	-0.09	1.00					
10. Thrill seeking	0.51	-0.09	0.03	0.31	-0.38	0.34	-0.25	-0.21	-0.04	1.00				
11. Confrontive coping	0.52	0.04	0.09	0.17	-0.26	0.67	-0.14	-0.21	0.01	0.38	1.00			
12. Task-focused coping	-0.43	0.05	-0.12	-0.07	0.40	-0.37	0.08	0.48	0.01	-0.43	-0.44	1.00		
13. Emotion-focused coping	-0.06	0.46	0.17	-0.45	0.09	0.17	0.65	-0.02	0.28	-0.13	-0.02	0.09	1.00	
14. Reappraisal coping	-0.08	0.17	0.01	-0.02	0.19	-0.10	0.07	0.27	-0.04	-0.06	-0.07	0.41	0.17	1.00
15. Avoidance coping	0.03	0.02	-0.07	0.14	-0.04	-0.01	-0.13	0.02	-0.03	0.05	0.05	0.14	0.00	0.29

r's ≥ 0.08; $p < 0.05$ (one-tailed); r's ≥ 0.11; $p < 0.01$ (one-tailed).

Driver Behaviour and Training

Table 6.3 **Summary of hierarchical multiple regression analysis for predicting speeding (N = 400)**

Variable	B	SE B	β	sr²	95% CI Lower	Upper
Worry and concern	0.09	0.04	0.11*	0.01	0.02	0.16
Likelihood of accident	0.25	0.10	0.09*	0.01	0.04	0.46
Efficacy	−0.01	0.06	−0.01	0.00	−0.13	0.10
Aversion to risk-taking	−0.29	0.04	−0.31†	0.07	−0.37	−0.21
Aggression	0.02	0.02	0.08	0.00	−0.01	0.06
Dislike of driving	−0.02	0.02	−0.07	0.00	−0.05	0.02
Hazard monitoring	−0.02	0.01	−0.06	0.00	−0.05	0.01
Fatigue proneness	0.00	0.01	0.01	0.00	−0.02	0.02
thrill seeking	0.05	0.01	0.23†	0.03	0.03	0.06
Confrontive coping	0.06	0.01	0.25†	0.03	0.04	0.09
Task-focused coping	−0.01	0.01	−0.04	0.00	−0.04	0.02
Emotion-focused coping	−0.01	0.01	−0.05	0.00	−0.04	0.01
Reappraisal coping	0.01	0.01	0.05	0.00	−0.01	0.04
Avoidance coping	−0.01	0.01	−0.02	0.00	−0.03	0.02

*$p < 0.05$; † $p < 0.01$.

Additional standard regression analyses were conducted for the following subgroups: drivers less than or equal to 20 years old (N = 108), males (N = 79), and drivers who are less frequent drivers (N = 105). These subgroups represented no more than 27 per cent of the overall sample and therefore the results from the overall analysis may not reflect the importance of aversion to risk-taking as a predictor of speeding for these drivers.

The results of the three standard multiple regression analyses were all significant with the overall model explaining 58 per cent of the variance in speeding for younger drivers ($R^2 = 0.58$), with $F(14, 93) = 9.17$, $P < 0.001$. For males, the overall model explained 56 per cent of the variance in Speeding ($R^2 = 0.56$), with $F(14, 64) = 5.84$, $P < 0.001$, while for less frequent drivers, the overall model explained 49 per cent of the variance in speeding ($R^2 = 0.49$), with $F(14, 90) = 6.15$, $P < 0.001$.

There is not a great deal of difference between these results and the results for the overall sample with the overall R^2 values being higher for younger drivers and for males. However, the unique contribution of aversion to risk-taking differed for these three subgroups. For younger drivers, aversion to risk-taking was still the strongest unique predictor (sr^2 = 0.07) with a beta weight of –0.36 (t = –4.05, $P < 0.001$). For the males, aversion to risk-taking was not a significant predictor (sr^2 = 0.01) with a beta weight of –0.16 (t = –1.34, *ns*), while for less frequent drivers, aversion to risk-taking was the second strongest unique predictor (sr^2 = 0.04) with a beta weight of –0.25 (t = –2.57, $P < 0.05$) after thrill seeking (sr^2 = 0.06). Therefore, it needs to be recognised that the importance of aversion to risk-taking in predicting speeding does depend on the characteristics of the group with a greater proportion of females and, to a lesser degree, a greater proportion of more frequent drivers serving to strengthen the importance of this measure of risk perceptions.

Discussion

The overall conceptual model of predictors of speeding was able to predict 50 per cent of the variance in speeding which indicates that it is a very well-specified model. We were able to demonstrate that there are several unique predictors of speeding, including three risk perception variables (worry and concern, likelihood of oneself having an accident and aversion to risk-taking), one personality variable (thrill seeking), and one coping strategy (confrontive coping). It might be tempting to conclude that the other variables are not important when predicting speeding but this is not true. The results show which variables can contribute uniquely to the prediction of speeding after all of the other predictors have been controlled. Even though aversion to risk-taking contributed an additional seven per cent of the variance, that means that the other predictors together accounted for 43 per cent of the variance. Therefore, the results suggest that at least three and perhaps as many as five predictors should be included in the conceptual model and that these include personality and coping variables in addition to measures of risk perceptions.

The additional analyses examining the potential moderating effects of age, gender and frequency of driving showed that this unique contribution of aversion to risk taking varies between 1 and 7 per cent in the three subgroups tested. The unique contribution of the other significant predictors, such as thrill seeking and confrontive coping, was not reported in the results. However, these predictors demonstrated similar variation to aversion to risk-taking in that they were not consistently significant unique predictors of speeding in the subgroups. Therefore, researchers must consider the possible moderating effects of these factors when specifying models that link individual attitudes, perceptions and attributes to risky driving behaviours.

The current study extends the results of a previous study by Machin and Sankey (2008) by including a wider range of ages in the sample and also expanding the range of predictor variables to include drivers' coping strategies. While we have not specified a structural equation model examining whether risk perceptions mediate the influence of personality characteristics, we have extended the previous study by including age, gender and driving frequency as potential moderators of the importance of aversion to risk-taking.

One implication of these results is that research into risky driving behaviours has developed very strong conceptual models which explain a great deal of the variance in speeding. These models can be simplified so that we only need to consider a small number of predictor variables, say between three and five, which will capture the majority of the variance in speeding. It is always difficult to consider the simultaneous effects of 15 predictors so this is a definite advantage in researching risky driving behaviours such as speeding. The role of risk perceptions such as aversion to risk-taking is quite important across both younger and older drivers, but less important for drivers who drive less frequently and not important for male drivers.

References

Aarts, L. and van Schagen, I. (2006). Driving speed and the risk of road crashes: A review. *Accident Analysis and Prevention*, 38(2), 215–224.

Dalziel, J.R. and Job, R.F.S. (1997). *Taxi Drivers and Road Safety. A report to the Federal Office of Road Safety*. Canberra, Department of Transport and Regional Development.

Deery, H.A. (1999). Hazard and risk perception among novice drivers. *Journal of Safety Research*, 30(4), 225–236.

Jonah, B.A. (1997). Sensation seeking and risky driving: A review and synthesis of the literature. *Accident Analysis and Prevention*, 29(5), 651–665.

Lam, L.T. (2003). Factors associated with young drivers' car crash injury: comparisons among learner, provisional, and full licensees. *Accident Analysis and Prevention*, 35(6), 913–920.

Machin, M.A. and Sankey, K.S. (2008). Relationships between young drivers' personality characteristics, risk perceptions, and driving behaviour. *Accident Analysis and Prevention*, 40(2), 541–547.

Mast, M.S. et al. (2008). Masculinity causes speeding in young men. *Accident Analysis and Prevention*, 40(2), 840–842.

Matthews, G. (2001). A transactional model of driver stress. In P. Hancock and P. Desmond, eds, *Human Factors in Transportation: Stress, Workload, and Fatigue*. Majwah, NJ, Lawrence Erlbaum Associates, pp. 133–163.

Matthews, G. et al. (1996). Validation of the driver stress inventory and driver coping questionnaire. Paper presented at the International Conference on Traffic and Transport Psychology, May 1996, Valencia, Spain.

Matthews, G. et al. (1997). A comprehensive questionnaire measure of driver stress and affect. In T. Rothengatter and E.C. Vaya, eds, *Traffic and Transport Psychology: Theory and Application*. Oxford, Pergamon, pp. 317–324.

Rundmo, T. and Iversen, H. (2004). Risk perception and driving behaviour among adolescents in two Norwegian counties before and after a traffic safety campaign. *Safety Science*, 42(1), 1–21.

Steiner, D.L. (2003). Starting at the beginning: an introduction to coefficient alpha and internal consistency. *Journal of Personality Assessment*, 80(1), 99–103.

Ulleberg, P. and Rundmo, T. (2003). Personality, attitudes and risk perception as predictors of risky driving behaviour among young drivers. *Safety Science*, 41(5), 427–443.

Yagil, D. (1998). Gender and age-related differences in attitudes toward traffic laws and traffic violations. *Transportation Research Part F*, 1(2), 123–135.

Chapter 7

Young Drivers: Investigating the Link Between Impulsivity and Problem Driver Status

Fearghal O'Brien, Simon Dunne and Michael Gormley

Introduction

Young drivers are massively over-represented in crash statistics. Within the US, road traffic accidents account for about 40 per cent of all deaths among teenagers (Shope and Bingham, 2008) and in 2004 alone, almost 12,000 people between the ages of 10 and 24 died in motor vehicle accidents (D'Angelo and Halpern-Felsher, 2008). The statistics in many other countries are broadly similar to this with a recent OECD report (2006) finding road traffic accidents to be the single greatest killer of young people in all OECD countries. Among young drivers, males are involved in a disproportionate number of road traffic accidents. In Ireland young males are involved in about five times as many fatal accidents as young females (CSO, 2007). This is not a phenomenon unique to Ireland: males are involved in more accidents than females across all OECD countries (OECD, 2006) – a difference that is particularly pronounced among young drivers. Özkan and Lajunen (2006) suggest that three main factors underlie the problem: *exposure, driving skill* and *driving style.*

Exposure refers both to the amount of time spent on the road and to the kinds of driving conditions engaged in by drivers. Clarke et al. (2006) have found that younger people are more likely to be involved in accidents at night and Shope and Bingham (2008) report that younger people often drive vehicles that are less safe than those of their older counterparts. These conditions under which young people choose to drive may predispose them to a higher risk of accident involvement. However, time spent on the road is unlikely to be the only cause of higher crash rates among young males with findings from the Insurance Institute for Highway Safety (2005) indicating that even when distance travelled is taken into account, young males are still involved in more crashes than any other group. Driving skill is also likely to contribute to crash statistics among young people. Support for this point of view comes from findings that seem to indicate that some driver training programmes can be effective in reducing crash involvement (Senserrick, 2007). While there is no doubt that the relative inexperience of young drivers contributes to their involvement in road accidents, inexperience cannot solely account for their increased risk. Firstly, if it were simply a matter of inexperience, then we would not expect to find a difference in crash rates between young males and females. Secondly, a comparison between 20-year-olds and 30-year-olds with the same level of driving experience found that the younger drivers still had a higher crash likelihood than the older drivers (Groeger, 2006).

The significance of driving style becomes apparent when one considers the important role that human factors play in a large number of accidents. In a large-scale study, Streff (1991) examined police records for almost two million vehicles that had been involved in a crash. He found that there was at least one instance of unsafe driving involved in 80 per cent of these cases. Examples of these unsafe driving behaviours include speeding (in 16.9 per cent of cases), following too close (10.3 per cent) and failure to yield (19.3 per cent). Harrington found that speed was the most common violation associated with accidents that resulted in fatal injury (as cited in Clarke et al. 2002, p. 3). The popular perception of young drivers, particularly young males, driving dangerously is supported by a number of studies. Young people do break the speed limit more frequently and more excessively than older drivers (Fuller et al., 2008) and they drive more aggressively in general (Ho and Gee, 2008). Åberg and Rimmo (1998) found using the Driver Behaviour Questionnaire that young drivers reported more road violations than older drivers for both males and females, and that males reported more violations than females in all age groups. If experience cannot account for these differences, as discussed above, then how can these levels of risky driving among young people, and young males in particular, be explained?

There is now a significant amount of research to suggest that neurodevelopmental factors may play an important role in high levels of risk-taking behaviours among young people, and Steinberg (2008) has developed a comprehensive theory to explain this. He suggests that increased risk-taking during adolescence has very little to do with risk perception and risk appraisal as others have suggested (e.g., Deery, 1999). He supports this claim with references to a recent review of adolescent decision-making by Reyna and Farley (2006) in which they highlight a number of counter-intuitive findings. Importantly, they found that not only are young people similar to adults in their estimation of risks, but young people are actually more likely to overestimate some risks, such as those associated with HIV or lung cancer. Groeger (2006) found that older drivers, more so than young drivers, have a more positive view of their own driving than is warranted. Steinberg instead suggests that adolescent risk-taking can be best understood with reference to the development of two separate brain systems (2008).

Firstly, he suggests that risk-taking increases during early adolescence because of changes in the brain's socio-emotional network caused mainly by a decrease in autoreceptor levels in the dopamine system. According to Steinberg, this leads to an increase in sensation-seeking behaviour among adolescents, particularly in the presence of peers. He supports this claim with reference to a number of brain imaging studies which have found that social stimuli and regular rewards both elicited a similar neural response in adolescents but not in adults (Galvan et al.,

2005, as cited in Steinberg, 2008). This sensation-seeking behaviour is curbed by the neurological development of a number of brain areas, especially those associated with action monitoring (Hogan et al., 2005), which leads to a decrease in risk-taking behaviour towards the end of adolescence and into early adulthood. A number of neuroanatomical findings corroborate this description. The caudate nucleus, a subcortical structure associated with goal-directed behaviours (Grahn et al., 2008), does not reach adult levels of maturation until after adolescence (Giedd, 2004). The dorsolateral prefrontal cortex, a cortical structure associated with planning and regulation, is another brain region that does not fully mature until the early 20s (Paus, 2005). Steinberg (2008) suggests that adolescents are particularly prone to risk-taking because their socio-emotional systems are maturing but neural systems associated with impulse control have not yet developed.

There are, however, some complications with the application of Steinberg's theory of risky behaviour to risky driving among young people. Firstly, most of the studies that Steinberg discusses are concerned with the comparison of adolescents and adults, with this younger group typically having an average age of about 15 or 16 years. However, within driver behaviour research, young drivers are often those aged 17–21, or even 17–24 years of age. Secondly, Steinberg does not account for higher levels of risky behaviours among males in general, though he does note that males seem to be more susceptible to the influence of peers. Nonetheless, this theory may be helpful in explaining some of the observed differences between young and older drivers, especially when one considers that sensation-seeking and impulse control have previously been linked to problematic driving.

Sensation-seeking has been associated with risky driving in a large number of studies. Jonah (1997) examined about 40 studies and found that in a large majority of them there was a positive relationship between sensation-seeking and risky driving. Arnett and colleagues (1997) found that there was a correlation of 0.34 between sensation-seeking and frequency of speeding, and that there was a correlation of 0.23 between sensation-seeking and frequency of driving while intoxicated. Jonah and colleagues (2001) split their sample into low sensation-seekers and high sensation-seekers on the basis of a median split of their scores on a sensation-seeking scale. They found that high sensation-seekers were more likely to drive without a seatbelt and to have committed a driving violation in the last two years.

Impulsiveness has not been studied in the context of driver behaviour as frequently as has sensation-seeking. It is important to distinguish these two concepts from one another with sensation-seeking referring to one's preference for novel experiences while impulsivity refers to self-control over one's own actions (Dahlen et al., 2005). Dahlen and colleagues did find a correlation of 0.35 between self-reported risky driving and impulsivity (measured using the BIS-11) and a correlation of 0.23 between aggressive driving and impulsivity. Impulsivity has been related to problematic driving in past research predominantly through the use of self-report measures.

The present study seeks to investigate the relationship between impulsivity and risky driving using the Go/NoGo task – a task that has been frequently used as a measure of response inhibition. Response inhibition refers to a number of separate, but related, skills – the ability to withhold a pre-planned response, to interrupt one that has already begun or to ignore interference (Tamm et al., 2002). Response inhibition is often linked with impulsivity with Horn and colleagues (2003) identifying the former as a key component of the latter. Response inhibition has been used in research investigating issues that are thought to be related to problems of impulsivity such as alcohol abuse (Nigg et al., 2006) and clinical conditions like ADHD (Smith et al., 2006). Furthermore, response inhibition is conceptually related to problematic driving as self-regulation would seem to be an important component of both safe driving and resisting the influence of one's peers. A number of studies lend support to this conceptualisation as they have found that adults outperform adolescents on tasks of response inhibition (Johnstone et al., 2005; Rubia et al., 2006; Tamm et al., 2002).

If the underdevelopment of brain systems plays a role in dangerous driving, it would be expected that problematic young drivers should differ from non-problematic young drivers in a cognitive measure that is related to these systems. As discussed, the Go/NoGo task has been used to demonstrate differences between adolescents and older participants in the past and so it may have utility in distinguishing between risky and non-risky young drivers.

Method

Participants

Sixty undergraduate students (30 male and 30 female) from Trinity College, Dublin took part in the research. The results of four participants (three males and one female) were removed from analysis as these participants did not complete the experimental task according to the instructions provided. All participants had either a full or provisional driving licence for a car. Participants from the School of Psychology were given course credits for their participation while students from other schools in the college were rewarded with a chocolate bar. The average age of the participants was 19.2 years (SD 0.82).

Questionnaires

Driver experience form   This form was used to obtain basic information about driving experience and past accident involvement. Participants were required to provide details relating to how long they had been driving, what type of licence they held (full or provisional), the size of their car engine and the distance they had travelled by car in the last year. Furthermore, they were required to list how many penalty points they had accumulated, the number of accidents they had been involved in, the outcome of these accidents (serious injury or property damage) and the number of accidents for which they were at fault.

Problem Driver Questionnaire (PDQ) The PDQ (Gormley and Fuller, 2008) is an 11-item questionnaire which focuses on the past behaviours of participants rather than on their attitudes and opinions. The questionnaire focuses predominantly on speeding behaviour though there are also questions relating to drink-driving, dangerous overtaking, racing other drivers and close following. Each question asks participants about the frequency of their involvement in the activity and all questions are answered on a six-point *Likert* scale with a choice ranging from 'most days' to 'never' and with an additional option of 'I don't know'. Participants are not just asked how often they speed but also the degree to which they break the speed limit and the nature of the roads on which they speed. Unlike a number of other measures of driver behaviour, the score associated with each answer (e.g., 'most days') is dependent on the question being asked. A participant who admits to driving at 35km/h in a 30km/h zone 'most days' and at 40km/h in a 30km/h zone 'most days' will receive a higher score for the higher speed (where a high score is indicative of problematic driving). The validity of the questionnaire was supported by Gormley and Fuller's study (2008) as they found that problematic drivers reported more accidents in the previous three years than non-problematic drivers.

Dula Dangerous Driving Index (DDDI) The DDDI (Dula and Ballard, 2003) is divided into three subscales: aggressive driving, negative emotions while driving and risky driving. Participants are required to indicate the frequency of their involvement in 28 different dangerous driving behaviours using a five-point likert scale ranging from 'never' to 'always'. Willemsen and colleagues (2008) found the DDDI to have good test–retest reliability and the validity of the test was supported by a number of results. They found that females and older drivers scored lower on the test – which is to be expected since these groups differ markedly in rates of dangerous driving.

Materials

The driver experience form, PDQ and DDDI were presented on A4 paper. The Go/NoGo task was presented on 20 inch Apple iMac using Superlab 4.0.8 or a Sony VAIO laptop using Superlab 4.0.7.b. The two stimuli used in the task were the letters 'X' and 'Y' which were presented in uppercase and in font size 350. The font colour used was white on a black background.

Procedure

Participants all completed the driver experience form, the DDDI and the PDQ before they were presented with the response inhibition task. The task used two different stimuli – the letters 'X' and 'Y', which were presented in an alternating pattern. Participants were instructed to press the response key (the space bar) as soon as each stimulus was presented unless that stimulus had been presented in the preceding trial. For example, if participants were presented with the sequence X, Y, X, Y, Y; they would press the response key for the first four trials but would withhold their response on the fifth trial. Each trial lasted for 1000ms regardless of the participant's response. Each trial consisted of a blank screen presented for 300ms and one of the two stimuli for 700ms. There was a total of 317 trials in the experiment and 20 of these were lures – trials which required participants to withhold their response. Lures were followed by at least two regular Go trials.

Data analysis

Participants were assigned to the problematic driving group and the non-problematic driving group on the basis on their score in the PDQ. A median split of the scores was carried out separately for both males and females to obtain similar numbers in all four groups – problematic males, non-problematic males, problematic females and non-problematic females. Comparisons between these groups were conducted for four different measures within the Go/NoGo task. These measures were reaction time to Go trials, number of omission errors (failed Go trials), number of commission errors (failed NoGo trials) and reaction time to failed NoGo trials. The DDDI was also used to designate drivers as problematic and non-problematic and a separate analysis of the four experimental measures was completed on the basis of this alternative grouping.

Results

Preliminary analysis

There were no differences between males and females in the number of accidents that they had been involved in, the number of penalty points they had accumulated, their age, their driver experience or the distance they had travelled in the last year. No difference was found between problematic drivers and non-problematic drivers, as designated by the PDQ or the DDDI, on any of these variables either. It is also worth noting that none of the 56 participants had received any penalty points and none had been involved in an accident that led to injury or death. In fact, only eight of the participants had ever been involved in an accident. On average, participants had been driving for 20 months (SD 13.3) and in the last year they had travelled 3,587km (SD 8420).

Overall participants had an average score of 23.4 (SD 15.1) on the PDQ – with 0 and 66 being the lowest and highest possible scores respectively. The average score for males on the PDQ was 28.6 (SD 16.1) while the average score for females was 18.7 (SD 12.6) and this difference was found to be statistically significant

t (54) = 2.57, P < 0.05, two-tailed, d = 0.68. Males and females did not differ in their scores on the DDDI.

The results of all of the Go/NoGo measures were normally distributed with the exception of the number of omission errors which was found to be positively skewed. A log transformation was carried out which led to a normal distribution of omission errors.

Questionnaires

There was a positive correlation between participant scores on the PDQ and the DDDI, r = 0.5, P (two-tailed) < 0.001. When the DDDI was used to assign participants to problematic and non-problematic groups, no difference between these groups was found on any of the four Go/NoGo measures. Accordingly, the results reported below all relate to problematic and non-problematic drivers as determined by the PDQ.

Four separate factorial ANOVAs were conducted – one for each of the Go/NoGo measures. Each ANOVA had two IVs – gender and problem driver status, and each of these IVs had two levels – male and female, and problematic and non-problematic respectively. Table 7.1 below gives a summary of the findings.

Table 7.1   Mean scores for all groups on the four Go/NoGo measures

	Males		Females	
	Problematic	**Non-problematic**	**Problematic**	**Non-problematic**
C errors	11.5	10	13	13.5
Go RT (ms)	340	375	345	342
O errors	32	8	23	7
No/go RT (ms)	311	347	312	325

C errors, commission errors; O errors, omission errors.

For both the number of commission errors variable and the reaction time to No/Go trials variable there was no significant main effect of gender or problem driver status nor was there an interaction between these two independent variables.

Go reaction time

There was a non-significant main effect of gender and problem driver status on reaction time to Go trials. Although the interaction between these variables was found to be non-significant, it did come very close F (1, 52) = 3.42, P = 0.07. The cause of this potential interaction can be seen in Figure 7.1 with non-problematic females having a faster reaction time of 342ms (SD 25.7) compared with non-problematic males who had an average reaction time of 375ms (SD 33.4). This difference was reversed for problematic drivers with males responding faster than females – 340ms (SD 50) and 345ms (SD 38.7) respectively. Post hoc *t*-tests were carried out to determine if these differences were statistically significant.[1] Non-problematic females were found to respond faster than non-problematic males *t* (25) = 2.84, $P < 0.01$, two-tailed, $d = 1.09$. The difference between problematic males and problematic females was not found to be statistically significant.

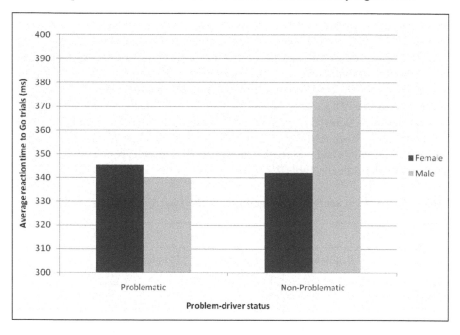

Figure 7.1   Reaction time to Go trials for both male and female problematic and non-problematic drivers

Omission errors

There was a non-significant main effect of gender on the number of omission errors and the interaction between gender and problem driver status was not found to be significant. However, a significant main effect of problem driver status was found $F (1, 51) = 8.08$, $P < 0.01$. Figure 7.2 illustrates the difference in omission errors between problematic and non-problematic drivers and it can be seen that this difference exists for both males and females.

---

1   A Bonferroni correction was applied to account for four simultaneous *t* tests. Accordingly, a *P* value of < 0.013 was needed to indicate statistical significance.

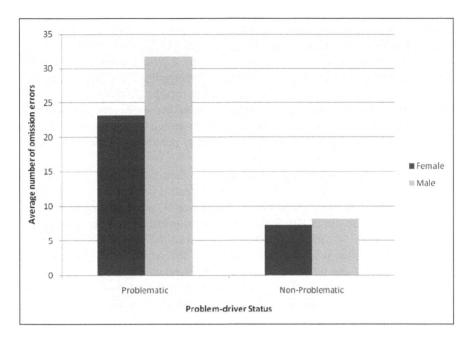

Figure 7.2   Number of omission errors committed by both male and female problematic and non-problematic drivers

Levene's test for Homogeneity of Variance was found to be significant for this ANOVA. Accordingly, a Mann–Whitney U test was carried out to test if there was a difference between problematic drivers and non-problematic drivers in omission errors which proved to be significant $U = 219.5$, $z = -2.83$, $P < 0.01$.

Discussion

The findings of this study were not entirely as predicted. Although some differences were found between problematic and non-problematic drivers, the implication of these differences is unclear at this stage. If it were simply the case that problematic drivers were deficient at inhibiting responses, then it would be expected that they would have made more commission errors – incorrect responses on NoGo trials. This is one of the differences that Rubia et al. (2006) found between adults and younger participants which they interpret as an indication of poorer inhibitory functions in younger people. Yet in the present study no significant difference in the number of commission errors was found between problematic and non-problematic drivers and there was no difference between the two driver groups in their reaction time for these commission errors.

The most striking difference between the two groups is the higher level of omission errors for problematic drivers. The significance of this is unclear, though

Johnstone et al. (2005) suggest that since the task is predominantly concerned with the inhibition of prepotent responses, any observed errors should be seen as a failure in inhibitory functioning. In their own study, a difference in total omission errors was found between younger and older participants. The most unexpected finding in this study was the difference in reaction time to Go trials between non-problematic males and females. Since all of these drivers were considered to drive in a relatively safe manner, it is not clear why there should be a difference between males and females. Further research is needed to determine if this is a robust finding and, if so, what its implications are.

The failure to find any significant results when participants were sorted according to their scores on the DDDI is of interest. Furthermore, while males scored higher on the PDQ, there was no difference between males and females on their responses to the DDDI. Since a medium correlation was found between participants' scores on each questionnaire, and since the questions used in both are reasonably similar, it is surprising that these differences exist. One possible explanation is that the more advanced scoring mechanism used for the PDQ (see Results section) makes the questionnaire more suited than the DDDI to distinguishing between problematic and non-problematic drivers.

There were a number of limitations in the present study. The sample variance was not ideal for an analysis based on a median split since there was a relatively low variability in the PDQ scores. This problem could be addressed in two ways. Firstly, participants could be recruited from a number of different backgrounds – the present study used only college students. Secondly, the PDQ could be administered to a much larger sample with a follow-up analysis only focusing on those participants with very low and very high scores.

The above findings can be related back to Steinberg's (2008) theory of adolescent risk-taking. He proposed that adolescents generally take more risks than adults do because of their higher levels of impulsivity due to underdeveloped cognitive control mechanisms. The Go/NoGo task is a measure of response inhibition and accordingly it is an indication of levels of impulse control (Tamm et al., 2002). The problematic drivers in this study were those who reported exhibiting more risky driving behaviours. Since these drivers also made more mistakes in the Go/NoGo task, this suggests that Steinberg's theory of adolescent risk-taking may apply to dangerous driving among young people. The fact that these significant results were found even with limitations of variability in PDQ scores lends support to the utility of this approach in studying problematic driving. An investigation into the relationship of problematic driving and performance on a Go/NoGo task among adults would be necessary in order to determine if impulse control is a contributing factor in all risky driving behaviour or if this is just the case among young drivers as Steinberg's theory would predict.

References

Åberg, L. and Rimmo, P.-A. (1998). Dimensions of aberrant driver behaviour. *Ergonomics*, 41(1), 39–56.

Arnett, J.J. et al. (1997). Reckless driving in adolescence: 'state' and 'trait' factors. *Accident Analysis and Prevention*, 29(1), 57–63.

Clarke, D.D. et al. (2002). *In-depth Accident Causation Study of Young Drivers*. Road Safety Research Report No. 54. London, Department for Transport.

Clarke, D.D. et al. (2006). Young driver accidents in the UK: the influence of age, experience, and time of day. *Accident Analysis and Prevention*, 38(5), 871–878.

CSO (2008). *Transport 2008*. Dublin, Central Statistics Office.

D'Angelo, L.J. and Halpern-Felsher, B.L. (2008). from the exam room to behind the wheel: can healthcare providers affect automobile morbidity and mortality in teens? *American Journal of Preventive Medicine*, 35(3, Supplement 1), S304–S309.

Dahlen, E.R. et al. (2005). Driving anger, sensation seeking, impulsiveness, and boredom proneness in the prediction of unsafe driving. *Accident Analysis and Prevention*, 37(2), 341–348.

Deery, H.A. (1999). Hazard and risk perception among young novice drivers. *Journal of Safety Research*, 30(4), 225–236.

Dula, C.S. and Ballard, M.E. (2003). Development and evaluation of a measure of dangerous, aggressive, negative emotional, and risky driving. *Journal of Applied Social Psychology*, 33(2), 263–282.

Fuller, R. et al. (2008). *The Conditions for Inappropriate High Speed: A Review of the Research Literature from 1995 to 2006*. Road Safety Research Report No. 92. London, Department for Transport.

Giedd, J.N. (2004). Structural magnetic resonance imaging of the adolescent brain. *Annals of the New York Academy of Sciences*, 1021, 77–85.

Gormley, M. and Fuller, R. (2008). *Investigation of High-risk Behaviour in Irish Young Male Drivers*. Ballina: Road Safety Authority.

Grahn, J.A. et al. (2008). The cognitive functions of the caudate nucleus. *Progress in Neurobiology*, 86(3), 141–155.

Groeger, J.A. (2006). Youthfulness, inexperience, and sleep loss: the problems young drivers face and those they pose for us. *Injury Prevention*, 12(suppl_1), i19–24.

Ho, R. and Gee, R.Y. (2008). Young men driving dangerously: development of the Motives for Dangerous Driving Scale (MDDS). *Australian Journal of Psychology*, 60(2), 91–100.

Hogan, A.M. et al. (2005). Maturation of action monitoring from adolescence to adulthood: an ERP study. *Developmental Science*, 8(6), 525–534.

Horn, N.R. et al. (2003). Response inhibition and impulsivity: an fMRI study. *Neuropsychologia*, 41(14), 1959–1966.

Insurance Institute for Highway Safety (2005). *Fatality Facts 2005: Teenagers.* Available at www.iihs.org/research/fatality_facts_2005/teenagers.html.

Johnstone, S.J. et al. (2005). Development of inhibitory processing during the Go/NoGo task – a behavioral and event-related potential study of children and adults. *Journal of Psychophysiology*, 19(1), 11–23.

Jonah, B.A. (1997). Sensation seeking and risky driving: A review and synthesis of the literature. *Accident Analysis and Prevention*, 29(5), 651–665.

Jonah, B.A. et al. (2001). Sensation seeking, risky driving and behavioral adaptation. *Accident Analysis and Prevention*, 33(5), 679–684.

Nigg, J.T., Wong, M.M. et al. (2006). Poor response inhibition as a predictor of problem drinking and illicit drug use in adolescents at risk for alcoholism and other substance use disorders. *Journal of the American Academy of Child and Adolescent Psychiatry*, 45(4), 468–475.

OECD (2006). *Young Drivers: The Road to Safety.* Paris: OECD Press.

Özkan, T. and Lajunen, T. (2006). What causes the differences in driving between young men and women? The effects of gender roles and sex on young drivers' driving behaviour and self-assessment of skills. *Transportation Research Part F: Traffic Psychology and Behaviour*, 9(4), 269–277.

Paus, T. (2005). Mapping brain maturation and cognitive development during adolescence. *Trends in Cognitive Sciences*, 9(2), 60–68.

Reyna, V.F. and Farley, F. (2006). Risk and rationality in adolescent decision making: implications for theory, practice, and public policy. *Psychological Science in the Public Interest*, 7(1), 1–44.

Rubia, K. et al. (2006). Progressive increase of frontostriatal brain activation from childhood to adulthood during event-related tasks of cognitive control. *Human Brain Mapping*, 27(12), 973–993.

Senserrick, T.M. (2007). Recent developments in young driver education, training and licensing in Australia. *Journal of Safety Research*, 38(2), 237–244.

Shope, J.T. and Bingham, C.R. (2008). Teen driving: motor-vehicle crashes and factors that contribute. *American Journal of Preventive Medicine*, 35(3, Supplement 1), S261–S271.

Smith, A.B. et al. (2006). Task-specific hypoactivation in prefrontal and temporoparietal brain regions during motor inhibition and task switching in medication-naive children and adolescents with attention deficit hyperactivity disorder. *American Journal of Psychiatry*, 163(6), 1044–1051.

Steinberg, L. (2008). A social neuroscience perspective on adolescent risk-taking. *Developmental Review*, 28(1), 78–106.

Streff, F.M. (1991). Crash avoidence – new opportunities for behavior analysis. *Journal of Applied Behavior Analysis*, 24(1), 77–79.

Tamm, L. et al. (2002). Maturation of brain function associated with response inhibition. *Journal of the American Academy of Child and Adolescent Psychiatry*, 41(10), 1231–1238.

Willemsen, J. et al. (2008). The Dula Dangerous Driving Index: an investigation of reliability and validity across cultures. *Accident Analysis and Prevention*, 40(2), 798–806.

Chapter 8

Relationships Between Driving Style, Self-reported Driving Behaviour and Personality

S.M. Skippon, N. Reed, T. Luke, R. Robbins, M. Chattington,
and A.H. Harrison

Introduction

Driving style is an important factor in fuel efficiency. A recent study estimated that educating drivers in more efficient driving styles could reduce US cumulative transport CO_2 emissions from 2010 to 2050 by 1.17 billion tonnes (Cambridge Systematics, 2009).

Ask anyone to account for why another person drives in a particular way, and they will offer explanations that draw on aspects of the driver's personality. But if personality influences driving style, then interventions to educate drivers in fuel-efficient driving face a more nuanced task. To design the most effective interventions we must answer the question: how far does personality explain the way drivers tend to drive?

There is broad consensus that the 'Five Factor' model (FFM) captures the main independent dimensions of personality (Costa and McCrae, 1995; Goldberg, 1981; Liebert and Liebert, 1998; McCrae and Costa, 2003). In the FFM, personality is represented by the factors neuroticism, extraversion, openness to experience, conscientiousness and agreeableness (Costa and McCrae, 1995). The five main factors are each composed of six narrower 'facets'. Driving research using the FFM has focused on accident risk, rather than driving style. Sumer and colleagues use a Five-Factor measure, the Big Five Inventory (BFI) (Benet-Martinez and John, 1998) to study accident risk in a sample of 1001 Turkish drivers. BFI scores were compared with scores on the Driver Behaviour Questionnaire (Reason et al., 1990) and self-reported accident history. All five factors had indirect effects on accident risk, mediated by their effects on 'aberrant driving behaviour'. Low conscientiousness scores were associated with higher accident risk and more aberrant driving behaviours, as were low agreeableness scores. Extraversion and neuroticism had weak positive correlations with aberrant driving behaviours and indirectly with accident risk. Conscientiousness was also found to correlate negatively with accident frequency by Arthur and Graziano (1996), and agreeableness was negatively correlated with accident and driving penalty rate by Cellar et al. (2000). Benfield and colleagues (2007) found weak negative correlations between agreeableness and conscientiousness and various measures of driver anger and aggressive tendencies. All of these studies used self-reports rather than direct measures of driving.

Other studies, also focused on accident risk, have used narrower measures of personality traits. Ulleberg (2002) found that risky driving by young drivers was associated with two 'subtypes' of personalities: low altruism + low anxiety + high sensation-seeking; and high aggression + high anxiety + high sensation-seeking. In a later study (Ulleberg and Rundmo, 2003) the authors found that the combination of high aggression + high sensation-seeking was associated with high-risk attitudes to driving among more experienced drivers, while the combination of high anxiety + high altruism was associated with low-risk attitudes to driving. Sumer (2003) found that sensation-seeking was positively correlated with driving speed. Jonah (1997), in a review of 40 studies, found sensation-seeking correlated with various measures of 'risky driving' with correlation coefficients in the range 0.3 to 0.4.

There are other models of personality besides the FFM. Lajunen (2001) compared national means on the three factors of the Eysenck Personality Questionnaire (EPQ) (Eysenck and Eysenck, 1975), extraversion, neuroticism and psychoticism, with traffic fatality rates in 35 countries. Extraversion was positively correlated to the rate of fatalities, while neuroticism correlated negatively. Furnham and Saipe (1993) found that psychoticism was higher in a group of drivers with driving convictions than a group without. EPQ psychoticism is related to, though not identical with, low agreeableness in the FFM.

There is evidence, then, that personality traits relate to accident risk. However, accidents are distal consequences of driving behaviour, in the sense that most drivers, even risky ones, have them infrequently compared to their overall time spent driving. It seems plausible that driving *style*, the pattern of driving behaviour that drivers engage in constantly while driving, is more proximately related to personality. In this picture driving style mediates the relationship between personality traits and accident frequency.

Researchers have developed a number of instruments that measure self-reported driving behaviour. Parker and colleagues (1995), and Ozkan and Lajunen (2005), using the Driver Behaviour Questionnaire (DBQ) (Reason et al., 1990), found that self-reported violations were positively associated with higher accident rates, while non-volitional errors and lapses were not. The Driver Stress Inventory (DSI) (Matthews et al., 1997), developed from the Driver Behaviour Inventory (DBI) (Gulian et al., 1989), assesses aspects of driving behaviour that represent different vulnerabilities to stress. Matthews and colleagues (1998) compared DBI scores with driving behaviour in a simulator. High scores on DBI aggression were associated with driving at higher speeds, making control errors, and more frequent overtaking, while high scores on dislike of driving were associated with driving at lower speeds, less frequent overtaking, errors when following other vehicles and lateral control errors.

How do these self-report measures of driving behaviour relate to personality? Dorn and Matthews (1995) claimed that DBI factors, which measure relatively narrow behavioural traits, were better predictors of driving mood, and self-appraisal of driving skills, than measures of more broadly defined personality traits. This is consistent with Ajzen's (2005) view that narrowly defined traits are better predictors of behaviour than general traits. Matthews et al. (1991) identified correlations between dimensions of the EPQ and the DBI. Neuroticism was positively correlated with aggression and dislike of driving. Dorn and Matthews (1992) compared DBI scores with two different FFM measures and with Catell's 16PF questionnaire (Catell et al., 1970). Again neuroticism was positively correlated with aggression and dislike of driving. Both affection (which relates to the NEO PI domain agreeableness), and conscience (which relates to the NEO PI domain conscientiousness) were negatively correlated with DBI aggression. DBQ violations were found to correlate with sensation-seeking by Rimmo and Alberg (1999). Sumer and colleagues (2005) found a weak positive correlation between violations and neuroticism, and weak negative correlations between violations and agreeableness and conscientiousness.

So there is evidence that self-reported driving behaviour relates to personality. Self-report has advantages, in that it provides information covering an extended period of driving, and is logistically relatively straightforward to accomplish. However, it has disadvantages too: there is a risk of impression management, and it depends on accurate and full recall.

In this study FFM personality traits were compared directly with driving in a driving simulator. In an earlier paper we reported that some factors of the self-report instruments DSI, DCQ and DBQ (Skippon et al., 2008) correlated with direct measures of driving style. These self-report instruments are also compared here with FFM traits, providing a triangulation between methods.

Method

The simulator consisted of a medium-sized family hatchback surrounded by four screens giving 210° front vision and 60° rear vision, enabling the normal use of the vehicle's driving and wing mirrors. The motion system had three degrees of freedom (heave, pitch and roll). Engine noise, road noise and the sounds of passing traffic were provided. Participants walked up to and entered the car in a normal manner, setting up expectations of a close-to-normal driving experience. The simulated route consisted of 10.2km of rural single carriageway, with appropriate UK signage and markings, and roadside features such as trees, bushes, fences, walls and buildings. The route was designed to create repeated opportunities for acceleration and deceleration. Participants' driving style was measured by recording control actions and vehicle behaviour, such as: road speed; engine speed; accelerator pedal, brake pedal and clutch pedal depression; number of gear changes; lane position; time spent in opposite lane; and various specific measures associated with overtaking, hill climbs and bends.

Exploratory factor analysis showed that many of the driving style measures could be explained by a smaller number of hypothesised underlying factors.

However, such factors are not themselves behaviours; indeed to some extent they may be interpreted as representing dispositional tendencies towards behaviours. In this study, where we sought to identify relationships between personality and behaviours, it was not considered appropriate to reduce the actual behaviours to hypothesised underlying constructs in this way.

In normal driving, people experience variation in the performance of their vehicle (for instance, as passenger and load weight changes). Many drivers also have regular access to more than one vehicle. Four vehicle performance conditions were used to give measures of driving style over a range of available vehicle performance. Performance was varied in two ways: (a) the overall torque curve was modified from 'standard' to 'high performance', giving an 8 per cent difference in acceleration, approximately equivalent to changing engine size from a 1.4l to a 1.6l engine in a C class car; (b) overall driveability was modified from standard by introducing a delay between the accelerator pedal action and the vehicle response.

Driving style is also influenced by the driving situation, and the driver's immediate goals. To represent this source of variation in natural driving, experimental drives were completed in one of two 'driving situation' conditions. In the first, drivers were asked to complete the route driving normally. In the second, drivers were put under time pressure, and presented with repeated needs to overtake slower vehicles.

To control for potential confounding effects of learning and fatigue, participants completed drives in only four of the possible eight combinations of vehicle performance and driving situation. For each participant the mean of each driving style measure was calculated from all four drives. Across the participant pool all conditions were represented equally in a counterbalanced design.

After their experimental drives, participants completed several self-report questionnaires: The Driver Behaviour Questionnaire (DBQ); the Driver Stress Inventory (DSI) and the related Driver Coping Questionnaire (DCQ). Various versions of these questionnaires have been used in previous studies. Af Wahlberg and colleagues (2009) discuss four studies using versions of the DBQ with 50, 32 and 16 items, three with six-point Likert scales and one with a five-point scale. Factors have often been identified on an individual-study basis using factor analytic methods, so that the items contributing to the reported factors have not necessarily been consistent between studies. As the focus of the present study was on the relationships between existing measures, it was not appropriate to develop new solutions specific to the study sample, so the DBQ, DSI, and DCQ were treated as established psychometric questionnaires with factor structure, scoring scheme and norms determined by the original studies.

Participants later completed the NEO PI-R personality inventory at home after their driving session, via a web interface.

Forty-three participants were recruited, of whom all completed the experimental session and self-report driving questionnaires. Twenty-seven also completed the NEO PI-R. All of them were male (to control for possible gender effects), had a full UK manual driving licence, greater than five years driving experience, and drove 8,000 miles per year or greater. The mean age was 50.5 ± 15.2 years and the mean number of years of driving experience was 31.9 ± 13.5 so the sample represented older, experienced drivers who might be expected to have established driving styles.

Results

Comparison of driving style in the simulator with self-reported driving behaviour

DSI *thrill-seeking* was moderately correlated with many driving style measures, as shown in Table 8.1.

Table 8.1   Significant correlations between DSI factors and driving style measures (SD = standard deviation)

Factor	Aspect of driving style	Pearson *r* (N = 43)	*P* (2-tailed)
Aggression	Average speed (mph)	0.34	0.03
	Average RPM	0.38	0.01
	Average SD of lane position in 6 sharp curves	0.32	0.03
	Average time in neutral between gear changes(s)	−0.32	0.04
Dislike of driving	Average clutch position	0.34	0.03
	Duration spent out of lane (s)	−0.33	0.03
	Second overtake – time exposed to danger(s)	0.30	0.05
Thrill seeking	Time to complete(s)	−0.43	0.004
	Average speed (mph)	0.52	<0.001
	Maximum speed (mph)	0.59	<0.001
	Average RPM	0.51	0.001
	Average accelerator position	0.39	0.011
	Average brake position	0.45	0.003
	SD speed	0.53	<0.001
	SD RPM	0.33	0.029
	SD brake position	0.44	0.007
	Average speed in 12 normal curves (mph)	0.46	0.002
	Average SD of lane position in 12 normal curves	0.34	0.027
	Average speed in 6 sharp curves (mph)	0.43	0.004
	Average SD of lane position in 6 sharp curves	0.41	0.006

SD = standard deviation.

Participants with high thrill-seeking scores completed the route quicker, tended to drive faster, used both accelerator and brake more, showed greater variation in speed and showed greater variation in lateral position on bends, suggesting that they cut across corners to complete the route more quickly. DSI aggression was weakly correlated with four aspects of driving style. Participants who scored higher on aggression tended to drive faster, with higher engine speed and greater variation in lateral position in sharp curves, and made gear changes more quickly.

The confrontive coping factor of the DCQ was weakly correlated with six aspects of driving style, as shown in Table 8.2. Participants who scored higher on confrontive coping tended to drive faster, have higher average speed on bends, and show more variation in lateral position on bends.

Table 8.2   Significant correlations between DCQ factors and driving style measures

Factor	Aspect of driving style	Pearson r (N = 43)	P (2-tailed)
Confrontive coping	Time to complete(s)	−0.32	0.037
	Average speed (mph)	0.36	0.019
	Average RPM	0.31	0.049
	Average speed in 12 normal curves (mph)	0.33	0.033
	Average speed in 6 sharp curves (mph)	0.33	0.033
	Average SDLP in 6 sharp curves	0.38	0.013
Reappraisal coping	Duration spent out of lane(s)	0.43	0.004
	Average time in neutral between gear changes(s)	0.36	0.018
Avoidance coping	Duration spent out of lane(s)	0.37	0.017

The DBQ factor violations was correlated with nine aspects of driving style, as shown in Table 8.3. Participants who scored higher on the violations factor tended to drive faster, show more variation in speed, use the brake more and spend more time out of lane on bends. This set of behaviours resembles the set that correlated with thrill-seeking. In the case of violations there are slightly fewer correlations with specific behaviours, and the correlation coefficients tend to be smaller.

Personality and driving style in the simulator

Neuroticism correlated with three aspects of driving style in the simulator (Table 8.4), all related to average speed during the drive. The neuroticism facet angry hostility (tendency to experience anger and related states such as frustration and

Table 8.3 **Significant correlations between DBQ factors and driving style measures**

Factor	Aspect of driving style	Pearson *r* (N = 43)	*P* (2-tailed)
Violations	Time to complete(s)	−0.37	0.014
	Average speed (mph)	0.44	0.004
	Maximum speed (mph)	0.39	0.010
	Average RPM	0.39	0.010
	Average brake position	0.31	0.044
	SD speed	0.33	0.033
	Average speed in 12 normal curves (mph)	0.39	0.011
	Average speed in 6 sharp curves (mph)	0.36	0.016
	Average SD of lane position in 6 sharp curves	0.38	0.012
Errors	SD accelerator position	−0.31	0.044
Lapses	Maximum brake position	0.31	0.041

Table 8.4 **Significant correlations between driving style measures and scores on the domains of the NEO PI-R personality inventory**

NEO PI-R domain	Aspect of driving style	Pearson *r* (N = 43)	*P* (2-tailed)
Neuroticism	Time to complete(s)	−0.39	0.040
	Average speed (mph)	0.40	0.035
	Average speed in 12 normal curves	0.42	0.025
Openness	Average clutch position	0.41	0.032
	First overtake – time exposed to danger(s)	−0.39	0.040
Agreeableness	Time to complete(s)	0.52	0.005
	Average speed (mph)	−0.54	0.003
	Maximum speed (mph)	−0.43	0.023
	Average RPM	−0.40	0.035
	Average accelerator position	−0.41	0.031
	SD speed	−0.44	0.020
	Average speed in 12 normal curves (mph)	−0.47	0.012
	Average speed in 6 sharp curves (mph)	−0.40	0.036
Conscientiousness	Average accelerator position	−0.37	0.050
	Average speed in 12 normal curves (mph)	−0.38	0.045

bitterness) was the main contributor to the overall domain correlations. Angry hostility itself was positively correlated with other aspects of behaviour (higher speed, RPM, more accelerator and brake usage, greater variation in speed) that suggest a faster driving style. High agreeableness was associated with a slower driving style. Significant correlations with agreeableness appeared to have been driven most by correlations in the trust and modesty facets. Extraversion was not significantly correlated with any aspect of driving style. However, the extraversion facet excitement-seeking was weakly positively correlated with average speed, suggesting that it was associated with faster driving. The facet positive emotions was negatively correlated with a number of aspects of driving style: Average speed, RPM, Mean accelerator position, Mean brake position and the variations (standard deviations) of each of these. This suggests that positive emotions was positively associated with a slower and smoother driving style, involving lower overall speeds and less use of the controls.

Personality and self-reported driving behaviour

DSI aggression was positively correlated with neuroticism, and negatively correlated with agreeableness and conscientiousness (Table 8.5). Aggression also correlated positively with one specific facet of extraversion, excitement-seeking, but negatively with another, warmth. This accounts for the absence of a significant correlation with the overall extraversion domain. Thrill-seeking correlated negatively with agreeableness and with two of its facets, trust and compliance. It also correlated negatively with two facets of conscientiousness, dutifulness and deliberation. Thrill-seeking correlated positively with the extraversion facet excitement-seeking, and with two facets of neuroticism. Fatigue proneness correlated positively with conscientiousness and with three of its facets.

DCQ confrontive coping was positively correlated with neuroticism, and negatively correlated with agreeableness and conscientiousness (Table 8.6). There were significant correlations between confrontive coping and several facets of these domains. Confrontive coping also correlated positively with one specific facet of extraversion, excitement-seeking.

DBQ violations had multiple weak negative correlations with facets of openness, agreeableness and conscientiousness, and weak positive correlations with facets of neuroticism (Table 8.7). DBQ violations correlated positively with one facet of extraversion (excitement-seeking) and negatively with another (positive emotions).

Discussion

It seems reasonable to expect that measured driving style and self-reported driving behaviour would be associated to some degree. We found that measured driving style was indeed related to some factors in the DSI, DCQ and DBQ self-report

Table 8.5   Significant correlations between DSI factors and NEO PI-R domains and facets (facet names in this and subsequent tables are preceded by a letter indicating which domain they belong too: C for conscientiousness, A for agreeableness etc.)

Factor	NEO PI-R domain or facet	Pearson *r* (N = 27)	*P* (2-tailed)
	Neuroticism	**0.536**	**0.004**
	Agreeableness	**−0.623**	**0.001**
	Conscientiousness	**−0.414**	**0.032**
	N: Anxiety	0.500	0.008
	N: Angry hostility	0.618	0.001
Aggression	N: Self-consciousness	0.389	0.045
	N: Impulsiveness	0.428	0.026
	E: Warmth	−0.588	0.001
	E: Excitement seeking	0.439	0.022
	O: Openness to actions	−0.439	0.022
	A: Trust	−0.543	0.003
	A: Altruism	−0.665	<0.001
	A: Compliance	−0.411	0.033
	C: Dutifulness	−0.464	0.015
	C: Self-discipline	−0.409	0.034
Dislike of driving	N: Self-consciousness	0.495	0.009
	Agreeableness	**−0.578**	0.002
	N: Angry hostility	0.545	0.003
	N: Depression	0.397	0.041
Thrill seeking	E: Excitement seeking	0.632	<0.001
	A: Trust	−0.566	0.002
	A: Compliance	−0.417	0.030
	C: Dutifulness	−0.418	0.030
	C: Deliberation	−0.418	0.030
Hazard monitoring	E: Assertiveness	0.589	0.001
	O: Openness to ideas	0.389	0.045
	Conscientiousness	**0.489**	**0.010**
Fatigue proneness	C: Competence	0.402	0.038
	C: Achievement-striving	0.412	0.033
	C: Self-discipline	0.439	0.022

Facet names in this and subsequent tables are preceded by a letter indicating which domain they belong too: C for conscientiousness, A for agreeableness etc.

Driver Behaviour and Training

Table 8.6 **Significant correlations between DCQ factors and NEO PI-R domains and facets**

Factor	NEO PI-R domain or facet	Pearson *r* (N = 27)	*P* (2-tailed)
Confrontive coping	Neuroticism	0.588	0.002
	Agreeableness	−0.433	0.031
	Conscientiousness	−0.508	0.010
	N: Anxiety	0.551	0.004
	N: Angry hostility	0.793	<0.001
	N: Depression	0.482	0.015
	N: Impulsiveness	0.406	0.044
	E: Excitement seeking	0.478	0.016
	A: Trust	−0.526	0.007
	A: Compliance	−0.397	0.049
	C: Dutifulness	−0.461	0.020
	C: Self-discipline	−0.482	0.025
Reappraisal coping	E: Excitement seeking	−0.415	0.039
	O: Openness to values	−0.435	0.030
Avoidance coping	O: Openness to values	−0.439	0.028
Task-focused coping	E: Excitement seeking	−0.526	0.007
	C: Deliberation	0.449	0.025
Emotion-focused coping	N: Self-consciousness	0.412	0.041

questionnaires. For instance faster, more dynamic driving style in the simulator, involving higher speeds, more acceleration and braking, was positively associated with DSI thrill-seeking and aggression. These findings are partially consistent with those of Matthews et al. (1997), who reported associations between the aggression and dislike of driving factors of the DBI (predecessor of the DSI) and aspects of driving behaviour in a simulator. In our study these factors did correlate significantly with driving style, but not as strongly as did thrill-seeking. Only the violations factor of the DBQ was related to driving style. Parker et al. (1995), and Ozkan and Lajunen (2005) found that DBQ violations were positively correlated with accident risk. Our findings are consistent with this, if we assume that driving

Table 8.7   Significant correlations between the violations factor of the DBQ, and NEO PI-R domains and facets

Factor	NEO PI-R domain or facet	Pearson r (N = 27)	P (2-tailed)
Violations	N: Angry hostility	0.431	0.025
	N: Depression	0.407	0.035
	E: Warmth	−0.438	0.022
	E: Excitement seeking	0.403	0.037
	E: Positive emotions	−0.479	0.012
	O: Openness to fantasy	−0.435	0.023
	A: Trust	−0.394	0.042
	C: Self-discipline	−0.385	0.048

style mediates accident risk. These findings lend support to the validity of the self-report measures, particularly the DSI.

Turning now to personality, we return to the question posed in the introduction, how far does personality 'explain' the way drivers drive? There was a very clear negative association between agreeableness and faster, more dynamic driving style. People who scored lower on agreeableness adopted a faster driving style; people who scored highly on agreeableness adopted a slower, smoother driving style. A similar association was found with self-reported driving: agreeableness correlated negatively with DSI aggression and thrill-seeking, and DCQ confrontive coping. These findings are in agreement with that of Dorn and Matthews (1992) that affection (related to agreeableness) correlated with DBI aggression.

There were moderate associations between neuroticism and faster, more dynamic driving. Supporting this is the finding that, although the domain extraversion as a whole did not correlate with aspects of driving style, the facet positive emotions was negatively correlated with measures of faster driving style. Thus people who experience more negative emotions, and people who experience less positive emotions, tended to drive faster. Dorn and Matthews (1992) found that neuroticism correlated positively with DBI aggression and dislike of driving. In our study the correlation with DSI aggression was stronger, and although neuroticism itself was not significantly correlated with DSI dislike of driving, its facet self-consciousness was.

Conscientiousness was associated with slower, more careful driving styles, but the effects were weaker than those for agreeableness. Conscientiousness was also negatively associated with DSI aggression and DCQ confrontive coping. This is in agreement with Dorn and Matthews' (1992) finding that conscience (which relates to conscientiousness) correlated with DBI aggression. Sumer and colleagues (2005) found that conscientiousness correlated negatively but weakly with all three DBQ factors, and Arthur and Graziano (1996) found that it correlated negatively with accident risk.

Two driving style measures correlated with openness, but these were an idiosyncratic combination and we can offer no useful interpretation, beyond the possibility that it reflects a chance result. Openness was not consistently associated with any of the self-report factors.

No measures of driving style correlated with extraversion. Initially this is perhaps surprising, but it seems that two facets of extraversion, excitement-seeking and positive emotions, had oppositely directed associations: Excitement-seeking was positively associated with faster driving, while positive emotions was negatively associated. A somewhat similar pattern was found when comparing extraversion scores with self-reported driving questionnaire behaviour. Extraversion itself was not significantly correlated with any of the self-reported factors. This is in agreement with the findings of Dorn and Matthews (1992) for the DBI and Sumer and colleagues (2005) for the DBQ. However, excitement-seeking was positively correlated with DSI aggression and thrill-seeking, DCQ confrontive coping and DBQ violations, while positive emotions was negatively correlated with DBQ violations.

In the psychometric literature the FFM has been criticised on the basis that only the extraversion and neuroticism domains are fully independent (Kline, 2000). The results obtained here tend to support the view that openness is independent of conscientiousness and agreeableness, since the latter domains were associated with driving style and self-reported driving behaviour while openness was not. To a lesser extent the results support the independence of agreeableness and conscientiousness, as agreeableness was more strongly related to driving style and self-reported driving behaviour than was conscientiousness.

Figure 8.1 summarises the relationships between driving style, self-reported driving behaviours and personality. Returning to our original question, driving style is indeed related, in part, to personality. Few previous studies have related personality directly to driving style, as opposed to more distal measures.

This was a correlational study, so it is not possible to draw the formal conclusion that these aspects of personality *cause* faster and slower driving styles. However, given the stability of personality traits in adulthood, the reverse direction of causality, that driving style causes these aspects of personality, certainly seems less plausible. Given the large number of variables it is reasonable to expect some correlations might have occurred by chance, so some caution is needed in interpretation. The weak correlations between openness and two measures of driving style may best be explained as chance associations. On the other hand, where there were multiple associations (such as those between agreeableness and measures of driving style) we can be confident of real effects.

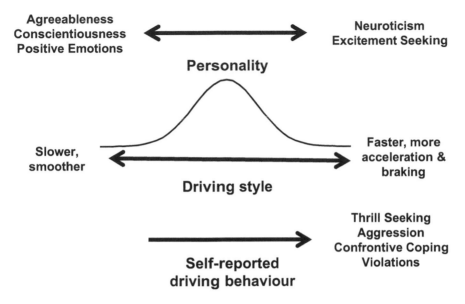

Figure 8.1   **Significant relationships between driving style, self-reported driving behaviours and domains and facets of personality**

Conclusions

Faster, more dynamic, less fuel-efficient driving is associated with high neuroticism, high excitement-seeking, low agreeableness and low conscientiousness. The more any of these are combined, the more likely it is that the person will have a more dynamic and less fuel-efficient driving style. Slower, smoother and more fuel-efficient driving is associated with the opposite combination of personality factors: low neuroticism, low excitement-seeking, high agreeableness and high conscientiousness. Of course there are other determinants of driving style too: for instance goals, moods, experience and situation. The observed correlations with personality in this study tended to be in the range $r = 0.4$ to 0.5: so while personality influences driving style, it by no means fully determines it. This suggests that there remains plenty of scope for interventions to improve fuel efficiency via changes to driving style.

References

Arthur, W. and Graziano, W.G. (1996). The five-factor model, conscientiousness, and driving accident involvement. *Journal of Personality*, 64, 3, 593–618.
Ajzen, I. (2005). *Attitudes, Personality and Behaviour*, 2nd edn. Maidenhead, Open University Press.

Benet-Martinez, V. and John, O.P. (1998). Los Cinco Grandes across cultures and ethnic groups: multitrait – multimethod analyses of the Big Five in Spanish and English. *Journal of Personality and Social Psychology*, 75, 729–750.

Benfield, J.A. et al. (2007). Driver Personality and anthropomorphic attributions of vehicle personality relate to reported aggressive driving tendencies. *Personality and Individual Differences*, 42, 247–258.

Cambridge Systematics Inc. (2009). *Moving Cooler. An Analysis of Transportation Strategies for Reducing Greenhouse Gas Emissions.* Washington, DC, Urban Land Insitute.

Catell, R.B. et al. (1970). *Handbook for the Sixteen Personality Factor Questionnaire.* Champaign, IL, IPAT.

Cellar, D.F. et al. (2000). The five-factor model and driving behaviour: personality and involvement in vehicular accidents. *Psychological Reports*, 86, 454–456.

Costa, P.T. and McCrae, R.R. (1995). Domains and facets: hierarchical personality assessment using the revised neo personality inventory. *Journal of Personality Assessment*, 64, 1, 21–50.

Dorn, L. and Matthews, G. (1992). Two further studies of personality correlates of driver stress. *Personality and Individual Differences*, 13, 8, 949–951.

Dorn, L. and Matthews, G. (1995). Prediction of mood and risk appraisals from trait measures: two studies of simulated driving. *European Journal of Personality*, 9, 25–42.

Eysenck, H.J. and Eysenck, S.B.G. (1975). *The Eysenck Personality Questionnaire* London, Hodder and Stoughton.

Furnham, A. and Saipe, J. (1993). Personality correlates of convicted drivers. *Personality and Individual Differences*, 14, 329–336.

Goldberg, L.R. (1981). Language and individual differences: the search for universals in personality lexicons. In L. Wheeler, ed., *Review of Personality and Social Psychology*, Vol. 2. Beverley Hills, CA, Sage, pp. 141–165.

Gulian, E. et al. (1989). Dimensions of driver stress. *Ergonomics*, 32, 585–602.

Huguenin, R.D. and Rumar, K. (2001). Models in traffic psychology. In P. Emmanuel-Barjonet, ed., *Traffic Psychology Today.* Norwell, MA, Kluwer.

Jonah, B.A. (1997). Sensation seeking and risky driving: a review and synthesis of the literature. *Accident Analysis and Prevention*, 29, 651–665.

Kline, P. (2000). *Handbook of Psychological Testing*, 2nd edn. London, Routledge.

Lajunen, T. (2001). Personality and accident liability: are extraversion, neuroticism and psychoticism related to traffic and occupational fatalities? *Personality and Individual Differences*, 31, 1365–1276.

Liebert, R.M. and Liebert, L.L. (1998). *Liebert and Speigler's Personality: Strategies and Issues.* Pacific Grove, CA, Brookes/Cole Publishing Company.

Matthews, G. et al. (1991). Personality correlates of driver stress. *Personality and Individual Differences* 12, 535–549.

Matthews, G. et al. (1997). A comprehensive questionnaire measure of driver stress and affect. In T. Rothengatter and E. Carbonell Vaya, eds, *Traffic and*

Transport Psychology: Theory and Application. Amsterdam, Pergamon, pp. 317–324.

Matthews, G. et al. (1998). Driver stress and performance on a driving simulator. *Human Factors*, 40, 136–149.

McCrae, R.R. and Costa, P.T. (2003). *Personality in Adulthood: A Five-Factor Theory Perspective*. New York, Guilford Press.

Özkan, T. and Lajunen, T. (2005). A new addition to DBQ: Positive Driver Behaviours Scale. *Transportation Research Part F*, 8, 355–368.

Parker, D. et al. (1995). Behavioural characteristics and involvement in different types of traffic accident. *Accident Analysis and Prevention*, 27, 4, 571–581.

Reason, J.T. et al. (1990). Errors and violations on the roads. *Ergonomics*, 33, 1315–1332.

Rimmo, P.-A. and Aberg, L. (1999). On the distinction between violations and errors: sensation seeking associations. *Transportation Research F*, 2, 151–166.

Skippon, S.M. et al. (2008). Questionnaire measures of attitudes and driving behaviour: their relationships to driving behaviour measured in a driving simulator. Driving Simulation Conference 2008 Asia-Pacific, Seoul.

Sumer, N. (2003). Personality and behavioural predictors of traffic accidents: testing a contextual mediated model. *Accident Analysis and Prevention*, 35, 949–964.

Sumer, N. et al. (2005). Big five personality traits as the distal predictors of road accident involvement. In G. Underwood, ed., *Traffic and Transport Psychology: Theory and Application*. Amsterdam, Elsevier, pp. 215–227.

Ulleberg, P. (2002). Personality subtypes of young drivers. Relationship to risk-taking preferences, accident involvement, and response to a traffic safety campaign. *Transportation Research F*, 4, 279–297.

Ulleberg, P. and Rundmo, T. (2003). Personality, attitudes and risk perception as predictors of risky driving behaviour among young drivers. *Safety Science*, 41, 427–443.

Wahlberg, af A.E. et al. (2009). The Manchester Driver Behaviour Questionnaire as a Predictor of Road Traffic Accidents, Theoretical Issues in Ergonomic Science.

Chapter 9

Public Perception of Risk of Being Caught Committing Traffic Offences

Isah Noradrenalina, M. Maslina and L. S. Kee

Introduction

This chapter attempts to examine the change in perception levels of road users on enforcement visibility and the probability of being caught for specific traffic offences during Ops Sikap XVI Chinese New Year 2008. Ops, also known as operations, is a series of enforcements which take place during the peak season of certain festivities. The specific objectives are to analyse road users' perception of being caught before during and after implementation of the Ops, determine factors that influenced road users' perception of being caught, and identify the effectiveness of overt and covert enforcement approaches.

In Malaysia, during every festive season lives are lost in an increasingly alarming number of accidents. What is supposed to be a joyful period of celebrations turns into sorrow and tragedy. The Malaysian tradition of going back to their hometown during this festival period, better known as *Balik Kampung*, should not turn into horrific memories. Tragedies and accidents seem to be the norm for every festive season – *Hari Raya Aidilfitri*, Chinese New Year, Christmas and *Deepavali*. The Chinese New Year celebration is the second most celebrated festival based on the percentage of celebrants: 30 per cent of the entire population who are ethnic Chinese. However, due to the long duration of the festive holiday, the rest of society regardless of race takes time off to be with their families and loved ones in their respective home towns. The number of deaths per day increases tremendously during the festive season compared with normal days as people travelled home. This statement is supported by statistics of Ops Sikap Chinese New Year obtained five years ago (see Table 9.1). The data showed an increase in the number of accidents every year.

This study will only focus on road users' perceptions during Chinese New Year 2008. The Ops Sikap XVI Chinese New Year 2008 started on 31 January until 9 February 2008 nationwide. As a result, the government launched an integrated safety campaign for the Chinese New Year by combining several programmes which are:

- lowering of the speed limit on state and federal roads from 90kph to 80kph during the annual festive traffic operations this Chinese New Year

Table 9.1   Ops Sikap Chinese New Year report year 2003–2007

Year	No. of accidents	No. of fatal crashes	No. of fatalities	Average daily fatalities
2007 11 to 25 February	14930	192	207	13.8
2006 23 January to 6 February	13153	205	226	15.1
2005 2 February to 16 February	11462	161	188	12.5
2004 15 January to 29 January	11481	169	174	11.6
2003 25 January to 8 February	10241	153	168	11.2

Source: Road Safety Department, Ministry of Transport.

- the prohibition of freight vehicles from roads
- overt and covert enforcement operations
- increased frequency of campaign advertisements in the media
- increased frequency of emergency services on roads.

To ensure the Ops were carried out according to plan, the relevant authorities conducted enforcement and monitoring during the festive season. A total of 6,562 enforcement personnel from the police and Road Transport Department (JPJ) were deployed nationwide for the Ops. Royal Malaysian Police conducted monitoring through 19 observation towers, 16 on federal roads and 3 on expressways, to detect vehicles committing violations. Enforcement by overt means was conducted by means of deploying enforcement personnel to book those committing serious traffic offences. This overt enforcement activity was supported by covert operations which employed the use of unmarked vehicles equipped with video cameras and speed detectors. Furthermore, Road Transport Department personnel went undercover as express bus passengers on randomly selected express buses to ensure bus drivers abide by the rules. Police also supported enforcement on these bus drivers by conducting spot checks in uniform at bus lay-bys.

In terms of penalties and fines, a mandatory compound of RM300 was charged to private vehicle owners for six types of main offences which are speeding, queue-jumping, running traffic lights on stop, using the emergency lane unlawfully, dangerously cutting into traffic and overtaking on double lines. For motorcyclists and their pillion riders, the focus was on serious offences such as improper use of safety helmets, running traffic lights, dangerous riding and weaving in traffic and dangerous changes of lanes by not signalling.

As for the express bus, tour bus and additional buses, enforcement was conducted to cover serious offences such as driving in excess of 8 hours every 24 hours whereby each driver has to rest every 2 hours on each journey, no co-driver for journeys in excess of 300km or 4 hours drive, failure of driver to log their movements in the log book and not complying to the terms of their licence.

For increased frequency through media campaigns, special road safety adverts tailor-made for the festive seasons were broadcast starting from 20 January until 17 February 2008 through television channels and radio as well as mainstream newspapers. A special address on road safety by the Prime Minister of Malaysia and the Minister of Transportation was also published in the major newspapers. All the above programmes were conducted in order to control and reduce the ever-increasing rate of accidents and fatalities on roads during every festive season.

Rationale for Traffic Enforcement

Traffic enforcement is predicated on deterrence as the mechanism to achieve behaviour change. The underlying principle of deterrence is that human behaviour can be modified by making people fearful of the consequences of their illegal actions (Gibbs, 1975; Ross, 1982; Homel, 1988; Bjornskau and Elvik, 1992, all as cited in Zaal, 1994). Road users are assumed to make rational decisions on whether or not to offend (Corbett and Simon, 1992), and, according to Palmer (1977) as cited in Zaal (1994) they will be law-abiding if the expected consequences of offending are more onerous than those of complying. Homel (1988) as cited in Zaal tested this theory using data from random breath testing and found that the actual chance of getting caught is important but that the perception of the likelihood of getting caught plays an equally important role (Vanlaar, 2007).

There are two types of deterrence, specific or simple deterrence and general deterrence. Simple deterrence is the short-term mechanism by which people react to the threat of personally experiencing penalties for traffic offences they committed. General deterrence is the long-term mechanism of habit formation and moral education which follows from exposure of the population to deterrence, over time. Traffic law enforcement is traditionally based on both types of deterrence.

Overt and Covert Enforcement Methods

Studies show that the method of traffic enforcement is an important factor in helping to modify road user behaviour and reduce crash casualties and fatalities. Highly visible (i.e. overt) enforcement acts to remind motorists that enforcement is present and potentially increases the actual and perceived risk of detection (Fildes and Lee, 1993 as cited in Zaal, 1994). Non-visible (i.e. covert) enforcement is said to heighten road users' sense of uncertainty about detection risk and to prevent them adapting their traffic offence behaviour at specific times and locations when enforcement is visibly being carried out (Ostvik and Elvik, 1990 as cited in Zaal, 1994).

Greater reductions in casualty crashes have been reported with covert rather than overt enforcement (Diamantopoulou and Cameron, 2002). Reductions were also evident when a mix of overt and covert radar enforcement was used (Diamantopoulou and Cameron, 2002).

As in previous Ops, overt and covert enforcement activities were also used during Ops CNY 2008 to reduce rule violation and crashes. Examples of overt enforcement in Malaysia are police roadblocks, patrols in marked police cars, police warning signs of speed camera presence and manned police towers. Covert enforcement would employ methods such as hidden speed cameras, unmarked vehicles, disguised police officers in buses and the camera capture of queue jumping.

The Time Halo Effect

Based on previous research, periods of actual and perceived enforcement increases the perception of the risk of detection and reduces traffic offences for certain amounts of time. This is the 'time halo effect' which is a form of the perception of the risk of detection and is specifically the length of time that the effects of the awareness of enforcement on motorist behaviour continues after the enforcement operations have ended.

What is less clear is the nature of the relationship between the amount of time that enforcement is carried out and the time halo effect. Studies have produced mixed findings. Variables which may affect results range from geographical locations (eg. urban and rural) (Nilsson and Sjogren, 1981 as cited in Cameron et al., 2003; Diamantopoulou et al., 1998) to the enforcement methods, whether it is overt or covert, or both. Time halos of ten days, following six days of enforcement have been reported (Nilsson and Sjogren, 1981, as cited in Cameron et al., 2003) as well as one to four days after enforcement (Diamantopoulou et al., 1998). However, the study only covers the effect of enforcement on casualty crashes reduction and not the effect of enforcement on the perception of being caught.

Although studies on most of the previous Ops Sikap and Ops Bersepadu did not report time halo effects, it is expected that such an effect will be seen from the perception study of Ops Sikap Chinese New Year 2008. Information on time halo effects can be useful in several respects, such as planning traffic enforcement strategies and programmes, determining the effectiveness and cost-effectiveness of traffic enforcement methods, the optional days of effective enforcement and developing or improving interventions for change in road user behaviour. The halo effect also includes distance, which is not dealt with in this research.

Method

A series of cross-sectional surveys on alternate days for six weeks were carried out before (17 to 30 January), during (31 January to 14 February) and after (15 to

28 February) the launch of the Ops. Studies were conducted at six locations: three locations were chosen from federal roads and the rest were from expressways with a total of 10,800 respondents involved in this study. Respondents were chosen based on the ratio of Malaysian population – Malays are the highest population at 60 per cent – followed by Chinese (30 per cent) and Indian (10 per cent). Locations at Sg Buloh, Genting Sempah and Seremban represent samples from expressways while Klang (Selangor), Muar (Johor) and Sg Petani (Kedah) represent samples from federal roads. For expressways, one Rest Area was randomly selected to represent each region: Sg Buloh Rest Area (northern), Genting Sempah Rest Area (east coast) and Seremban Rest Area (southern) region. For federal roads, one state was randomly selected to represent each region: Selangor (central), Johor (southern) and Kedah (northern). The locations were chosen based on high accident-prone areas. All respondents are licensed drivers and they were informed about the study and its procedures to participate in the study. The respondents were asked to fill a three-page self-administered questionnaire which consisted of Likert scale questions on their perception of enforcement activities and type of enforcement.

Results and Discussion

Sample characteristics

Table 9.2 summarises the characteristics of the sample in this study. A total of 8360 respondents (78 per cent) were male. A high percentage (54 per cent) of the respondents were between 26 and 45 years of age. The average age of the sample was 34 years. The minority group was slightly over sampled with 72 per cent Malay, 18 per cent Chinese and 10 per cent Indian and others. All respondents were fully licensed drivers. A majority of the respondents surveyed (64 per cent) frequently used a car for travelling. This is followed by 29 per cent motorcycles, the remainder were 7 per cent van drivers and others. Nearly half of the respondents (47 per cent) have experienced being summonsed. Another interesting finding was that a large number of respondents that have experienced being summonsed (90 per cent) had received between one to five summonses.

Perception study analysis

In order to observe the pattern of perception of being caught (POBC), the collected data were categorised into three main-time periods; before Ops (17 to 30 January), during Ops (31 January to 14 February) and after Ops (15 to 28 February). The percentage collected was from 0 to 100 per cent where 0 per cent indicates perception of not being caught at all while 100 per cent would indicate a sure perception of being caught. The study was done at locations in expressways and federal roads throughout Peninsular Malaysia.

Table 9.2   Respondents characteristics

Variables	Frequency (N)	%
Gender (N = 10766)		
Male	8360	77.7
Female	2406	22.3
Age of the respondents (N = 10743)		
16 to 25	3179	29.6
26 to 35	3396	31.6
36 to 45	2389	22.3
46 to 55	1383	12.9
More than 55	389	3.6
Mean 34		
Range 16–85		
Ethnicity (N = 10788)		
Malay	7748	71.8
Chinese	1975	18.3
Indian and others	1065	9.9
Type of licence (N = 10796)		
D	8637	80.0
B/B1/B2	6991	64.7
E/E1/E2	698	6.5
Vehicle frequently used (N = 10764)		
Car	6883	63.9
Motorcycle	3064	28.5
MPV/SUV/4WD	444	4.1
Van	302	2.8
Others	71	0.7

Table 9.2   Respondents characteristics *concluded*

Variables	Frequency (N)	%
Experience of being summonsed (N = 10799)		
No	5763	53.4
Yes	5036	46.6
No. of summonses received (N = 5036)		
1–5	4543	90.2
6–10	410	8.1
More than 10	83	1.6
Mean 3		
Range 1–40		

Just before the Ops started, there was a slight fluctuation with an average of 40 per cent for both expressways and federal roads. At the beginning of the Ops, the POBC was still low at an average of 46 per cent. This might be due to low awareness that Ops Sikap had started. The percentage peaks to an average of 57 per cent during the middle of the Ops and increases steadily straight after the Ops. This is shown in Figure 9.1 and demonstrates the time halo effect of having the Ops. Towards the end of the Ops, the POBC percentage started to drop down to 52 per cent which was about the same percentage as just before the Ops started. The average POBC percentage after the Ops fluctuates around 42–52 per cent.

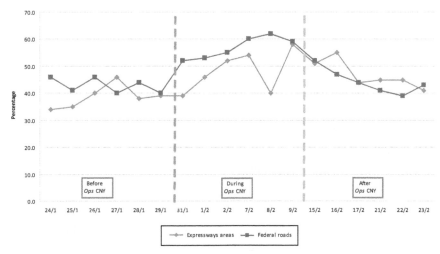

Figure 9.1   Perception of being caught at all locations surveyed

Expressway locations

For expressway travellers, overall POBC was increased from 39 per cent before the Ops to 54 per cent during the Ops in respect of all offences. This was a good indicator showing the awareness of road users of Ops being launched and taking effect. Of all the offences, the POBC for speeding was at a record high of 48 per cent before the Ops and increased during the Ops to 60 per cent. The POBC for speeding percentage rose again to 67 per cent as the Ops was nearly over before it decreased to an average of 50 per cent after the Ops period. The overall perception in expressways increased to 58 per cent just after the Ops ended. This was probably due to the expectation that enforcement was still being implemented even after the Ops ended. Results of POBC for expressway travellers are shown in Figure 9.2.

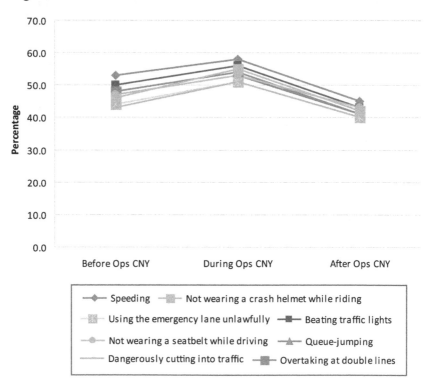

Figure 9.2   Percentage of perception of being caught on expressways

Federal roads

For federal roads, overall POBC was increased from 40 per cent before the Ops to 60 per cent during the Ops in all offences. This was a good indicator showing the awareness of road users of Ops being launched and taking effect. Of all the offences, the POBC for speeding was at a record high of 48 per cent before the Ops and increased significantly during the Ops to 58 per cent. The POBC for speeding percentage increased again to 62 per cent as the Ops was nearly over before it decreased to an average of 44 per cent after the Ops period. The overall perception in expressways slightly decreased to 59 per cent just after the Ops ended. This was probably due to the expectation that enforcement would decrease after the Ops ended. Figure 9.3 below shows the POBC on federal roads.

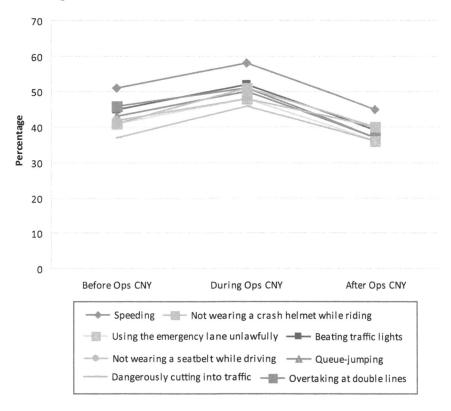

Figure 9.3   Percentage of perception of being caught on federal roads

Traffic offences

Although the perception of being caught increased overall during the Ops (see Table 9.3), road users felt that the probability of being caught was higher for some offences than for others. These data were obtained by respondents ranking the probability of detection for nine offences on the basis of their experience and knowledge. Overall, in descending order, the offences of speeding, beating traffic lights and overtaking on double lines were perceived as posing the highest risk of detection (see Table 9.3).

Table 9.3   Traffic offences according to possibility of being detected among respondents

Items	Percentage		
	Before	**During**	**After**
Speeding	48	58	52
Beating traffic lights	42	55	47
Queue-jumping	41	52	44
Overtaking at double lines	43	53	45
Using the emergency lane unlawfully	38	50	42
Not wearing a front seatbelt while driving	40	50	46
Not using a crash helmet while riding	38	53	47
Dangerously cutting into traffic	36	48	43

Factors that influenced their perception

It is important to bear in mind that in general, most self-report instruments (i.e. questionnaires) cannot reveal the causes of a phenomenon. Surveys can however reveal whether perceptions are associated with identified research variables. Understanding which variables are associated with the sharp rise in motorist perception of being caught during the Ops would help in the improvement of current interventions and the development of new behaviour modification programmes. Basically, four major factors affect the public's perception of being caught. These are: (a) visibility of enforcement activities, (b) media exposure and road safety campaign, (c) experience of being caught, and (d) traffic punishment. The factors which respondents on expressways and federal roads (respectively) said most influenced their perception of being caught were traffic punishment (67–69 per cent), media exposure and road safety campaigns (64–65 per cent), followed by the visibility of enforcement activities (55–57 per cent). See Figures 9.4 and 9.5 for details.

Overt and covert approaches

The survey also investigated the effectiveness of both the overt and covert approaches in influencing road user perception of being caught. On expressways, the perception of being caught by covert enforcement methods was slightly higher in all the periods surveyed – before 51 per cent, during 54 per cent and after 55 per cent – than the perception of being caught by overt enforcement activities: before 45 per cent, during 53 per cent and after 51 per cent (see Figure 9.6). While this

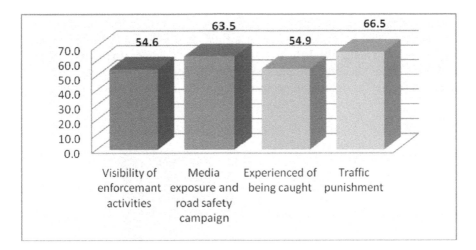

Figure 9.4   **Factors associated to the perception of being caught – expressways**

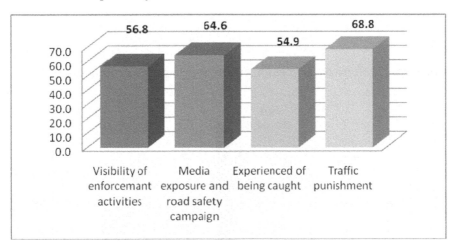

Figure 9.5   **Factors associated to the perception of being caught – federal roads**

greater perception has not yet been mapped against crash statistics during those time periods, this pattern of higher perception risk from covert approaches appears consistent with findings in other countries where covert enforcement produced greater reduction in casualty crashes than did overt methods (Diamantopoulou and Cameron, 2002). Results from federal roads showed the same pattern as expressways. From Figure 9.7, it is apparent that the perception of the risk of being caught was higher from overt enforcement – before 52 per cent, during 53 per cent and after 46 per cent the Ops – compared with covert methods during the same period of time – 46 per cent before, during 49 per cent and 48 per cent after Ops. Note however that the differences are marginal. In conclusion, for all locations, the perception of being caught by covert enforcement methods was slightly higher in all the periods surveyed (48 per cent) than the perception of being caught by overt enforcement activities (46 per cent). Again the differences are marginal.

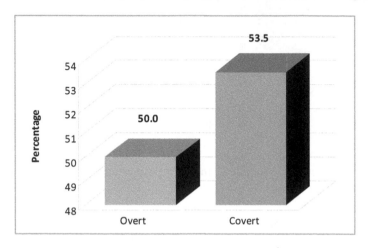

Figure 9.6   Effectiveness of enforcement approaches – expressways

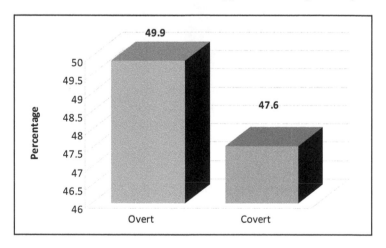

Figure 9.7   Effectiveness of enforcement approaches – federal roads

Conclusion

Ops Sikap Chinese New Year 2008 executed by all relevant agencies show a reduction in fatality, 190 deaths compared to 226 deaths in 2006. A reduction of

17 death or 8.2 per cent throughout Ops Sikap XVI compared to the previous Ops shows that the effort by traffic officers was effective. However, the number of accidents increased to 14,991 cases compared to 14,930 cases during the previous Ops. This shows that although enforcement was thorough during that period accidents still occurred. Nevertheless, along with support with other initiatives such as media exposure on road safety campaigns, visibility of enforcement, and experience of being summonsed, it contributed to a reduction in fatality rate during this study. It is an indicator that all agencies must collaborate in increasing the perception risk of being caught rate among the public to produce a more safety-oriented driving culture in Malaysia.

References

Bjornskau, T. and Elvik, R. (1992). Can road traffic law enforcement permanently reduce the number of accidents? *Accident Analysis and Prevention*, 24(5), 507–520.

Cameron, M. et al. (2003). *Interaction between Speed Camera Enforcement and Speed-related Mass Media Publicity in Victoria*. Report No. 201, Monash University Accident Research Centre, Victoria, Australia. http://www/monash. edu.au/muarc/reports/muarc201.pdf>, accessed 1 June 2008.

Corbett, C. and Simon, F. (1992). Police and public perceptions of the seriousness of traffic offences. *British Journal of Criminology, Delinquency and Deviant Social Behaviour*, 31(2), 153–164.

Diamantopoulou, K et al. (1998b). *Evaluation of Moving Mode Radar for Speed Enforcement in Victoria 1995–1997*. Report No. 141, Monash University Accident Research Centre, Victoria, Australia. http://www/monash.edu.au/ muarc/reports/muarc187.pdf>, accessed 3 May 2008.

Diamantopoulou, K. and Cameron, M. (2002). *An Evaluation of the Effectiveness of Overt and Covert Speed Enforcement Achieved through Mobile Radar Operations*. Report No. 187, Monash University Accident Research Centre.

Fildes, B.N. and Lee, S. (1993). *The Speed Review: Road Environment, Behaviour, Speed Limits, Enforcement and Crashes + Appendix of Speed Workshop Papers*. Canberra, ACT, Federal Office of Road Safety FORS/Rosebery, NSW, Road and Traffic Authority of New South Wales RTA, Road Safety Bureau RSB, Report No. CR 127/127A (FORS) Consultant Report; CR 3193/CR 3193A (RSB).

Gibbs, J.P. (1975). *Crime, Punishment and Deterrence*. New York: Elsevier.

Homel, R. (1988). *Policing and Punishing the Drinking Driver. A Study of General and Specific Deterrence*. New York, Springer Verlag.

Nilsson, E. and Sjogren, L.O. (1981). Relationship between enforcement, traffic speeds and traffic accidents. Paper presented at the International Symposium on the Effect of Speed Limits on Traffic Accidents and Fuel Consumption, Ireland.

Ostvik E. and Elvik, R. (1990). The effects of speed enforcement on individual road user behaviour and accidents. In *Enforcement and Rewarding: Strategies and Effects: Proceedings of the International Road Safety Symposium in Copenhagen, Denmark, September 19–21, 1990*, pp. 56–59.

Palmer, J. (1977). Economic analyses of the deterrent effect of punishment: A review. *Journal of Research in Crime and Delinquency*, 14, 4–21.

Ross, H.L. (1982). *Deterring the Driving Driver: Legal Policy and Social Control*, revised and updated edition. Lexington, MA, Lexington Book.

Vanlaar, W. (2007). Less is more: The influence of traffic count on drinking and driving behaviour, *Accident Analysis and Prevention*, 40, 1018–1022.

Zaal, D. (1994). *Traffic Law Enforcement: A Review of the Literature*. Report No. 53, Monash University Accident Research Centre. http://www/monash.edu. au/muarc/reports/muarc053.pdf>, accessed 21 July 2008.

Chapter 10

Rear Seatbelt Wearing in Malaysia: Public Awareness and Practice

Norlen Mohamed, Muhammad Fadhli Mohd Yusoff
and Isah Noradrenalina

Introduction

Road traffic fatalities are expected to increase as the world continues to motorise; similarly, Malaysia is a country undergoing rapid motorisation. A steady increase in the number of new vehicles registered each year has been reported (see Figure 10.1). In 2007, a total of 1,034,418 new vehicles were registered including 16,825,150 vehicles registered in Malaysia of which 44.3 per cent were cars (Road Transport Department, 2007). The World Health Organization report estimates that traffic accidents are responsible for approximately 1.2 million deaths per year and 50 million injured people all over the world (WHO, 2004). Based on the current trend, projections estimated that these accidents will be the third most common cause of death by 2020 in the world (Murray and Lopez, 1996). In Malaysia, road traffic accidents were the third leading cause of death amongst males, and seventh most common cause of death amongst females in the year 2000 (Institute for Public Health, 2004). In 2005, of 6,200 road traffic deaths registered, car drivers and passengers made up about 20 per cent of the deaths which is second to the number of deaths involving motorcycles accidents (57.8 per cent) (Royal Malaysian Police, 2005). These alarming situations resulted in the formulation of the Malaysian government's Road Safety Plan 2006–2010.

Promotion of seatbelt wearing

The Malaysian Road Safety Plan 2006–2010 was established to achieve the following goals:

- Reduce the number of road traffic deaths per 10,000 registered vehicles by 52 per cent from 4.2 per cent in 2005 to 2 per cent in 2010.
- Ten deaths per 100,000 population compared to 23 deaths per 100,000 population in 2005; and ten deaths per one billion vehicles per kilometre travelled compared with 18 deaths per one billion vehicles per kilometre travelled in 2005.

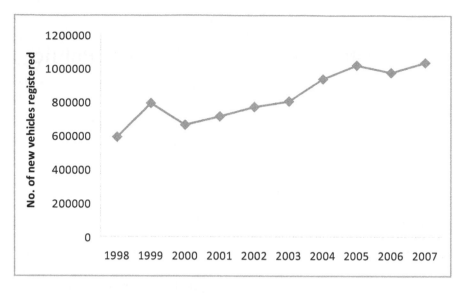

Figure 10.1   New vehicle registrations in Malaysia, 1998–2007 (Road Transport Department, 2007)

Nine strategies have been formulated and documented in the Plan as a framework for the government to work with stakeholders to achieve the set goals by 2010 (Road Safety Department, 2006). Strategy No. 7 focuses on high risk road users most frequently involved in accidents, that is, motorcyclist, pillion riders, car drivers and passengers and pedestrians. With regards to drivers and passengers safety, seatbelt wearing is one of the interventions initiated by the government to reduce death and severely injured cases. To support the Malaysian Road Safety Plan 2006–2010, a Strategic Road Safety Interventions and Potential Fatality Reduction 2007–2010 was estimated by a road safety expert (Radin, 2007). Under this strategic interventions plan a target of 80 per cent of rear seatbelt wearing has been set for Malaysia by 2010. The achievement of this target would be expected to save 84 lives of back-seat passengers per year; the estimation was made based on the average number of back-seat passenger deaths per year (350) and the potential reduction of 30 per cent of rear seatbelt effectiveness in preventing death (Table 10.1). This is an underestimation of the real potential of rear seatbelt effectiveness as a few scientific studies have shown a higher effectiveness rate of rear seatbelt in preventing death of rear passengers involved in a crash. For instance, Shimamura and colleagues (2005) showed a reduction of 45 per cent in the killed or severely injured when previously unbelted passengers began to use seatbelts. Morgan (1999) reported that back-seat lap belts and shoulder belts are 32 per cent (95 per cent confident limit: 23 to 40 per cent) and 44 per cent (95 per cent confident limit: 35 to 50 per cent) effective in reducing fatalities compared to unrestrained back seat occupants.

Table 10.1   Strategic Road Safety Interventions and Potential Fatality Reduction 2007–2010 (Radin, 2007)

Program	% Intervention coverage				Potential reduction %	No. of deaths	Expected No. of fatality reduction			
	2007	2008	2009	2010			2007	2008	2009	2010
AES										
Speed cameras	20	60	100	100	30	1400	84	252	420	420
Red light cameras	20	60	90	90	40	150	12	16	54	54
Lane discipline	0	20	60	80	20	450	0	18	54	72
Helmet program	30	65	100	100	50	1500	225	488	750	750
Airbags	10	20	40	60	30	400	12	24	48	72
Driver training	10	30	50	60	10	300	3	9	15	18
RSE and CBP	10	20	50	80	20	400	8	16	40	64
Motorcycle lanes	10	20	30	40	80	500	40	80	120	160
Black spots	10	20	30	40	30	500	15	30	45	60
Others	10	20	30	40	20	350	7	14	21	28
						6300	427	1009	1630	1782
Deaths/10,000 vehicles							3.45	2.94	2.45	2.21

Implementation of rear seatbelt policy: A feasibility study

Prior to the government decision to implement seatbelt wearing, a preliminary study on rear seatbelt availability and accessibility in Malaysia was carried out by Malaysian Institute of Road Safety Research (MIROS), and the result was recommended to the government for implementation. The study aimed to provide the government with more information to better understand the rear seatbelt scenario in Malaysia before devising road safety strategies and policy. Based on the study, almost 90 per cent of the passenger vehicles carried only three passengers or fewer, and 10 per cent of road users who carry additional passengers will have difficulty complying with the law. A figure of 64.2 per cent of passenger vehicles observed in rural areas, 75 percent in urban areas, and 87 per cent on long trips were fitted with three rear seatbelts. The study concluded that most Malaysians cars are fitted with rear seatbelts and that most Malaysians have access to rear seatbelts (Rahmat et al., 2007). The following measures were recommended to the government: (1) passenger vehicles that have rear seatbelts should comply with the law; (2) passenger vehicles that have no rear seatbelts but are able to be retro-fitted with seatbelts should be given a grace period to do so; (3) passenger vehicles that have no rear seatbelts and are unable to be retro-fitted with seatbelts should be exempted.

Rear seatbelt policy Effective on 1 January 2009, the rear seatbelt regulation is enforced on all types of private passenger vehicles except as follows:

1. vehicles registered before 1 January 1995 (not compulsory but are encouraged to retro-fit and use, if without any technical constraint)
2. vehicles registered after 1 January 1995 that do not have anchorage points
3. commercial vehicles including taxis and rental vehicles
4. passenger vehicles with more than eight seats including the driver's seat
5. goods vehicles with permissible laden weight of more than 3.5 tonnes.

For vehicles registered after 1 January 1995 without rear seatbelts fitted but with anchorage points, vehicle owners are given a grace period of three years to retro-fit. This grace period started from 1 January 2009. If the number of passengers in the vehicle is more than the number of seatbelts available, the driver of the vehicle concerned is responsible for ensuring that all the seatbelts are used. Enforcement action would be meted out if they fail to comply.

Advocacy campaign to promote seatbelt wearing

The policy of rear seatbelt wearing was implemented using a stepwise approach to avoid it being viewed as 'hastily implemented' by the public. The government launched a national advocacy campaign beginning in June 2008, i.e., six months prior to the effective date of enforcement. The first three months of the campaign focused heavily on raising public awareness of the benefit of wearing a rear seatbelt and the upcoming effective date of rear seatbelt law enforcement, while the elements of stern advice and warnings by enforcement officers were included in the second three-month advocacy period. By 1 January 2009, law enforcement was implemented in addition to the existing advocacy activities (Figure 10.2). For the campaign, the government adopted a theme 'click it front and back, you can make a difference', to motivate the desired behaviour change. A combination of message delivery methods were used during the campaign which included paid mass media, campaigns by the roadside, community forum and delivery of pamphlets to road users. For a mass media method of delivery, before airing to the public, message pre-testing was conducted among a group of people to ensure that they would understand the messages presented.

Evaluation of the policy implementation

The evaluation process was strategised in phases in line with the stepwise policy implementation as shown in Figure 10.2. It served as a feedback mechanism for the next phase of the stepwise policy implementation. The objective of this report was to present the achievements of the first three-month advocacy programme (Phase 1) and highlight some of the recommendations that should be included in the next step of advocacy activities. Public awareness level and rear seatbelt wearing rates was the outcome of interest in this evaluation.

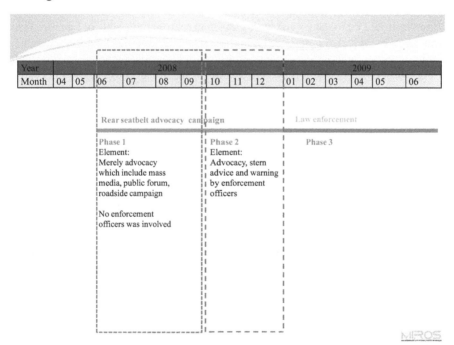

Figure 10.2 A stepwise approach of the implementation of rear seatbelt policy

Method

Observational survey on rate of rear seatbelt wearing

Seven states were selected to represent different regions in Malaysia for the survey on rear seatbelt wearing in the Malaysian population. The survey was conducted in three states that represent non-expressway road and three main expressways to represent the expressway road in Malaysia. Two districts were selected from each state – one urban and one rural. The survey targeted all occupants of light-duty vehicles, which included car, light truck, van, multipurpose vehicles (MPV) and sport utility vehicles (SUVs). Each observation period was for a period of two hours and took place during daylight hours (between 09:00 and 18:30). Over a three-month period, a total of 18,741 rear occupants were observed during the course of the survey.

Survey on public awareness

The survey was conducted on a total of 1,767 respondents in the states of Selangor and Penang area. Two districts were selected from each state. The districts of Shah Alam and Kuala Selangor were selected to represent the state of Selangor while Butterworth and Permatang Pauh were chosen to represent the state of Penang. The locations were selected based on the highest (Penang) and the lowest (Selangor) fatality rates per total registered vehicles by state. The respondents were informed of the study objectives and procedures, and on receiving consent, the interview was conducted.

Results and Discussion

Observational survey on rate of rear seatbelt wearing

Based on the observations made in the seven states from June to August 2008, generally the rate of rear seatbelt use in this country was still low. The overall rate was 2.9 per cent for June and July and 2.5 per cent for August, as shown in Figure 10.3. The first two months after the launch of the advocacy programme, the seatbelt wearing rate showed an increasing pattern though the extent of the change was small. However, in August, all localities exhibited a decreasing trend except in rural areas. The study has shown that urban areas had a higher seatbelt wearing rate compared to rural area. By type of road, rear seatbelt wearing rate was higher on expressway as compared to non-expressway. As the overall rear seatbelt wearing rate was low, the breakdown figure by gender and type of vehicle was not analysed. There was no specific reason to explain the downward trend that was observed at all localities in August 2008.

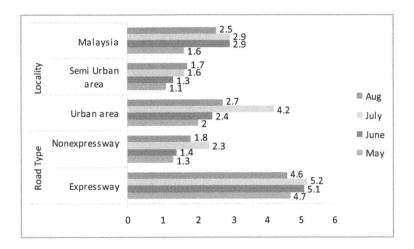

Figure 10.3   Rear seatbelt wearing rate by type of road and locality, May–August, 2008

Survey on public awareness of rear seatbelt wearing

The respondents were predominantly male (60 per cent) as compared to female with 85 per cent of them being 26–55 years old. Twelve per cent of the respondents belong to the younger age group (17–25 years old), and only three per cent of respondents were from the older age group (more than 55 years old).

Public awareness on the upcoming rear seatbelt law

Figure 10.4 shows that public awareness level on the upcoming mandatory use of rear seatbelts is very low. Almost 80 per cent of the respondents stated that they

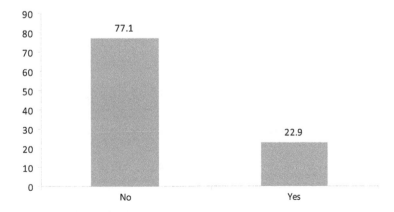

Figure 10.4   Public awareness of impending enforcement of rear seatbelt law

were unaware of the impending rear seatbelt law. This finding highlights that the public had not received adequate messages on the effective date of rear seatbelt law implementation. Based on the principles of justice and human rights, the public must be well informed about any law that is to be implemented before a fine can be imposed for violating the law. More advocacy activities were needed to address this sufficiently to avoid public panic and complaints on implementation of the law.

Knowledge on benefits and attitude toward rear seatbelt wearing

More than 70 per cent of the respondents were aware that using rear seatbelts can save lives and reduce serious injuries, as shown in Figure 10.5. To some extent, advocacy programmes had been successful in educating the public about the benefits of using rear seatbelts. However, more than 60 per cent of the respondents stated that they would only use a seatbelt if it is made mandatory. These findings highlight the importance of the enforcement component to influence compliance behaviour to rear seatbelt use among the public. It is well known that as enforcement activities increase, the compliance rate to traffic rules also increases (Kulanthayan, 2001). The same finding was recorded by Kulanthayan et al. (2004) who found that car drivers and front-seat passengers would be twice as likely to comply with seatbelt use if enforcement activities were in place. Other studies also indicated similar findings and confirmed that enforcement is a powerful tool to change road user behaviour (Kaye et al., 1995; Rothengatter, 1990). This explains the low rate of seatbelt use reported in our study and justifies the need for inclusion of enforcement activities in addition to the existing advocacy programmes.

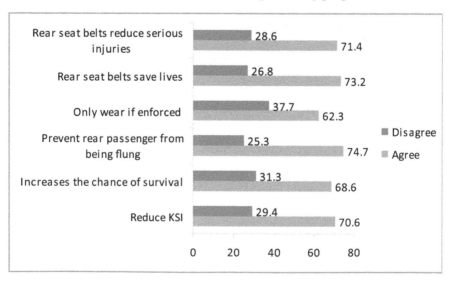

Figure 10.5   Public awareness on benefit and attitude towards rear seatbelt wearing

Reasons for not wearing rear seatbelt

Figure 10.6 shows various reasons given by the respondents for not wearing a rear seatbelt. The five main factors were related to negative attitude which include; it is troublesome; the rear seatbelt is unduly constraining; do not know that they should wear it; it is uncomfortable; and laziness. Nearly 30 per cent of respondents did not wear a rear seatbelt because of wrong perceptions and feeling safe without a rear seatbelt. More than 10 per cent responded that they did not wear a rear seatbelt because of seatbelt issues which included: the seatbelt was not functioning, seatbelts were not fitted in the car, or not enough seatbelts for rear passengers. The issue of not enough seatbelts for rear passengers is very sensitive as the average number of children per family in Malaysia is more than three. Even though the culture of staying in a nuclear family is preferred by most Malaysians, staying in the extended family is still widely being practiced. Because of the sensitivity of the situation the government had passed the message to the public and the enforcement agencies that those who fail to wear a rear seatbelt because there are not enough seatbelts in the car should not be fined.

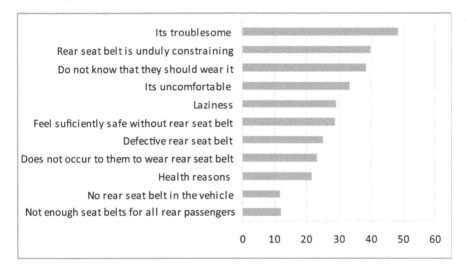

Figure 10.6 Reasons for not wearing rear seatbelts

Conclusion

The advocacy programme had been successful in making more than 70 per cent of the public aware of the importance of rear seatbelt use, including how rear seatbelts can help in reducing the risk and severity of injury. However, this evaluation has found no significant improvement in terms of the use rate despite the three-month advocacy effort by the government. It is clear that the message had reached the public through the advocacy programme but was not translated into practice because of an attitude problem. This is supported by data that more than 60 per cent of the respondents stated that they would only use the rear seatbelt if the law is enforced even though they know that seatbelt use saves lives. With regard to the impending law that the mandatory use of rear seatbelts would begin on 1 January 2009, the message had not received adequate attention from the public.

The study served as a feedback mechanism for the improvement of the stepwise advocacy campaign initiated by the government. The findings of this evaluation suggested the need for the inclusion of a visible enforcement element in the next step of the advocacy programme for a greater impact on changing public behaviour towards the use of rear seatbelt, as well as the advocacy campaign needing to address the issues highlighted by the public as the law of rear seatbelt use comes into force.

References

Institute for Public Health (2004). *Malaysian Burden of Disease and Injury Study*. Ministry of Health Malaysia.
Kaye, B.K. et al. (1995). Increasing seat belt use through PI&E and enforcement: the thumbs up campaign. *Journal of Safety Research*, 26, 235–245.
Kulanthayan, K.C.M. (2001). Modelling of compliance behaviour of motorcyclists to proper usage of safety helmet in Malaysia. Doctoral Thesis. Universiti Putra Malaysia.
Kulanthayan S. et al. (2004). Seatbelt use among car users in Malaysia. *IATSS Research*, 28, 19–25.
Morgan, C. (1999). *Effectiveness of Lap/Shoulder Belts in the Back Outboard Seating Position*. NHTSA Report Number DOT HS 808945. Springfield, Virginia, NHTSA.
Murray C.J. and Lopez A.D. (1996). *The Global Burden of Disease: A Comprehensive Assessment of Mortality and Disability from Diseases, Injuries and Risk Factors in 1990 and Projected to 2020*. Boston, MA, Harvard School of Public Health.
Rahmat et al. (2007). *An Assessment of Rear Seatbelt Availability and Accessibility in Malaysia: A Preliminary Study*, MRR 04. Kuala Lumpur, Malaysian Institute of Road Safety Research.
Radin Umar, R.S. (2007). Integrated approach to road safety in Malaysia. The 7th Malaysian Road Conference, Sunway 2007.
Road Safety Department (2006). *Malaysia Road Safety Plan 2006–2010*. Putrajaya, Department of Road Safety.
Road Transport Department (2007). *Malaysia's New Car Registration, 1998–2007*. Putrajaya, Road Transport Department of Malaysia.
Rothengatter, T. (1990). Behaviour change on the road. *Journal of International Association of Traffic and Safety Sciences*, 14(1), 102–105.

Royal Malaysian Police (2005). *Statistical Report on Road Accidents in Malaysia – 2006*. Bukit Aman, Kuala Lumpur, Traffic Branch, Royal Malaysian Police.

Shimamura, M. et al. (2005). Method to evaluate the effectiveness of safety belt use by the rear passengers on the injury severity of front seat passengers. *Accidents Analysis and Prevention*, 37, 5–17.

WHO (2004). *World Report on Road Traffic Injury and Prevention*. Genvea, World Health Organization.

Chapter 11

The Continuous Evaluation of Driver Rehabilitation Programmes in Austria

Julia Bardodej, Franz Nechtelberger and Martin Nechtelberger

Introduction

Driver rehabilitation programmes are systematic road safety interventions originally developed in the United States in the 1950s and 1960s. Like other road safety measures, driver rehabilitation programmes are intended to reduce the number of road accidents, injuries and fatalities but they are aimed specifically at drivers who have already committed an offence, in particular driving while impaired by alcohol and speeding. Rehabilitation courses are designed to influence attitude and potentially the behaviour of this high-risk group of drivers given that the punishment of withdrawal of driving licences alone has proved to be insufficient for the prevention of repeated traffic offences. Therefore, traffic offenders will usually have to participate in a rehabilitation course in order to regain or keep their licence to drive (Bartl et al., 2002).

In Austria, the development of driver rehabilitation programmes first started in 1973. In 1997 the attendance of rehabilitation courses became compulsory for certain types of traffic offences, such as driving with a blood alcohol level of 120 millilitres (1.2 per cent) or above. The attendance of a drink-drive rehabilitation course also became compulsory for drivers holding a probationary licence who get caught driving with a blood alcohol concentration of 0.1 per cent or higher, for drivers who refuse to take a breath test when asked to do so by police authorities, and for drivers who have committed at least two offences within the Demerit Point System if at least one of those offences was alcohol-related. In 2002 the basic contents and structure of driver rehabilitation courses were clearly defined by the official Austrian Rehabilitation Course Regulation (Republik Österreich, 2002). The rehabilitation programme thereby has to consist of 15 to 18 units of 50 minutes each and has to comprise at least four course sessions, or five sessions for repeated offenders, and be completed within 22 to 40 days. There are 6 to 12 participants allowed on each course. Only in exceptional cases may the programme also be conducted as an individual training. The courses are only to be offered by officially registered course providers and must be conducted by a registered traffic psychologist. As driver rehabilitation programmes are costly measures both in their development as well as in their implementation and the very purpose of the courses is defined through their effectiveness, the Austrian Rehabilitation Course

Regulation recommends that programmes should be evaluated on a regular basis in order to ensure the highest possible quality of the courses and allow for the integration of up-to-date research findings into the course design.

In order to maintain a high quality standard of their provided services, Austrian Applied Psychology Ltd (AAP) are conducting regular evaluations of their driver rehabilitation courses in cooperation with the Institute of Economic Psychology, Educational Psychology and Evaluation of the University of Vienna. The AAP's summative evaluations are based on the evaluation model by Kirkpatrick (1996), which is a hierarchical model consisting of four levels. The first and lowest level is entitled 'reactions' and comprises the programme participants' acceptance and satisfaction with the programme, the second level, 'learning', includes the acquisition of knowledge, attitudes and skills and is required in order to reach the next and third level, 'behaviour'. The third level refers to actual changes in behaviour and the transfer of the acquired knowledge into real-life situations. The fourth and final level is called 'results', meaning results on an organisational or institutional level such as reduction in recidivism or crashes. Figure 11.1 shows Kirkpatrick's model applied to the evaluation of driver rehabilitation courses. Evaluations of the AAP's courses focus on the first two levels of this model.

The first evaluation was conducted in 2003 and focused on changes in attitudes and knowledge throughout the driver rehabilitation course for drink-drive offenders as well as the general acceptance of the measure and its related legislation (Schickhofer, 2003). The course participants took part in a survey before ($n_{t1} =$ 248) and after ($n_{t2} = 221$) the course as well as six weeks after completion of the course ($n_{t3} = 67$). The majority of participants, 83.6 per cent, were male and 59.6

Stage 1

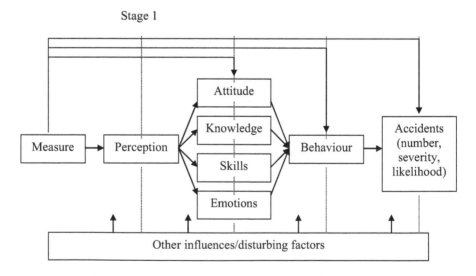

Figure 11.1   The four stages of effect measurement (Utzmann, 2008)

per cent were between 24 and 44 years old. The participants' educational level approximately represented that of the general Austrian population with 56.6 per cent having completed compulsory school, with an apprenticeship or professional training as their highest education. The majority of participants (66.5 per cent) were employed as workers or civil servants. The results showed a significant increase in offence-related knowledge as well as a more positive attitude towards laws concerning driving under the influence of alcohol and the measure itself. After the course, a significantly greater number of participants approved of a lower legal alcohol limit when driving a motorised vehicle (see Figure 11.2). Furthermore, participation in the course led to a greater awareness and understanding regarding the severity of the committed offence and a more realistic view regarding own abilities.

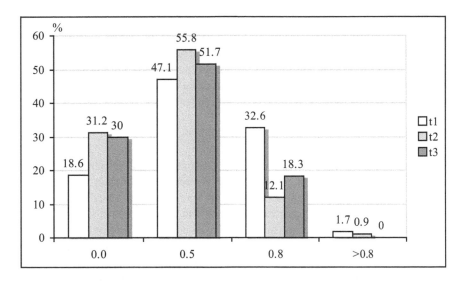

Figure 11.2   Blood alcohol concentration level which participants of the first evaluation (in percentages) believed to be the adequate legal limit at three different times of measurement

For the second evaluation three main course goals were defined (Lüftenegger, 2006). The primary goal was the achievement of changes in the participants' attitudes, which may lead to positive changes in their behaviour. The secondary goals were the enhancement of self-efficacy and internal attribution of their drink-driving behaviour. The third goal was the transfer of offence-related knowledge to actual behaviour as a wide number of alcohol-related offences are based on a lack of knowledge about effects of alcohol. Additionally, it was expected that effects of the courses might differ between first-time offenders and repeated offenders. Therefore, the purpose of the study was to investigate if the rehabilitation course for drink-drive offenders leads to changes in participants' attitudes and/or an increase in offence-related knowledge and if there are any differences in these effects between first-time offenders and offenders who have already attended one or more rehabilitation courses. A secondary purpose was to investigate how the participants rate the courses and trainers.

Methods

Participants

The evaluation was conducted in eight different branches of the AAP across Austria. In total, 285 course participants took part in the evaluation, 248 of which were male (87 per cent) and only 37 were female (13 per cent). The age of the participants varied between 18 and 74 years with an average of 37.19 (SD = 12.08) years. For the majority of drivers (37.2 per cent), the registered blood alcohol concentration which had led to the compulsory participation in the rehabilitation measure had been 160 millilitres (1.6 per cent) or higher. The questionnaires were administered to 268 drivers and could be used for further statistical calculations. Within this sample, 58 had already taken part in a rehabilitation course at least once.

Measures

The data were collected using a standardised questionnaire administered by course trainers. Successful completion of the course is necessary to regain their driving licence, so it was decided that participants remain anonymous to reduce the influence of social desirability on their responses to the questionnaire. This was achieved through a special coding system. The evaluation was based on a pre-course and post-course design; therefore the data collection took place before the first and after the last course unit. Individual attributional styles were assessed using Krampen's Questionnaire for Locus of Control and Competence Beliefs (FKK) (Krampen, 1991). This questionnaire consists of four primary scales with eight items each. The scales 'self-concept of own abilities' and 'internality' provide information on internal attributions, while the scales 'socially caused externality' and 'fatalistic externality' offer insights into external attributions. General self-efficacy beliefs were assessed with a scale by Schwarzer and Jerusalem (1999), which investigates how efficient participants believe themselves to be at dealing with everyday problems. Additionally, specific self-efficacy was also measured using a scale inspired by Kases (2002) and Bandura's concept of self-efficacy (Bandura, 1986). The idea behind this scale was to determine how likely people were to resist driving under the influence of alcohol under certain circumstances, such as friends expecting a lift home or not having much money left for a cab. The amount of knowledge acquired throughout the course was measured with six multiple-choice items regarding effects of alcohol and legal measures for driving while intoxicated. Finally, three items were included in the post-test in which the participants could rate the course and the trainer on a five-point scale.

Data analysis A multivariate analysis of variance with repeated measurements (MANOVA) was conducted to find out if the course leads to changes in the participants' attitudes or an increase in knowledge and if these findings differ between first-time offenders and offenders who had already attended one or more driver rehabilitation courses. Univariate analyses were also conducted for each scale.

Results

The univariate analyses resulted in significantly higher values for post-course compared with pre-course responses for four out of seven scales. The largest difference was found for offence-related knowledge. For pre-course, an average of 1.75 (SD = 1.07) out of six items were answered correctly, for post-course an average of 3.18 (SD = 1.03) correct answers was achieved. The scales 'concept of own abilities', 'internality' and 'specific self-efficacy' also resulted in significant differences between the pre- and post-course (see Table 11.1). The number of attended rehabilitation courses was shown to have an effect on the acquisition of knowledge, as first-time participants clearly differed from repeated offenders in the pre-course (M = 1.59, SD = 1.02 versus M = 2.34, SD = 1.05 for repeated offenders) but on average scored almost equally high in the post-course (M = 3.19 for first-time offenders and M = 3.16 for repeated offenders). The trainers and courses scored good ratings with an average of 1.2 for the trainers and 1.72 for the courses on a five-point scale. The multivariate analysis showed significant differences regarding the number of attended courses with $F(7,260) = 2.082$, $P = 0.046$ and the time of measurement, that is before and after the course, with $F(7,260) = 27.719$, $P < 0.001$. The interaction was also significant with $F(7,260) = 2.994$, $p = 0.005$.

Table 11.1   Results of the univariate analyses of the second evaluation (Lüftenegger, 2006)

Scale	F-test (F(1,267))	Sig. (P)	Eta²	M (SD) pre-test	M (SD) post-test
Self-concept of own abilities	8.226	0.004	0.030	4.50 (0.66)	4.57 (0.68)
Internality	5.907	0.016	0.022	4.28 (0.68)	4.41 (0.70)
Socially caused externality	2.418	0.121	0.009	2.77 (0.75)	2.71 (0.71)
Fatalistic externality	0.588	0.444	0.002	2.71 (0.75)	2.70 (0.80)
General self-efficacy	3.720	0.055	0.014	3.15 (0.45)	3.20 (0.42)
Specific self-efficacy	28.199	< 0.001	0.096	5.18 (1.09)	5.54 (0.75)
Knowledge	154.936	< 0.001	0.368	1.75 (1.07)	3.18 (1.03)

Discussion

The primary goal of influencing the course participants' relevant attitudes was achieved immediately after the course. Significant changes were found in the scales 'concept of own abilities', 'internality', 'specific self-efficacy' and 'knowledge'. Thus, the programme appeared to strengthen the participants' self-efficacy and encourage internal attribution, while external attributions remained unaltered throughout the course. Therefore, after the course participants may be more aware of the influence of their own actions on their lives and believed to have a higher level of control over their actions. It is not clear how long-lasting this effect may be. Further research would need to consider long-term effects of this intervention. The most significant effect of the course was the increase in offence-related knowledge, which is also a tertiary goal for the course. This is a very important achievement as a large number of traffic offences are based on ignorance or a lack of knowledge. These findings are also concurrent with the results of previous research work by Kases (2002) and Schickhofer (2003). A comparison of first-time offenders and repeated offenders showed that first-time attendees knew less about the offence at the beginning of the course but reached the same level of knowledge as the repeated attendees by the end of the course. Therefore, whilst an increase in knowledge was reported for both groups, the increase was much greater for first-time offenders than for participants who had already taken part in a rehabilitation course. Finally, the courses and trainers received a very positive rating by the course participants, which shows that there is a high acceptance of the intervention among traffic offenders. This is especially important as the general acceptance of the courses and trainers is a vital precondition for influencing participants' attitudes and the acquisition of knowledge.

Outlook – the third evaluation

A third evaluation of the AAP's driver rehabilitation programmes is currently being conducted in all branches across Austria. As in previous evaluations, this is an investigation of whether drink-drive courses influence participants' attitudes towards their offence and whether there is an increase in offence-related knowledge. Additionally, further research will consider whether these effects differ between native German-speaking course participants and non-native German participants, a steadily growing participant group that has not received much special attention in evaluations of driver rehabilitation courses so far. In order to avoid the loss of data through drop-outs caused by language problems, the German questionnaires were also translated into four other languages. Using scientific translation standards, the questionnaires were translated into Serbian, Croatian, Turkish and Polish and then re-translated into German and compared with the original questionnaires in order to ensure the correspondence of all questionnaires as regards content and meaning. The acceptance of the measure and its trainers will be part of the assessment as it is an essential influence on openness to change and knowledge acquisition.

The goal of this study is the quality control and assurance of the AAP's driver rehabilitation courses with the best possible inclusion of all course participants and an optimisation of the courses based on current research findings, with special consideration being given to non-German speaking participants.

Summary

In summary, the driver rehabilitation courses provided by the AAP were found to be effective in influencing participants' attitudes and increasing their knowledge in both evaluations conducted so far. However, further research using a control group of offenders that do not take part in the course would serve as a comparison for a full interpretation of the effectiveness of the intervention. Further research would also need to establish the long-term effects of the course on drink-drive behaviour.

References

Bandura, A. (1986). *Social foundations of thought and action.* Engelwood Cliffs, NJ, Prentice-Hall.

Bartl, G. et al. (2002). *EU-Project 'Andrea'. Analysis of Driver Rehabilitation Programmes.* Wien, Kuratorium für Verkehrssicherheit.

Kases, M. (2002). Einstellungs- und Verhaltenstraining für alkoholauffällig gewordene Kraftfahrer. Eine Überprüfung der Wirksamkeit unter sozialpsychologischen Aspekten. Unpublished diploma thesis, Universität, Wien.

Kirkpatrick, D. (1996). Great ideas revisited. Revisiting Kirkpatrick's four-level model, *Training and Development*, 50(1), 54–57.

Krampen, G. (1991). *Fragebogen zu Kompetenz- und Kontrollüberzeugungen (FKK).* Göttingen, Hogrefe.

Lüftenegger, M. (2006). *Evaluation von verkehrspsychologischen Nachschulungskursen.* Unpublished diploma thesis, Universität, Wien.

Republik Österreich (2002). 357. Verordnung, Nachschulungsverordnung FSG_NV. In Republik Österreich, ed., *Bundesgesetzblatt für die Republik Österreich, Jahrgang 2002, Teil II.* Wien, Republik Österreich, pp. 2597–2601.

Schickhofer, E. (2003). *Die Evaluierung von Nachschulungskursen für alkoholauffällige Verkehrsteilnehmer.* Unpublished diploma thesis, Universität, Wien.

Schwarzer, R. and Jerusalem, M. (eds) (1999). *Skalen zur Erfassung von Lehrer- und Schülermerkmalen. Dokumentation der psychometrischen Verfahren im Rahmen der Wissenschaftlichen Begleitung des Modellversuchs Selbstwirksame Schulen.* Berlin, Freie Universität Berlin.

Utzmann, I. (2008). Zur summativen Evaluation von Maßnahmen der Verkehrserziehung und aufklärung. *Zeitschrift für Verkehrssicherheit*, 1, 25–31.

Chapter 12

Perceptions of the Spanish Penalty Point Law

Maria Eugènia Gras, Sílvia Font-Mayolas, Mark J.M. Sullman,
Mònica Cunill and Montserrat Planes

Introduction

In Spain, as in most parts of the European Union, traffic accidents are one of the main causes of death amongst young people. Over 5,000 people aged between 16 and 24 years old die every year in the EU, with more than 700 of these deaths occurring in Spain (European Road Safety Observatory, 2008). One of the more recent attempts by the Spanish government to reduce the number of road deaths and injuries has been through the introduction of a penalty point licence law. This measure was introduced in July 2006, and was motivated by the experience of countries like the United Kingdom, France, Germany, Italy and Luxemburg. However, this law was not a popular measure amongst young people, who doubted it would increase safe driving, and agreement with the law was higher amongst older drivers (Sánchez-Martín, 2004).

Irrespective of its popularity, Spanish government records indicate that since the introduction of the law there has been a sustained decrease in accidents, deaths and serious injuries (DGT, 2008). Official statements attribute the cause of this reduction to the effect of this law on driving behaviour. The penalty point law may have caused a reduction in the number of crashes by decreasing engagement in risky behaviours in the same way found in other countries (Lenehan et al., 2005; Zambon et al., 2008). This explanation is supported by the fact that young Spanish drivers self-reported positive changes in driving behaviour since the introduction of the law (Font-Mayolas et al., 2008; Gras et al., 2008). A similar pattern of results were also found in a sample from the general population of Spanish drivers (Montoro et al., 2008).

Previous research has also shown that changing traffic laws can be an effective method for changing attitudes, which can in turn lead to positive changes in driving behaviour (Constant et al., 2008). The role of attitudes in behavioural change has been postulated by the Theory of Planned Behaviour (TPB) (Ajzen, 1991) and has received strong support from many studies on driving behaviour (Paris and Van den Broucke, 2008; Pelsmacker and Janssens, 2007; Simsekoglu and Lajunen, 2008). According to the TPB, behaviour is predicted by behavioural intentions, which are in turn predicted by attitudes, social norms and perceived behaviour control. This means that if drivers had a positive attitude towards the penalty point law this should decrease their intention of engaging in risky driving behaviour, and eventually their actual risky driving behaviour.

The present research aimed to investigate how young people evaluate the penalty point law and the perceived effects of this law on driving behaviour. The impact of gender and previous loss of points were also investigated, along with whether the perceived effectiveness of the law was related to attitudes towards the law.

Method

The sample consisted of 1,850 young people (51.6 per cent female) with ages ranging from 17 to 30 years old (Mean = 20.8; SD = 2.7). The participants were all undergraduate students at the University of Girona (Spain).

In addition to demographic variables (gender, age), participants were also asked whether they had a car or a motorcycle driving licence, whether they supported the penalty point law and if they had lost any points from their licence. The participants were then asked to report whether they believed the following behaviours had changed in frequency since the introduction of the penalty point law: obeying the speed limit when driving in the city and motorways, using a mobile phone while driving, drinking and driving, using a seatbelt in the rear seats and when driving in the city, and using a helmet (for motorcycles). Finally the participants were were asked to rate their agreement with the following statements, *'Since the introduction of the penalty point law, drivers are more compliant with the traffic rules'* and *'Since the introduction of the penalty point law, you are more likely to be fined if you do not obey the traffic rules.'* In all cases the scales ranged from 0 (I do not agree at all) to 10 (I totally agree).

The questionnaires were administered to all students present during normal class time. Participants were assured of their anonymity and the confidentiality of their responses.

Results

The great majority of participants (84.5 per cent males/80.1 per cent females) were drivers. Amongst females, 73.2 per cent drove a car and 26 per cent a motorcycle, while the percentage of male car drivers was 80.2 per cent and 45.7 per cent drove a motorcycle. Significantly more males than females reported driving cars ($P < 0.05$) and motorcycles ($P < 0.01$).

Seventy-seven drivers (5 per cent of drivers) had lost between two and twelve points from their driver's licence (Mean = 3.7; SD = 1.7). Significantly more males (7 per cent), than females (3 per cent), had lost points from their licence ($P < 0.01$).

The majority of participants (70 per cent) supported the introduction of the penalty point law, but there were differences by gender and also according to whether they had lost any points from their licence. Figure 12.1 shows the percentage of participants who supported the penalty point law, by gender and by loss of points (drivers only). Chi-squared tests showed that significantly more females than males and more drivers who had not lost any points from their licence supported the introduction of this law ($P < 0.01$).

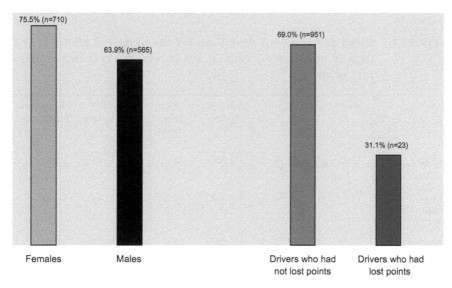

Figure 12.1 Percentage of participants who supported the penalty point law, by gender and loss of points

The means and standard deviations of the individual items measuring the perceived effects of the penalty point law, by gender, loss of points and support of the law are presented in Table 12.1. Mobile phone use was the traffic infringement that people reported having reduced the least following the introduction of the penalty point law, while using a seatbelt when driving in the city was the behaviour that they reported as having increased the most.

Furthermore, the results of t-tests showed that females were more positive than males about the effect of the penalty point law on driving behaviour ($P < 0.01$). More females reported an improvement in compliance with traffic rules, such as obeying speed limits, seatbelt/helmet use and not drink-driving. However, there were no gender differences in the perceived use of mobile phones while driving after the introduction of the law. In addition the perceptions of changes in the individual driving behaviours did not differ according to whether or not they had lost points from their licence.

Perceptions of changes following the penalty point law were also related to attitudes towards the law: participants who supported the introduction of the law perceived more positive changes in driving behaviour than those who did not support it ($P < 0.01$). Furthermore, females and drivers who had not lost any points from their licence reported general driving behaviour to be more compliant after the penalty point law was introduced ($P < 0.01$). A positive attitude towards the introduction of the law was also related to a perception of positive changes in general driving behaviour. Moreover, females perceived a higher risk of being fined, as did those who supported the introduction of the law. However, there was no difference according to whether or not they had lost points from their licence.

Table 12.1   Mean and standard deviations for the perceived effects of the penalty point law by gender, lost points (drivers) and support for the law

	Females	Males	Not lost points	Lost points: points (N = 74)	Positive evaluation	Negative evaluation
More use of seatbelt (city)	7.3* (2.4)	6.8*(2.7)	7.0 (2.6)	6.7 (3.0)	7.5* (2.2)	6.2* (3.0)
More use of seatbelt (rear seats)	7.0* (2.4)	6.6* (2.7)	6.8 (2.6)	6.3 (2.9)	7.2* (2.2)	5.9* (3.0)
More use of helmet (motorcyclists)	7.0* (2.4)	6.4* (2.9)	6.7 (2.7)	6.2 (3.3)	7.2* (2.3)	5.7* (3.2)
Less drink driving	6.4* (2.6)	6.1* (2.9)	6.2 (2.7)	5.8 (3.1)	6.6* (2.4)	5.3* (3.2)
Obey speed limits (motorway)	6.3* (2.5)	6.0* (2.8)	6.2 (2.7)	5.8 (3.1)	6.6* (2.3)	5.0* (3.1)
Obey speed limits (city)	6.2* (2.4)	5.7* (2.8)	5.9 (2.7)	5.8 (3.1)	6.5* (2.3)	4.8* (3.0)
Less use of mobile phone	5.5 (2.6)	5.3 (2.8)	5.3 (2.7)	5.3 (2.9)	5.7* (2.5)	4.6* (2.9)
More compliant driving behaviour	6.2* (2.5)	5.6* (2.9)	6.0* (2.7)	5.3* (2.8)	6.5* (2.4)	4.6* (2.9)
More probability being fined	7.5* (2.5)	6.9* (3.0)	7.2 (2.8)	7.4 (3.0)	7.4* (2.5)	6.7* (3.2)

* $P <0.01$.

Discussion

Most of the young Spanish adults supported the introduction of the penalty point law, while three out of every ten young people did not. Proportionally more men than women had negative attitudes towards the law. These results are in agreement with another Spanish survey (SARTRE 3) which also found that young men were the group who agreed the least with the penalty points law (Sanchez-Martín, 2004). It seems likely that this is related to the fact that young male drivers are the group which will be most affected by the imposition of the penalty point law, as they are the most likely to engage in aberrant driving behaviours and to be involved in a crash (DGT, 2008; Gras et al., 2006; Twisk and Stacey, 2007).

A positive attitude towards the penalty point law was strongly related to whether or not they had previously lost points from their licence. Only three out of every ten young drivers who had lost points supported the introduction of the law. Unfortunately the cross-sectional nature of the present study does not allow us to determine if they had negative attitudes towards the penalty point law because they had lost points or if their attitude was negative before losing the points. Further research is needed in order to investigate whether a loss of points results in a change in their attitude towards the law, as this has important implications in terms of changing behaviour.

More females reported an increase in compliant driving behaviours in general and in most individual driving behaviours, such as more frequent use of seatbelts in the city and for the rear seats, more helmet use by motorcycle riders, less drink-driving and obeying the speed limits on motorways and in the city. The only exception was using a mobile phone while driving, where there were no gender differences. Females also perceived a higher risk of being fined following the introduction of the law. These results can be explained by means of the frequency of engaging in these behaviours, as young male drivers are much more likely to not use a seatbelt or helmet, drink and drive, speed and to have been fined, while males and females tend to use a mobile phone while driving at a similar frequency (Sánchez-Martín, 2004). In addition, mobile phone use was perceived as the behaviour that had changed the least following the penalty point law. Montoro and colleagues (2008) found similar results in a sample from the general population of Spanish drivers. These authors reported that the only behaviour that was found to decline less than using a mobile phone while driving, following the implementation of the law, was parking in a prohibited area.

Previously losing points from their licence did not seem to be related to perceived changes in the individual driving behaviours measured here. However, amongst those who had not lost points, there was a perception that in general driving behaviour was more compliant since the introduction of the penalty point law. Perhaps losing points is related to changes in actual behaviour, rather than the perception of changes in other drivers' behaviours.

A positive attitude towards the penalty point law was strongly related to the perceived effects on general driving behaviour. A positive attitude towards the law was also associated with greater perceived changes in driving behaviour. This finding was not surprising, given that if we support the introduction of the law then also perceiving it to have had a positive effect surely follows.

This study had a number of weaknesses, which included the possibility of sampling bias, as all of the participants were university students. This means it is not possible to generalise these results to the general population of young Spanish drivers. The research also suffers from the normal limitations associated with self-report measures and cross-sectional studies.

Conclusion

The large changes in drivers' behaviour reported by the participants in this study following the implementation of the law appears to be in agreement with the substantial reduction of crashes in Spain during this time period. The present study also showed that a large number of young people do not support the introduction of the penalty point law. Further research is necessary to help improve the attitudes towards the law, mainly amongst males and those who had lost points from their licence. Changing attitudes towards the penalty point law could be an important countermeasure to increase safe driving, as is predicted by the TPB.

References

Ajzen, I. (1991). The theory of planned behaviour. *Organizational Behaviour and Human Decision Processes*, 50, 179–211.

Constant, A. et al. (2008). The recent dramatic decline in road mortality in France: How drivers' attitudes towards road traffic safety changed between 2001 and 2004 in the GAZEL cohort. *Health Educational Research*, 23 (5), 848–858.

DGT (2008). *Las principales cifras de la siniestralidad vial. España 2007*. Madrid, Dirección General de Tráfico.

European Road Safety Observatory (2008). *Annual Statistical Report 2008*. Available at http://ec.europa.eu.transport/wcm/road_safety/erso/data/Content/basic_facts.htm.

Fishbein, M. and Ajzen, J. (1975). *Belief, Attitude, Intention and Behaviour: An Introduction to Theory and Research*. Reading, MA, Addison-Wesley.

Font-Mayolas, S. et al. (2008). Is new penalty point law perceived useful to reduce drug consumption for young Spanish drivers? A paper presented at the International Conference on Traffic and Transportation Psychology, Washington, DC.

Gras, M.E. et al. (2006). Spanish drivers and their aberrant driving behaviours. *Transportation Research Part F*, 9, 129–137.

Gras, M.E. et al. (2008). Helmet use by young Spanish motorcycle riders: The impact of the penalty point law. A paper presented at the International Conference on Traffic and Transportation Psychology, Washington, DC.

Lenehan, B. et al. (2005). Immediate impact of 'penalty point legislation' on acute hospital trauma services. *Injury, International Journal of the Care of the Injured*, 36, 912–916.

Montoro, L. et al. (2008). Influencia del permiso de conducción por puntos en el comportamiento al volante. *Psicothema*, 20(4), 652–658.

Paris, H. and Van der Broucke, S. (2008). Measuring cognitive determinants of speeding: An application of the theory of planned behaviour. *Transportation Research Part F*, 11, 168–180.

Pelsmacker, P. and Janssens, W. (2007). The effect of norms, attitudes and habits on speeding behaviour: scale development and model building and estimation. *Accident Analysis and Prevention*, 39, 6–15.

Sánchez-Martín, F. (2004). *Actitudes sociales frente al riesgo vial en Europa. SARTRE 3*. Madrid, Dirección General de Tráfico.

Simsekoglu, O. and Lajunen, T. (2008). Social psychology of seatbelt use: A comparison of theory of planned behaviour and health belief model. *Transportation Research Part F*, 11, 181–191.

Twisk, D.A.M. and Stacey, C. (2007). Trends in young drive risk and countermeasures in European countries. *Journal of Safety Research*, 38, 245–257.

Zambon, F. et al. (2008). Sustainability of the effects of the demerit points system on seatbelt use: A region-wide before-and-after observational study. *Accident Analysis and Prevention*, 40, 231–237.

PART 3
Powered Two-wheeler Behaviour

Chapter 13

The Role of the Psychologist in a Moped Rider Training Programme

Patrícia António and Manuel Matos

Introduction

Road traffic injuries kill an estimated 32,000 people under 25 years of age every year and are the leading cause of deaths globally among 15–19-year-olds, while for those in the 10–14 and 20–24 age brackets it is the second leading cause. In Europe (EU27) injuries due to road traffic accidents among young people aged 15–24 account for 50 per cent of all young people's fatal injuries, including violence and suicide (Kumpula and Paavola, 2008). Among drivers, young males are almost three times more likely to be killed than their female counterparts, with exposure to risk being distributed differently according to mode of transport (WHO, 2007, 2007a).

Moped and motorcycle injuries among adolescents continue to represent a significant source of injury-related mortality and morbidity, especially in populations where mopeds and motorcycles are popular means of transport among teens such as in Portugal and other Mediterranean countries (e.g. France, Greece, Italy and Israel). This requires stronger political will and advocacy, a more proactive role of the health sector and financial investment in prevention efforts targeting young people.

In Portugal, a number of teenagers use mopeds and motorcycles without a licence or any kind of formal training. This itself represents an example of reckless behaviour. Police enforcement is not effective. Since 1999, the Portuguese Road Safety Association has tried to address this problem and reduce risk by easing beginner riders into a traffic environment via a moped rider training programme (FJC Programme) throughout the country. It is an educational programme providing an opportunity for 14–15-year-olds to obtain a special driver's licence for mopeds with engine volumes of 50 cm³ and includes in-class education (theory), in-vehicle training (practice in closed training ground and in public areas) and a psychological intervention (two hours of group dynamics and one-hour psychological assessment) led by a psychologist. The aim of the FJC programme is to enhance the theoretical and practical knowledge of young moped riders in several areas (e.g. traffic laws and legal issues, protective equipment, riding in traffic, road safety, adolescence and risk-taking behaviour). Whenever the trainer and psychologist identify risk-taking behaviours or fears amongst the participants that predict future reckless behaviour and accident involvement, specific interventions are undertaken to delay licensure. Serious risk cases are referred to specific psychological interventions such as individual, family or group therapy in public or private mental health care settings (António et al., 2005).

In 2006, we studied the effectiveness of the FJC Programme and the results of our research showed that there are positive effects of the programme. FJC participants showed higher rule-following riding/driving behaviour, self-awareness of physical and psychological features associated with the riding task and were less likely to commit traffic offences when compared with young riders in the control group. With respect to moped/motorcycle crash experience, contrary to expectations, more than half of FJC participants revealed they had been involved in a crash experience during the four years after the programme. Regarding car crash involvement, rates were similar for both groups.

Our combined results demonstrated that theoretical and practical knowledge about riding a moped is important and useful. Psychomotor and physiological aspects are essential for safe driving and must be a part of the training process. However, according to our study these factors do not necessarily prevent crash involvement and do not sufficiently address FJC participants' risk-taking. Our findings suggest that despite the vehicle used, accidents are thought to be caused by unconscious motivations and actions to release tension, adolescents may not be able to express what they are feeling and may have a limited ability to develop an awareness of their own mental and emotional sates and find meaning in their own and other people's behaviour. Therefore, accidents can be a reflection of psychopathology, namely borderline disorders, and further investigation is needed (António, 2006; António and Matos, 2008).

Considering these findings and the first author clinical practice as a psychologist in the FJC Programme, the purpose of this paper is to discuss the important role of the psychologist in a moped rider training programme aimed at addressing the psychological and emotional attitudes of young people towards road safety and their level of responsibility. Specific psychological interventions are discussed.

What Puts Adolescents and Young Adults at Risk?

Many factors impact the incidence and severity of injuries among young people. Adolescence is a period of rapid development when young people acquire new capacities (e.g. increased physical maturity, education and training) to fulfil a useful role in adult society but also face new risks which can lead to injuries. This rapid development does not always correspond to psychological development.

The pubertal phase may introduce discontinuity into the psyche of the adolescent. This discontinuity may threaten the young person's narcissistic foundations or object relations, and there is a risk of major breakdown (e.g. disorientation; depression; anxiety; impulsivity; committing self-destructive acts or other-oriented violence). Internal and external adaptations to these changes are required and teens need room to experiment and to experience the results of their own decision-making in many different situations (APA, 2002; Jeammet, 2002; Le Breton, 2004; Marcelli and Braconnier, 2005; Matos, 2005; António, 2006; António and Matos, 2008). Their perspective is *testing* – exhibiting themselves, checking the firmness and strength of external limits, and *experiment* – acting, taking their own decisions, feeling and recognise their self-reality.

Challenges and risks emanate also from social changes and social inequality. Rapid social change may overwhelm the individual, media influences, interpersonal issues and economic interdependency may mean that the adolescent has less emotional space for feelings, internal reflections and words. Previously accepted settings for internal reflection (e.g. the stability of family structure, the ritual of religion) may also be threatened. The cacophony of stimulation and the lack of familiar structures may interfere with time to focus on and develop the individual's internal world.

Today's adolescents need one thing that adults seem to have the least surplus of – time. Recent findings indicate that not enough time spent with parents is one of teens' top problems and also that those adolescents with supportive home and school environments have lower relative odds of engaging in risk-taking behaviours and lower relative odds of injury (Shope et al., 1996; Murray, 1998; APA, 2002; Marcelli and Braconnier, 2005; António, 2006; Löwe et al., 2008).

On the other hand, injuries are related to lifestyles given that adolescence and young adulthood is a time for exploration, and testing limits that may involve taking risks. With diminishing parental influence, the influence of peers and youth cultures may grow and sensation-seeking may be gratifying. In this sense, risk-taking may be an important way that adolescents shape their identities, try out their new decision-making skills, and develop realistic assessments of themselves, other people and the world. However, sometimes they overestimate their capacities to handle new situations, and these behaviours can pose real threats to their health and lives.

Between 15 and 24 years of age almost two-thirds of all fatalities are due to injuries, both unintentional (e.g. road traffic injuries, sport injuries and leisure injuries) and intentional (e.g. injuries due to violence, self-harm and suicide). Injury also contributes significantly to morbidity rates and lifelong disability. For risky driving, young people are in hazardous situations without adequate experience in handling them. This applies particularly to boys, who also drive motorcycles and cars more than girls and are more likely to have serious or fatal crashes. Boys are more likely than girls to be exposed to risk by driving at higher speeds, under the influence of alcohol and the use of other recreational drugs and not using helmets or seatbelts. Inappropriate risk judgement, problems in accurately perceiving their own vulnerability, the need for peer approval or to avoid peer rejection, and eagerness to take risk, play a significant role in crash causation (Biermann et al., 2005; Clarke et al., 2005; Gregersen, 2005; Weinberger et al., 2005; WHO, 2007, 2007a; Kumpula and Paavola 2008; Löwe et al., 2008).

Moreover, a large body of research on the neurological development of teens confirms that the brain is still growing and maturing during adolescence (Gogtay et al., 2004; Weinberger et al., 2005; Isler et al., 2008). Contrary to long-held ideas that the brain was mostly fully developed by the end of childhood, it is now clear that adolescence is a time of profound brain growth and change and the brain's frontal lobe is not fully developed until age 25, meaning that young drivers/riders may be more at risk for almost ten years.

An important part of the front lobes and one of the last areas of the brain to fully mature is the prefrontal cortex (PFC), which is responsible for such skills as setting priorities, organising plans and ideas, forming strategies, controlling impulses and allocating attention. So, impulse control, planning and decision-making are largely PFC functions highly related to driving performance and are still maturing during adolescence. These findings suggest that although the pliability and changeability of the adolescent brain are extremely well suited to meet the demands of teen life, the parts related to emotions and decision-making are still developing and for that reason teenagers are particularly vulnerable to risk-taking behaviour, such as drink-driving or driving too fast. Guidance from parents, other adults and institutions are essential while decision-making skills are being developed (Gregersen, 2005; Weinberger et al., 2005; Isler et al., 2008).

Risk-taking Behaviour

The difficulty when working with concepts of risk-taking behaviour is that there is no common definition in scientific literature. Previous studies mostly refer to risk behaviour which is a broader definition, including all behaviours and habits developed which may lead to impairment and damage to health. On the other hand, different theories give different answers as to what risk-taking is and where risk-taking in young people comes from (APA, 2002; Löwe et al., 2008).

Involvement in risky behaviour is always a trade-off between short-term gains and potential long-term consequences and the decision-making process depends on the capabilities and knowledge of the individual. For example, involvement in risky driving such as speeding might seem attractive to an adolescent who wants to impress their peers despite the possible long-term effects of a crash or injury. What is important to take into account is that the risk perception and assessment is bound to be biased and subjective (e.g. depending on past experiences, motivations, the present mood and emotions) and that human beings are not always able to make accurate objective calculations of situations. Some people may try to avoid risks because of some unpleasant past experiences and some people may want to keep a certain level of thrill in their lives by accepting risk as part of their daily lives (Broughton, 2005; Löwe et al., 2008).

Young people are often aware of the dangers of risky behaviours. However, they are more prone to engage in them than adults, perhaps due to a heightened need for stimuli. Risk-taking behaviour tends to start in early adolescence, involving experimentation of different 'new' behaviours, sensation-seeking, need to be perceived as adults, peer pressure and individual's perceptions of risk-taking behaviour and the social environment have been associated with heightened risk-taking among young people. Therefore, one of the main reasons for young people being over-represented in risk-taking activities such as road traffic crashes seems to be the tendency to involve themselves in 'risky' behaviour, which is age-related and has 'developmental sources' (SWOV, 2002; Arnett, 2002; Löwe et al., 2008).

From a psychodynamic point of view, adolescents may be more vulnerable to risk-taking due to a sense of helplessness and the need to fill this void with arousing thoughts to avoid feelings of depression and anxiety. Risk-taking is often a way of testing out personal determination and a sense of being alive. This serves to maintain self-cohesiveness, develop identity and maintain one's very existence, while at the same time destroying what is reprehensible about the self or a threat to a fragile sense of coherence. By playing with the hypothesis of dying of their own accord, adolescents are stimulating a feeling of freedom and conquering their fear, persuading themselves that at any moment they have an escape door. Abandonment and family indifference, but also over-protection, especially on the part of the mother, can be the root causes of such acts. Discredited paternal authority is also a common theme. Sometimes the causes may be found in violence, parental disharmony, or hostility shown by a stepfather or mother in a reconstructed family (Laufer, 2000; Le Breton, 2004; Marcelli and Braconnier, 2005).

So, the term risk-taking, as applied to adolescents, is used to designate a series of different behaviours (e.g. drug-taking, eating disorders, reckless driving) whose common feature lies in exposing themselves to the significant probability of being injured or dying. Such risk-taking attitudes stem from an intention. Some are adopted as a lifestyle (where risky driving is a significant component of a deviant lifestyle); others a reaction to specific circumstances. Others can involve unconscious motivations: a tension-releasing act (acting-out), which replaces verbalising their feelings, suggesting a rift between inner and outer worlds (António, 2006; António and Matos, 2008).

To sum up, according to some researchers reckless behaviour becomes virtually a normative characteristic of adolescent development. On the other hand, reckless behaviour may in some cases be a reflection of psychopathology. Therefore, the main point of interest for professionals working with adolescents is the determination and separation of factors that are features of a developmental stage, and those factors that can be a reflection of psychopathology (e.g. depression, anxiety, borderline disorders, mood disorders, eating disorders and addictions).

Intervention should not focus on eliminating risk-taking behaviour as a whole but perhaps introduce safer risk-taking. The important thing is to know how to cope with risk and avoid actions that involve a high element of danger. In this sense, injury prevention must be applied to adolescents' behaviours, must assure their protection, and attention must be paid to the risk-taking fundamental characteristic – *communication*. The key is to provide guidance in decision-making and encourage the youngsters to channel the positive development aspects of their energy into less dangerous and more constructive pursuits (APA, 2002; António and Matos, 2008; Löwe et al., 2008).

When Risk-taking Behaviour Becomes Problem Behaviour

Knowledge about the possible consequences of risky behaviour does not necessarily diminish risk-taking behaviour. As mentioned earlier, although adolescents do consider the risk before indulging in risky behaviour, risks appear to play only a secondary role in their risk assessment.

In fact, for some teens risk-taking behaviour may be a problem that can threaten their well-being in both the short and long term. Individuals who experience early trauma may defensively inhibit their capacity to think ahead to avoid having to consider their caregiver's wish to harm them. For example, some characteristics of severe borderline personality disorder may be rooted in development pathology association to this inhibition. In this sense, it is extremely important that professionals working with adolescents (e.g. teachers, trainers, psychologists) are able to differentiate between normal experimentation and signs of troubled or high-risk youths (psychopathology) so they can make appropriate referrals to mental heath care services when necessary.

Psychologically, special concern is warranted when high-risk behaviours begin as early as ages eight or nine and usually occur within a social context of peers who engage in the same activities. Additionally, it may be an important sign that an adolescent is in serious trouble and needs professional help if they are engaged in multiple risk behaviours. Research has found that serious problems tend to cluster in the same adolescents and that youths who are at greater risk for serious negative outcomes tend to engage in multiple problem behaviour such as drink-driving, alcohol and drug use, dropping out of school, unprotected sexual intercourse or delinquency, and usually have several antecedent risk factors in common, such as poor school performance and low self-esteem (APA, 2002; WHO, 2007; 2007a; Löwe et al., 2008).

Many factors including parenting and family management characteristics that include lack of monitoring and supervision, unclear expectations of youth behaviour or no (or only rare) rewarding of positive behaviour, exposure and susceptibility to peer pressure, living on a low income, socially disorganised communities, exposure to violence in the home and in the community, are powerful risk factors for increasing the chances that a youth will engage in high-risk problem behaviours (APA, 2002; Marcelli and Braconnier, 2005; António, 2006).

The Role of the Psychologist in Road Safety Programmes

Frequently in our society traffic offences and accidents are seen as the result of an action, and this is seen as the explanation. A deeper understanding would require recognising alternative underlying motivations and beliefs to account for such observed behaviour. The psychologist is in the best position as a professional to remove the inhibition against conscious awareness of those underlying motivations and beliefs, and should take part in training programmes for young and novice drivers.

Over the past few years, research has led to the recognition that young drivers' goals and attitudes increases or decreases the risks of driving and there is a close connection between unconscious motives, attitudes and personality development (Mayhew and Simpson, 2002; Broughton, 2005; Gregersen, 2005). Traffic and road safety programmes for young drivers need to address these findings to be effective in reducing a substantial number of injuries.

Talking with teenagers about risk behaviours in general is a useful approach to get the injury prevention message across. Consequently, it is very important to know what young people experience and feel before getting a rider or driving licence (APA, 2002; Löwe et al., 2008).

Teenagers and young adults need special consideration as road users not yet experienced enough to respond appropriately to situations that put them and others at risk. The physical, mental and behavioural characteristics of teenagers and young people need to be taken into account in trying to understand why they are at risk on the roads and in developing preventive strategies. Specific educational measures are needed in order to change the attitudes related to risk-taking and injuries and to develop risk competence such as resilience and coping skills in risky situations. The psychologist's role is fundamental in achieving these goals. This information is extremely important for planning and targeting prevention to those most at risk.

Psychological Intervention

In the FJC Programme the psychological intervention consists of two hours of group therapy (with no more than ten adolescents) and one hour of psychological assessment. The main goals are to address the psychological and emotional attitude of young people towards road safety and their level of responsibility, to create an attitude of concern of personal security and to promote a self-awareness of the adverse consequences of risk-taking behaviours in traffic scenarios. The FJC Programme requires participants to discover the risks themselves by developing the need to take a moment to decide 'shall I do this or not…', so they become more aware of the possible consequences.

Therefore, active techniques such as role-playing, brainstorming and psychological exercises are used in the group therapy in order to get participants involved. The main perspective is to create an interpersonal context where understanding of mental states of self and others becomes a focus. These active techniques provide a base level of stimulation capable of activating processes that allow self-awareness through relational dimensions inside the group which facilitate the acquisition of new patterns of thinking and relating these to others before they take action.

Some of the techniques often used are: photo-language (pictures associations), geometric dialogue, 'risk-taking line', domino game, 'if I was a traffic signal...', balloons game, 'stroll of the slow ones...' and others. The psychologist must be able to create situations where making behaviour understandable, meaningful and predictable is possible.

The psychological assessment uses the Zung Anxiety Self-Evaluate Scale which is an anxiety questionnaire, the J. Stork Suicide Risk Scale which measures depression in both dimensions, 'acting out' and 'acting in', and enables us to assess and study likely suicidal tendency, and the Zulliger Projective Test which is a personality projective test. The first two measures are used mainly because Matos (1991) in a research study found a positive link between a constellation of very low anxiety and high suicidal risk and being involved in accidents in a sample of moped riders aged between 15–19 years old. The Zulliger Projective Test is used because this enables a 'quick prognosis' of the adolescent mental state, and also provides a diagnosis of psychopathology traits using a group format application. Considering our recent research findings new psychopathology measures may be introduced in the psychological assessment battery in order to improve the predictors of young rider risk (António, 2006).

As mentioned before, whenever trainer and psychologist observations and the psychological assessment identify problem behaviours that can threaten adolescent well-being in both the short and long term, specific interventions are undertaken. They usually take place in an individual clinical interview with the teens and their parents, which allows a more in-depth evaluation. Appropriate referrals to mental health care services are provided when necessary. From these interventions it is also decided whether or not the high-risk teens can keep their special driving licence at that moment. This might delay the granting of the special licence for a period of time.

Risk-taking behaviour prevention with adolescents means searching the logic and true essence in which each teen acts, behaves and injures themselves. Teens act to provoke a response from others and not only to affirm their existence and their identity. Overall, a psychologist enrolled in a moped rider-training programme, such as the FJC Programme, aims to provide each adolescent with a way to find an active role in growing up, capacity to transform their inner world, accepting the reality of emotions and feelings and arriving at solutions. The way youngsters drive a moped, a motorcycle or a car can be the reflex of an internal poor mental state.

Summary

The ability to plan, adapt to social environment and to imagine possible future consequences of an action or to appropriately measure their emotional significance is still developing throughout adolescence. Accordingly, road safety prevention should help adolescents to develop their 'thinking process', to increase awareness of their at risk behaviour, to temper risk-taking tendencies and vulnerability, rather than focus on vehicle-handling skills. This is in contrast to the common practice where educational measures are focused on dishing out advice and telling young people what to do. The impact of the role of the psychologist lies in their capacity to activate teens' ability to develop an awareness of mental and emotional sates and thus find meaning in their own and other people's behaviour.

To reach the target, moped riding training programmes need to be intensive and multifaceted in areas such as adolescence and risk-taking behaviours, use methods fitting to modern forms of communication and be embedded in the youth culture (e.g. visuals that appeal to youth, use group dynamics and new media and interactive Internet tools). In addition to general rider training, trainers and psychologists should develop a partnership with young people, to find solutions to addressing adolescent risk-taking behaviour.

References

APA (American Psychological Association) (2002). *Developing Adolescents: A Reference for Professionals.* Washington, DC, American Psychological Association. Retrieved 17 June 2004 from www.apa.org/pi/pii/develop.pdf

António, P. et al. (2005). Driving at fifteen: Assessment of moped rider training among teens. In L. Dorn, ed., *Driver Behaviour and Training. Volume II.* Aldershot, Ashgate, pp. 252–260.

António, P. (2006). 'Avaliação do Programa de Formação de Jovens Ciclomotoristas – *Licença Especial 50 cc* da Prevenção Rodoviária Portuguesa. Sinistralidade e relação com a psicopatologia' (An Evaluation of the Portuguese Moped Rider Training Programme from Portuguese Road Prevention Association. Road traffic accidents and psychopathology). Dissertação de Mestrado. Lisboa: Faculdade de Psicologia e de Ciências da Educação, p. 228.

António, P. and Matos, M. (2008). An evaluation of the Portuguese moped rider training programme. In L. Dorn, ed., *Driver Behaviour and Training. Volume III.* Aldershot, Ashgate, pp. 398–413.

Arnett, J. (2002). Developmental sources of crash risk in young drivers. *Injury Prevention,* 8(Suppl II), ii17–ii23. Retrieved 10 August 2004 from http://ip.bmjjournals.com/cgi/reprint/8/suppl_2/ii17.

Biermann, A. et al. (2005). Development and first evaluation of a prediction model for risk of offences and accident involvement among young drivers. In L. Dorn, ed., *Driver Behaviour and Training. Volume II.* Aldershot, Ashgate, pp. 169–178.

Broughton, P. (2005). Designing powered two-wheeler training to match rider goals. In L. Dorn, ed., *Driver Behaviour and Training. Volume II.* Aldershot, Ashgate, pp. 233–242.

Clarke, D. et al. (2005). Voluntary risk taking and skill deficits in young driver accidents in the UK. *Accident Analysis and Prevention*, 37, 523–529.

Gogtay, N. et al. (2004). Dynamic mapping of human cortical development during childhood through early adulthood. PNAS, 101(21), 8174–8179. Retrieved 10 May 2008 from www.loni.ucla.edu.

Gregersen, N. (2005). Driver education – a difficult but possible safety measure. In L. Dorn, ed., *Driver Behaviour and Training. Volume II.* Aldershot, Ashgate, pp. 144–153.

Isler, R. et al. (2008). *The 'Frontal Lobe' project. Final report.* Hamilton, Traffic and Road Safety Research Group, Psychology Department, University of Waikato, New Zealand.

Jeammet, P. (2002). *Réponses a 100 questions sur l'adolescence* (Answers to 100 questions about adolescence). Paris: Éditions Solar.

Kumpula, H., and Paavalo, M. (2008). *Injuries and Risk-taking among Young People in Europe. The European Situation Analysis. EU-Project AdRisk.* Helsinki, KTL (National Public Health Institute).

Laiufer, M. (2000). *O adolescente suicida* (Suicidal adolescent). Lisboa, Climepsi Editores.

Le Breton, D. (2004). The anthropology of adolescent risk-taking behaviours. *Body and Society*, 10(1), 1–15. Retrieved 22 October 2004 from www. sagepublications.com.

Löwe, U. Braun, E. and Kisser, R. (2008). *Tackling Injuries among Adolescents and Young Adults: Strategy and Framework for Action. EU-Project AdRisk.* Vienna, KfV (Austrian Road Safety Board).

Marcelli, D. and Braconnier, A. (2005). *Adolescência e psicopatologia* (Adolescence and psychopathology). Lisboa, Climepsi Editores.

Matos, M. (1991). 'Factores de risco psicológico em jovens condutores de motorizada e sua influência relativa na ocorrência de acidentes' (Psychological risk factors among motorcycle riders and their influence on accident involvement). Dissertação de Doutoramento. Lisboa: Faculdade de Psicologia e de Ciências da Educação, p.581.

Matos, M (2005). *Adolescência, representação e psicanálise* (Adolescence, representation and psychoanalysis). Lisboa, Climepsi Editores.

Mayhew, D., and Simpson, H. (2002). The safety value of driver education and training. *Injury Prevention*, 8 (Supl II), ii3–ii8. Retrieved 10 August 2004 from http://ip.bmjjournals.com/cgi/reprint/8/suppl_2/ii3.

Murray, A. (1998). The home and school background of young drivers involved in traffic accidents. *Accident Analysis and Prevention*, 30(2), 169–182.

Shope, J. et al. (1996). Alcohol-related predictors of adolescent driving: Gender differences in crashes and offences. *Accident Analysis and Prevention*, 28(6), 755–764.

SWOV (2002). *'Hardcore' Problem Groups among Adolescents. Their Magnitude and Nature and the Implications for Road Safety Policies.* T. Wurst, ed.

Leidschendam, SWOV. Retrieved 10 August 2004 from http://www.swov.nl/rapport/r-2002-25.pdf.

Weinberger, D., Elvevag, B. And Giedd, J. (2005). *The Adolescent Brain: A Work in Progress.* The National Campaign to Prevent Teen Pregnancy. Retrieved 10 May 2009 from http://www.thenationalcampaign.org/resources/pdf/BRAIN.pdf.

WHO (2007). *Youth and road safety in Europe. Policy Briefing.* D. Sethi, F. Racioppi, and F. Mitis, eds. Rome, WHO Regional Office for Europe.

WHO (2007a). *Youth and Road Safety.* T. Toroyan and M. Peden, eds. Geneva, World Health Organization.

Chapter 14

Interim Evaluation of the UK's National RIDE Scheme

Cris Burgess, Paul Broughton, Fiona Fylan and Steve Stradling

During the period from 1996 to 2003, the number of powered two-wheelers (PTWs) on the UK's roads increased by almost 50 per cent (Christmas et al., 2008). Although this trend has levelled off in recent years, in 2007, 24,381 riders of PTWs were injured in collisions on the UK's roads and 588 were killed, representing 22 per cent of all road traffic deaths in the UK (Department for Transport, 2008). Given that PTWs make up only 1–2 per cent of traffic per 1,000 miles travelled (Sexton et al., 2004), it is estimated that riders of PTWs are 51 times more likely to suffer a road traffic collision than car drivers per mile travelled (Christmas et al., 2008).

Excessive speed has been implicated in a substantial proportion of killed and seriously injured (KSI) crashes involving PTWs (e.g. Broughton, 2005; Clarke et al., 2004, 2007), as has riding after drinking alcohol (Lynam et al., 2001). Sexton and colleagues (2004) suggest that the most important behavioural contribution to collision involvement, after miles travelled, is the reported frequency of errors, but these authors go on to suggest that: 'these errors occur in a context that suggests they may be closely linked with riding styles involving carelessness, inattention and excessive speed – i.e. styles that might be termed "violational"' (Sexton et al. 2004, p. 1).

It seems clear that confronting such violational riding styles should contribute to a reduction in motorcycle casualties and in order to tackle this issue, the UK government, local government and police agencies are investigating alternative methods to traditional prosecution for dealing with road traffic offenders riding PTWs.

The National RIDE Scheme is a police-led diversion for motorcycle offenders facing potential prosecution for a Section 3 offence under the Road Traffic Act (1988), for:

- Section 3 offences 'driving or riding without due care or consideration for other road users
- Section 39 'failing to comply with road signs'
- Excess speed (non-camera).

In some constabularies PTW offenders are offered the option of course attendance by a police officer at the roadside, thus avoiding penalty (demerit) points and a fine. This offer method however may be decided by a Central Ticket Office decision-maker in other constabularies. Riders must decide at this point if they accept the offer of a course, although they may later change their mind. At this point, the rider must surrender their driving licence, which will be returned to them upon confirmation from the service provider that they have successfully completed the RIDE course (this procedure varies between constabularies according to internal procedures). Successful completion effectively comprises attending on the date given and participating to some extent in the group discussions. It is extremely rare for RIDE clients not to successfully complete. The cost of the course is borne by the offender, though to encourage course take-up, cost to the client for the RIDE course is comparable to the cost of fine and points. Current take-up rate is approximately 65 per cent. Having been identified by a police officer as a result of their inappropriate riding behaviour, it may be assumed that clients attending RIDE courses may be considered more behaviourally deviant than the 'average' PTW rider and more likely to be involved in a collision resulting in injury. Therefore, it is likely that these are the most vulnerable PTW riders, those most likely to contribute to future casualty statistics.

The main RIDE objectives are to increase awareness of current riding behaviour and engender a positive and responsible approach to motorcycling. Through a series of classroom-based theory sessions, the one-day course encourages clients to continue to ride their motorcycle, but to examine their individual attitudes and motivations, their approach to risk, to probe their beliefs surrounding inappropriate riding behaviour, to consider the positive effects and benefits of mindset change and then to maintain these positive changes after course completion. The course employs a mix of information exchange, demonstrations and facilitated group discussion to achieve these outcomes using psychological mechanisms, rather than the traditional skills-based training approach.

The National RIDE Scheme is based on a standardised course model and combines course content and methods of delivery from the earlier Rider Risk Reduction scheme initiated by Devon County Council and Devon and Cornwall Police, successfully running in Devon since 2004, and from the RIDE course developed by a national multi-agency steering group on behalf of the UK's Department of Transport involving inputs from behavioural psychologists, local authorities and motorcycle professionals. This scheme now operates in a number of local authorities around the country, allowing transparency for inter-county client referrals.

RIDE is now an Association of Chief Police Officers (ACPO)-approved model for local authorities and police services who wish to use a similar diversionary scheme for motorcycle offenders in their own areas. The National RIDE launch was in September 2007 at Lancashire Police HQ and a subsequent RIDE Checkpoint meeting in February 2008 attracted representatives from 27 of the 34 police forces in England and Wales, along with officers from Northern Ireland's Roads Police Department. Currently, eight police services employ the scheme, making referrals to service providers who deliver the courses.

This chapter reports the results of the first year of a two-year evaluation study, investigating the effects of course attendance on motorcycle offenders attending RIDE courses in three police areas during 2008 and 2009; Devon, Humberside and Kent. The study investigates changes in behavioural intentions, control beliefs, normative beliefs, knowledge and motivation to ride safely in a sample of RIDE course clients, compared with a control group who provided data over approximately the same period.

Method

The study reported here is a mixed design, with 'before' (T1) and 'after' (T2) measures collected from RIDE course clients and comparing them with those from a control group who did not experience any formal intervention between the time-points. All respondents completed two online questionnaires with approximately five to eight weeks between time-points.

Participants

A total of 126 clients from three RIDE course service providers contributed to the study; 44 from Devon, 49 from Humberside and 33 from Kent. Of these, 104 provided full sets of data which contribute to the analyses reported here. These individuals attended one of 12 courses in groups of between five and 16 individuals (median 8) between August and October 2008. A control group (N = 108) was recruited via several online motorcycle forums with a strong focus on motorcycle performance issues, and using motorcycle club mailing lists (with permission). These respondents completed the measures between June and October 2008.

The characteristics of both groups may be seen in Table 14.1. The control group was significantly older than the RIDE group (47.0 years vs 39.0 years, t_{210} = 5.6, $P < 0.001$) and, while the RIDE group was almost exclusively male, the control group had a significantly larger proportion of female motorcyclists (one per cent vs 10.2 per cent female, χ^2_1 = 8.4, $P = 0.005$). In terms of years since passing the motorcycle test, engine size of main PTW, annual mileage, offending and crash history, the groups did not differ significantly, though for those respondents reporting penalty points on their driving licences, the RIDE group reported more PTW-related offences than the control group (0.4 vs 0.2, t_{238} = 2.1, $P = 0.038$). Further, a slightly higher proportion of the control group reported knowing someone who had been killed in a motorcycle crash (81.0 per cent vs 60.8 per cent, χ^2_1 = 2.9, $P = 0.070$), though this difference did not achieve statistical significance. This last observation is most likely a result of the control group riders' greater age.

Combining the groups, the average respondent reports 7.2 years since passing his motorcycle test, 3.25 points on his driving licence (though 53 per cent of the sample report no convictions), riding a large-capacity PTW (1,000 to 1,100cc) and covering 2,001–3,000 miles each year on his PTW. Around one-third of the sample have been involved in a PTW crash during their riding careers and 12.6 per cent have been involved in such a crash in the previous two years. The majority report knowing someone who had been killed (65.3 per cent of total sample) or seriously injured (76.6 per cent of total sample) on a PTW.

At follow-up, 52 per cent of RIDE clients (N = 54) and 59 per cent of control group respondents (N = 64) completed the second online measure. There were no significant differences between clients in this follow-up sample and the original client group in terms of any of the variables reported above, and so the follow-up sample may be considered representative of the original groups of clients.

Table 14.1   Descriptive statistics, by group

Variable	RIDE course (N = 104)	Control (N = 108)			
	Mean (SD)	Mean (SD)	t	df	P
Age (years)	39.0 (10.7)	47.0 (9.9)	5.6	210	<0.001
Time since passing motorcycle test (years)	7.0 (2.1)	7.4 (1.9)	1.7	210	NS
Penalty points on licence	3.5 (5.6)	3.0 (4.9)	0.8	210	NS
Offences on motorcycle (ever)	0.4 (0.9)	0.2 (0.6)	2.1	268	0.038
Engine size (modal category)	1000cc	1100cc	1.2	210	NS
Annual mileage (modal category)	2001-3000	2001-3000	1.5	210	NS
	%	%	$\chi 2$	df	P
Gender (% male)	99.0	89.8	8.4	1	0.005
Offences (all vehicles – ever)	31.9	34.7	0.3	1	NS
Offences (all vehicles – in last 2 years)	7.1	6.3	0.1	1	NS
Offences on motorcycle (ever)					
Offences on motorcycle (in last 2 years)					
Motorcycle crash (ever)	31.9	34.7	0.3	1	NS
Motorcycle crash (in last 2 years)	13.5	10.4	0.6	1	NS
Knows someone who died in motorcycle crash	60.8	81.0	2.9	1	NS
Knows someone seriously injured in motorcycle crash	74.3	85.0	1.0	1	NS

Procedure

All measures were completed using a web-based survey, with respondents using their 'driver' number (a unique identifying code printed on their driving licence) to log in to the system, thus allowing T1 and T2 measures to be linked. All respondents were invited by email/post to complete the measures, though completion is presented to RIDE clients as a condition of course attendance, so response rates for this group at the pre-course (T1) time-point was 100 per cent. RIDE clients completed the T1 measures within a two-week period prior to course attendance. An email request to complete the follow-up (T2) measures was sent to RIDE clients four weeks after confirmation from service providers of successful course completion, while control respondents were asked to complete the T2 measures six to eight weeks after completion of the T1 measures. Thus, the period between measures was approximately six to eight weeks for both groups.

Measures

Respondents completed extensive online self-report questionnaires (170 items at T1, 118 items at T2). Demographic and background measures were included at T1, including annual mileage on a PTW, career collision involvement and traffic convictions. The questionnaires at both time-points included measures of recent riding history (crashes/convictions), measures of riding behaviour and a number of items from an earlier study (Mannering and Grodsky, 1995) measuring attitudes and perceptions of injury involvement, along with a series of items developed for this study assessing respondents' motorcycling-related attitudes and motivations. Riding behaviour measures included preferences for riding on public roads at over 100mph, pulling 'wheelies', riding within an hour of drinking alcohol and margins for exceeding the speed limit. Based on the Stages of Change model (also known as the Transtheoretical Model of Behaviour Change) (Prochaska and DiClemente, 1983), a single item asked respondents to what extent they had changed their behaviour or, if not, whether they intended to do so in the future (see Table 14.5 for a description of the response scale).

The attitudinal measures included assessment of behavioural beliefs (four items; e.g. 'obeying all speed limits and traffic laws is a sign of a good rider', higher score is more safety appropriate), motivation to ride safely (four items; e.g. 'I want to stay safe on the roads', higher score is more safety appropriate), knowledge (ten items; e.g. 'poor weather conditions contribute to most bike crashes', such items are recoded so that higher score is safety appropriate), control beliefs (four items; e.g. 'I could ride safely if I wanted to', higher score is more safety appropriate), susceptibility to collision involvement (six items; e.g. 'I can break a few road traffic laws and still stay safe on the road', lower score is more safety appropriate), norms (four items; e.g. 'My mates expect me to break a few road traffic laws', lower score is more safety appropriate) and thrill-seeking (three items; e.g. 'I sometimes like to frighten myself a little when I'm riding', lower score is more safety appropriate).

However, as may be seen in Table 14.2, the internal reliabilities of these attitudinal scales tended to be unacceptably low and so a factor analysis of the combined group data at T1 (N = 212) was carried out, using principal axis factoring and a varimax (orthogonal) rotation, with the solution converging in nine iterations. The scree test suggested a five-factor solution, accounting for 46.4 per cent of the variance in the data. Internal reliabilities for the scales derived from these factors are substantially higher than the original scales reported above (see Table 14.2) and so it is these measures that are employed in the later analyses.

Table 14.2   Scale reliabilities by time-point (combined sample)

Scale:	T1 (N = 212)	T2 (N = 118)
Behavioural beliefs	0.59	0.60
Motivation	0.68	0.68
Knowledge	0.59	0.58
Control beliefs	0.43	0.49
Susceptibility	0.40	0.40
Norms	0.65	0.66
Thrill	0.74	0.75
Deviant beliefs	0.82	0.85
Motivation–volition	0.72	0.62
Thrill-culture	0.77	0.79
Susceptibility-control	0.66	0.65
Norms (modified)	0.77	0.76

Table 14.3 shows the items loading on each factor in the solution. The first factor contained nine items, accounting for 16.2 per cent of variance and is similar to a combination of the original 'susceptibility' and 'behavioural beliefs' items, here named 'deviant beliefs' (see Table 14.3 for items loading on this factor), with higher scores indicating a greater tendency to ride in a high-risk manner. The second factor (four items, 10.5 per cent variance) is similar to the 'motivation' measure, though with one 'control' item[1] present, here named 'motivation–volition' with

---

1   Nordgren and colleagues (2007) draw a distinction between control over exposure to a risky behaviour (volition) and control over the negative consequences of that risky

Table 14.3   Factor solution for attitudinal/motivation items

Factor 1: 'deviant beliefs', 16.2% of variance (T1, $\alpha = 0.82$; T2, $\alpha = 0.85$)

Riding quickly on twisty roads is a sign of a good rider (0.35)

Obeying all speed limits and traffic laws is a sign of a good rider (–0.33)

You are a better rider if you obey all speed limits and traffic laws, even if you do not ride quickly around twisty roads (–0.34)

It doesn't matter if I break a few road traffic laws (0.69)

I get a real thrill out of riding fast (0.33)

I find it difficult to obey all the road traffic laws while riding (0.59)

It's okay for me to go faster than the speed limit as long as I ride carefully (0.78)

I can break a few road traffic laws and still stay safe on the road (0.75)

I don't want to take risks when I'm riding (–0.31)

Factor loadings in parentheses.

higher scores, indicating a greater motivation to ride safely. The third factor (four items, 7.7 per cent variance) includes all the 'thrill' items, plus one item from the 'norms' measure ('All my friends take risks when they're riding').

Consequently, this new measure is here named 'thrill-culture' to reflect the personal and social influences on the individual relating to risk, with higher scores indicating a greater tendency for thrill-seeking. The fourth factor (four items, 6.5 per cent variance) largely reflects the 'susceptibility' items, while including two 'control beliefs' items which reflect a belief in control over the negative consequences of risky behaviour. Hence this scale is named 'susceptibility-control', with higher scores indicating a greater sense of control. Finally, the fifth factor (three items, 5.5 per cent variance) is a shortened version of the original 'norms' measure and so the name is retained, 'norms (modified)'. Higher scores on this measure indicate a greater normative influence on the respondent for high-risk riding.

Results

Statistical analyses were performed using the statistical package for the social scientist version 15 (SPSSv15) to investigate the effects of RIDE course attendance, while controlling for variation across time using a control group of

behaviour (control). The control item that loads on Factor 2 is of the former type, hence the name for the new measure.

respondents. Differences in characteristics between the groups (RIDE vs control) will be investigated, then comparisons between the groups (RIDE vs control) across the two time-points will be reported. The data collected for this study are extensive and so this chapter will report specific features of the overall analysis, focusing first on preferred speed, then on general intentions to modify risky riding behaviour (as measured by the SoC item) and, finally, attitudinal and motivational measures related to respondents' experiences of risk-taking whilst motorcycling.

Self-reported behaviour

Two items asked respondents to report how much it is acceptable to exceed the speed limit in 30 mph restricted zones and 60mph zones. A 2 × 2 mixed-measures ANOVA (Table 14.4) with group and time-point as factors revealed a highly significant interaction between group and time-point ($F_{1,107}$ = 9.0, P = 0.003) showing that the groups differed in terms of the change in perceptions of 'acceptable' speeding across the time-points. RIDE course clients reported a significant decrease in how far above the 30 mph they felt was acceptable after course attendance, compared with before (2.1 mph at T1 vs 1.7mph at T2), while the controls reported almost no change over the same period (2.3 mph at T1 vs

Table 14.4   Comparison of speed preferences: group by time-point (ANCOVAs)

Measure	RIDE course (N = 104) mean (standard error)		Control (N = 108) mean (standard error)		Interaction (group by time-point)		
	T1	T2	T1	T2	F	df	**P**
When you are riding in a 30 mph speed limit, by how much do you think it is acceptable to exceed the speed limit? (mph over limit)	2.1 (0.1)	1.7 (0.2)	2.3 (0.1)	2.4 (0.2)	9.0	1,107	.003
When you are riding in a 60 mph speed limit, by how much do you think it is acceptable to exceed the speed limit? (mph over limit)	3.5 (0.3)	2.3 (0.3)	4.0 (0.3)	3.9 (0.3)	2.0	1,107	ns

Age is included as a covariate in these analyses.

Table 14.5   Comparison of stage of change measures, group by time-point

Stage of change	RIDE course (N = 104)		Control (N = 108)	
	T_1 n (%)	T_2 n (%)	T_1 n (%)	T_2n (%)
Maintenance: 'I have already changed the way I ride my bike recently as much as I can and I am now trying to keep it that way.'	14 (16.5)	15 (29.4)	10 (10.9)	6 (10.2)
Action: 'I have tried to change the way I ride my bike recently and I will be trying to change it even more over the next 6 months.'	39 (45.9)	26 (51.0)	51 (55.4)	25 (42.4)
Preparation: 'I have already tried in small ways to change the way I ride my bike recently and I am planning now to continue to do so over the next 6 months.'	24 (28.2)	10 (19.6)	17 (18.5)	17 (28.8)
Contemplation: 'I have not tried to change the way I ride my bike recently, but I am thinking of doing so over the next 6 months.'	2 (2.4)	–	4 (4.3)	3 (5.1)
Precontemplation: 'I have not tried to change the way I ride my bike recently and I am not thinking of doing so in the next 6 months.'	6 (7.1)	–	10 (10.9)	8 (13.6)
Total	**85**	**51**	**92**	**59**

2.4 mph at T2). Although a similar pattern of response was evident in 60 mph zones, these differences were not statistically reliable.

When considering self-reported high-risk behaviours, 'riding at over 100mph on public roads' or 'riding within an hour of drinking alcohol', incomplete sets of data and a high proportion of missing values resulted in only a small number of respondents contributing to these analyses. Consequently, Fisher's Exact test was used to analyse these data. Of those respondents who reported having exceeded 100 mph on public roads in the six weeks up to T1, three of the 15 in the RIDE group and five of the 13 in the control group reported having done so again between T1 and T2. However, this difference is not statistically reliable (χ^2_1 = 1.2, ns). Of those respondents who reported having *not* exceeded 100mph on public roads in the six weeks up to T1, none of the 13 RIDE group clients reported doing so again since attending the course, but seven of the 21 control respondents did report having done so between T1 and T2 (χ^2_1 = 5.5, P = 0.022).

Only seven respondents reported riding their motorcycle within an hour of drinking alcohol, two in the RIDE group and five in the control group. Of these, one from the RIDE group and three from the control group had repeated the behaviour between T1 and T2 ($\chi^2_1 = 0.06$, ns). Of those respondents reporting *not* having ridden within an hour of drinking alcohol, neither of the two in the RIDE group nor three of the nine in the control group had done so between T1 and T2 ($\chi^2_1 = 0.9$, ns). Neither of these differences was statistically reliable.

Stages of change

Table 14.5 reports responses to the SoC items in terms of percentage of each group falling into each response category. To analyse these items, a 2×2 mixed-measures ANOVA was carried out with group and time-point as factors, treating the SoC measure as an ordinal dependent variable (1 = precontemplation, 5 = maintenance). Results revealed a significant group by time-point interaction ($F_{1,88}$ = 6.5, $P = 0.013$). Figure 14.1 indicates the nature of this interaction, with control group respondents regressing towards the lower, less appropriate end of the scale, while RIDE course respondents report increasing scores, thus progressing towards a more safety-appropriate stage.

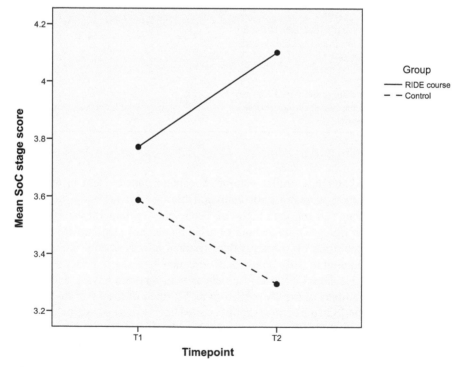

Figure 14.1   Stage of change stage score (group by time-point)

Table 14.6   Scales scores, group and time-point

Measure	RIDE course (N = 104) mean (standard error)		Control (N = 108) mean (standard error)		Interaction (group by time-point)		
	T1	T2	T1	T2	F	df	I*P*
Deviant beliefs	23.6 (0.8)	21.9 (0.8)	25.8 (0.7)	25.6 (0.8)	3.9	1,107	0.050
Motivation–volition	16.5 (0.4)	17.1 (0.3)	16.1 (0.3)	16.2 (0.3)	1.3	1,107	NS
Thrill-culture	14.6 (0.6)	14.4 (0.5)	14.4 (0.5)	14.2 (0.5)	0.1	1,107	NS
Susceptibility-control	17.5 (0.4)	18.2 (0.3)	16.9 (0.4)	17.3 (0.3)	0.5	1,107	NS
Norms (modified)	9.3 (0.4)	8.9 (0.4)	9.6 (0.4)	9.5 (0.4)	0.5	1,107	NS

Attitude and motivation measures

The RIDE and control groups were found to differ significantly in terms of age, with the control group older than the RIDE client group. Thus, the relationship between age and the attitudinal measures was investigated, in order to establish whether age should be controlled for in analyses involving these variables. Age was found to be significantly associated with a number of these measures and so differences in age between RIDE and control groups were controlled by incorporating age as a covariate in a series of two-way mixed measures ANCOVA procedures, taking as dependent variable in turn each of deviant beliefs, motivation–volition, thrill-culture, susceptibility and norms (modified).

Table 14.5 reports responses to these attitude/motivation measures. Results for the deviant beliefs scale show a significant group by time-point interaction ($F_{1,107}$ = 3.9, $P = 0.050$). Figure 14.2 shows the nature of this interaction, with RIDE clients reporting lower scores at T2 compared with T1 and control group respondents remaining relatively stable across time-points, thus indicating that RIDE clients have modified their beliefs relating to the link between their offending behaviour and collision involvement between time-points, compared with the controls.

Results for the other attitudinal measures; motivation–volition, thrill-culture, susceptibility-control and the modified norms measure reveal unreliable interactions, suggesting that these characteristics are not affected by course attendance.

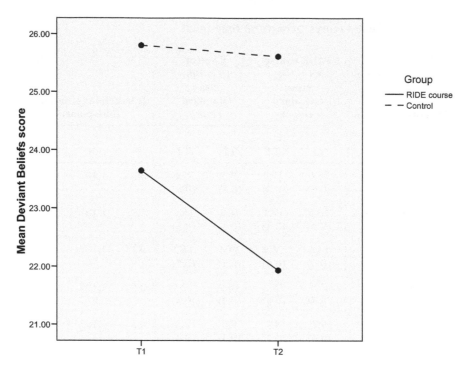

Figure 14.2   Deviant beliefs (group by time-point)

Discussion

This study reports some of the interim findings of an ongoing study which aims to assess the effects of attending a motorcycle offender course within the UK's national RIDE scheme on riders' self-reported behaviour and attitudinal measures. The results of these interim analyses indicate that RIDE course clients report desirable changes in a number of characteristics related to collision involvement, when compared with a control group who did not experience a formal intervention. It is important to note that the follow-up at T2 measures responses were provided approximately four weeks after successful course attendance. At this point, RIDE course clients will have been informed by service providers that they have successfully completed the course and there will be no further action taken over the incident that led to their course referral. Indeed, those riders who had been required to hand in their licence may have received it back from the police. Therefore, we can be reasonably confident that respondents feel able to respond honestly and without fear of reprisal.

The RIDE sample is almost exclusively male, riding large-capacity motorcycles and within the age-range reporting the highest involvement motorcycle crashes resulting in fatalities or serious injuries (Department for Transport, 2008). Just under one-third report existing penalty points on their driving licence and approximately two-thirds report previous crash involvement while riding their motorcycle. Of these crashes, three-quarters were reported as single-vehicle incidents. Further, the RIDE clients report significantly more penalty points on their driving licence arising from motorcycle-related offences than the control group. From these figures, it would appear that the 'right people' are being referred by the police onto the RIDE scheme.

In terms of self-reported behaviour, after course attendance RIDE clients reduced the margin over the speed limit in 30mph restricted speed areas that they felt acceptable compared with the controls. Of those RIDE clients who reported recently riding at excessive speeds (over 100mph on public roads) at T1, there was no effect of course attendance detectable at T2, but it is interesting to note that many of the control respondents who reported *not* having recently ridden excessively fast at T1, reported that they *had* done so by T2. Those in the RIDE group who reported not behaving in this way at T1 clearly did not feel tempted to do so after course attendance. A similar pattern of response is evident with riding soon after drinking alcohol, but is not statistically reliable. Given the very small numbers of cases involved in these analyses, these results should be considered with caution. Nonetheless, these two forms of motorcycling behaviour might both be considered particularly socially deviant, and Lawton and colleagues (1997) found evidence for a consistency in such forms of deviance. It is therefore possible that moderately socially deviant riders who have not engaged in such behaviour in the six weeks prior to course attendance (and so do not appear to do so regularly) are dissuaded from such deviance in the period following the course, an effect not present for those moderately deviant riders in the control group. For those riders who have ridden excessively fast in the period immediately prior to course attendance and so may be considered more socially deviant, course attendance appears not to effect changes in their excessive behaviour. Further investigation of the differential effects of course attendance will be possible with the larger dataset from the ongoing project and, with a longer follow-up period, analysis of the longevity of these effects will be possible.

RIDE clients are significantly more likely to report that they have thought about the manner in which they ride and are more likely to report attempts to change their riding to reduce their likelihood of collision involvement. For relatively high-risk riders, accepting that there is a need to change the way that they behave is clearly a requirement before such change can take place. However, as West (2006) and others argue, for those more risk-appropriate behaviours to be lasting is likely to require some longer-term influence. This issue will be discussed later.

When considering the effects of course attendance on the attitudinal measures, the reduction in the strength of deviant beliefs regarding the respondent's ability to ride outside of the rules and remain safe, indicates that clients have developed a better appreciation of the increased vulnerability that such behaviours produce. Part of the course focuses on a central function of traffic law in allowing road-users to predict others' actions. Making one's behaviour unpredictable to other road-users will inevitably lead to an increased likelihood that those other road-users will commit errors, failing to correctly anticipate the rider's behaviour. As Stradling and Parker (1996) suggest, the commission of an error whilst driving or riding combined with a violation is the best behavioural predictor of crash involvement on the road, and these two behaviours do not necessarily need to be committed by the same road-user. It would appear that RIDE clients are amenable to this message.

The longevity of the demonstrated effects of course attendance cannot be assessed from the present data. All that we know is that the effects are present four weeks after course attendance. Clearly, in order for an intervention to be effective, its influence must be lasting and there are some methods for achieving this goal, effectively keeping riders in the maintenance stage of Prochaska and DiClemente's (1983) model. As Prochaska and colleagues (1992) and, more recently, West (2006) and others point out, remaining in the maintenance stage is not a stable state and relapse is likely. Therefore, it is essential that methods are established to maximise the likelihood that new, more appropriate forms of behaviour are established, maintained over time and so become routine, forming a habitual element of an individual's behavioural repertoire. As an example of such a method, the formation of specific behavioural intentions (Gollwitzer, 1993) has been shown to improve the retention of specific safety-orientated behaviours (e.g. Elliott and Armitage, 2006). Indeed, there is an element of the RIDE course model that prescribes the development of such implementation intentions in course clients, but there are some issues surrounding consistency of course delivery that precludes the inclusion of this element in all courses. It is therefore likely that many respondents in the RIDE group did not make such explicit intentions and this may have influenced their future behaviour. A process of ensuring consistency of delivery is currently in hand and so it is anticipated that the behavioural intention exercise element will become more consistently applied in future.

Approximately one-quarter of riders in the RIDE group were apprehended by the police after the end of September, approximately the end of the 'traditional' motorcycling season. Therefore, they may consider themselves to be more experienced in riding in adverse weather and road conditions, and therefore 'better' riders than those who only ride in fine, dry, sunny weather. On this basis, it is likely that the course will have a weaker effect on these riders than those with less experience of adverse conditions. It is possible that these riders would have a more realistic view of the vulnerability of the 'average' rider, but would not consider themselves to be average. These considerations are important, as the two central intentions of the course are to increase the perceived risk of riding in an inappropriate manner and to increase the perceived vulnerability that experienced riders can lose after the repeated experience of *not* crashing. The larger sample resulting from the ongoing data collection is likely to resolve these issues, allowing comparisons between 'year-round' and 'summer only' riders. Furthermore, the differences in the change between time-points in attitudinal measures for the RIDE group compared with the control group are small and, although the deviant beliefs measure interaction is statistically reliable, it is possible that it is only one section of the RIDE clients' group that reports such a change. A larger sample will help to segment this group, in order to identify any differential effects of course attendance.

The RIDE course client group and the controls appear to be similar in all but their age and specifically motorcycle-related offending history, and so the comparisons made between these groups are considered reasonable. Of course, it is likely that there are differences between the groups that have not been measured in the current study and this possibility should be discussed. One of the main differences between the groups is that the RIDE clients have all had an encounter with an authority figure in the form of the police officer who stopped them at the roadside after observing them riding inappropriately. This encounter is likely to have caused the rider to consider their riding behaviour, even if this consideration was superficial and any need to change or moderate their behaviour quickly dismissed. However, it would seem sensible to control for these processes in future and, to that end, attempts are currently underway to recruit five new control groups: riders who have attended a police-run course with training based on traditional skills-based methods; riders who have engaged in an informal 'hearts and minds' conversation with police traffic officers (a policing tactic employed by several police services in the UK); riders who have been stopped by a police traffic officer as a result of inappropriate riding, but have only been issued with an informal warning; riders who have been stopped by a police traffic officer as a result of inappropriate riding, but have been issued with a fixed penalty notice (the traditional means of dealing with minor traffic offences, resulting in a fine and penalty points on the rider's licence); and riders who have been offered a place on the RIDE course but have declined the offer for some reason. In addition, it is important to investigate the longevity of any effects of course attendance, assessing the degree to which course clients develop permanent appropriate behavioural changes in their future riding. This study is ongoing, with a longer-term follow-up and the additional control groups discussed above.

Finally, an issue unrelated to the current research but one that is a matter for concerned interest is the familiarity respondents in both groups have with the mortality aspects related to riding a motorcycle. Roughly three-quarters of the combined sample know someone who has been killed on a motorcycle and an even larger proportion knows someone who has been seriously injured while riding. Perhaps, given the casualty statistics, this should come as little surprise but such rates are comparable with US armed forces personnel on active service (Cirillo, 2008). Anecdotal evidence suggests that motorcycling is frequently used as a form of therapy to temporarily 'filter out' negative aspects of the rider's life. The possible effects on cognitions and the behavioural consequences of such personally relevant encounters with death and serious injury deserve further investigation.

In summary, these results indicate that RIDE course attendance has a beneficial effect on clients' perceptions of risky riding and collision involvement while riding a motorcycle. The research project was completed in early 2010, and a more comprehensive report will be available. It is anticipated that many of the questions raised in this paper will then be answered.

Acknowledgements

This research is funded by Lancashire County Council, the Association of National Driver Improvement Service Providers (ANDISP), RSGB formerly LARSOA, and the Motorcycle Industry (MCIA). The data collection would not have been possible without the support of Devon County Council and the Devon Drivers Centre, Humberside Safer Roads Partnership and Kent County Council.

References

Broughton, J. (2005). *Car Occupant and Motorcyclist Deaths 1994–2002*. TRL Report TRL629. Crowthorne, Transport Research Laboratory.

Christmas, S. et al. (2008). *Understanding Motorcyclists' Attitudes to Safety, 'Passion and Performance'*. TRL Motorcycle Safety Research Report: PPRO 4/001/043 Crowthorne, Transport Research Laboratory.

Cirillo, V.J. (2008). Two faces of death – fatalities from disease and combat in America's principal wars, 1775 to present. *Perspectives in Biology and Medicine*, 51(1), 121–133.

Clarke, D.D. et al. (2004). *In-depth Study of Motorcycle Accidents*. TRL Road Safety Report No.54. Crowthorne, Transport Research Laboratory.

Clarke, D.D. et al. (2007). The role of motorcyclist and other driver behaviour in two types of serious accident in the UK. *Accident Analysis and Prevention*, 39, 974–981.

Department for Transport (2008). Road casualties in Great Britain 2007. Accessed 22 April 2009 from http://www.dft.gov.uk/pgr/statistics/datatablespublications/accidents/.

Elliott, M.A. and Armitage, C.J. (2006). Beyond the theory of planned behaviour: Testing a volitional intervention on drivers' compliance with speed limits. 26th International Congress of Applied Psychology, 16–21 July 2006, Athens, Greece.

Gollwitzer, P.M. (1993). Goal achievement: The role of intentions. *European Review of Social Psychology*, 4, 141–185.

Lawton, R. et al. (1997). Predicting road traffic accidents: The role of social deviance and violations. *British Journal of Psychology*, 88, 249–262.

Lynam, D. et al. (2001). *An Analysis of Police Reports of Fatal Accidents Involving Motorcycles*. TRL Report TRL492. Crowthorne, Transport Research Laboratory.

Mannering, F.L. and Grodsky, L.L. (1995). Statistical analysis of motorcyclists' perceived accident risk. *Accident Analysis and Prevention*, 27(1), 21–31.

Nordgren, L.F. et al. (2007). Unpacking perceived control in risk perception: The mediating role of anticipated regret. *Journal of Behavioral Decision Making*, 20, 533–544.

Prochaska, J.O. and DiClemente, C.C. (1983). Stages and processes of self-change of smoking: Toward an integrative model of change. *Journal of Consulting and Clinical Psychology*, 51(3), 390–395.

Prochaska, J.O. et al. (1992). In search of how people change: Applications to addictive behaviours. *American Psychologist*, 47(9), 1102–1114.

Sexton, B. et al. (2004). *The Accident Risk of Motorcyclists*. TRL Report TRL607. Crowthorne, Transport Research Laboratory.

Stradling, S.G. and Parker, D. (1996). Extending the Theory of Planned Behaviour: The role of personal norm, instrumental beliefs and affective beliefs in predicting driving violations. Paper given at the International Conference on Traffic and Transport Psychology, Valencia, Spain, May 1996.

West, R. (2005). Time for change: Putting the Transtheoretical (Stages of Change) model to rest. *Journal of Addiction*, 100, 1036–1039.

Chapter 15

Encouraging Rider Behaviour Change by Using Respected Communicators

Paul S. Broughton, Sandy Allan and Linda Walker

Introduction

Reaching your target audience with road safety messages is essential if you wish your message to be received and acted upon. A badly targeted or designed intervention will be ineffective and may even have a negative effect, for example being rejected by the target group as not being appropriate to them or patronising, therefore alienating members of the target group. This problem may be further exaggerated if the target group is, or its members feel that they are, part of a community (Broughton and Walker, 2009; Goldberg et al., 1997). This can be the case for motorcyclists who often believe they are members of the greater collective of 'bikers' (Broughton, 2007). This is an issue that needs to be considered at policy level, as it is at this level that general intervention principles are set; therefore policy makers need to be aware of potential message rejection by target groups.

Motorcyclists are vulnerable road users, with the relative risk of being involved in a killed or seriously injured crash (KSI) being 54 times higher than car drivers in 2006 (DfT, 2008), therefore it can be argued that it is even more important to get effective safety messages through that motorcyclists will act on. Safety messages may be designed to counter misheld beliefs, for example that most powered two-wheeler (PTW) KSI crashes are caused by other road users or faulty road infrastructure (Broughton and Walker, 2009). If messages are to be effective they have to appreciate the motivation of riders; for example half of riders ride because of the challenge that riding can give them, rather than the risk that the riding activity can provide (Broughton, 2008). With this in mind, policy makers need to implement methods to give riders accurate road safety knowledge. This information must be designed to affect a change in attitudes leading to safer riding behaviours.

The Lothian and Borders Police Force in south-east Scotland area have a plethora of roads that are attractive to leisure riders. These roads attract a large number of riders, especially in the summer months, with this increase in exposure often being reflected in the number of crashes in the area. In an attempt to mitigate this increase the police force developed a campaign that, rather than discouraging riders to come to the area, gives motorcyclists information to help them enjoy the area safely. The most innovative element of this intervention was that the safety messages would be delivered by police motorcyclists. Although it may seem counter-intuitive to have police, who normally enforce the law, giving out 'tourist information' to motorcyclists, its potential for success lay in the way that police riders are viewed within the motorcycle community. Generally police riders are firstly viewed as part of the motorcycle community; it is a secondary consideration that they are part of a law enforcement agency. They are also highly respected for their biking abilities. The 'Around the Corner' (AtC) intervention discussed in this paper, sought to utilise this respect for police riders to more effectively communicate safety messages to this vulnerable group.

The 'Around the Corner' intervention

The AtC campaign consists of three elements: a booklet, a dedicated website and police contact with riders. The booklet consists of a map of the area showing common biking routes, tea stops, etc.; it is also annotated with specific road safety messages such as indicating if particular corners are deceptive and may cause problems to riders. The booklet is designed to send out a welcoming message; providing riders with information they may want, rather than simply focusing on road safety information.

The website (http://www.aroundthecorner.org.uk) is also based around the principle of welcoming riders to the area, and states that: 'Our message is simple: enjoy what the Borders have to offer but recognise your own ability and read the road' (Around the Corner, 2009b).

Within the website are routes designed for riders to enjoy as well as top riding tips from police riders. As with the booklet, the website is designed to give motorcycle riders information that they want, not simply safety information.

The final element of the intervention is the use of police riders to speak to motorcyclists on a one-to-one basis. This element is described on the website:

> Throughout the season our advanced bikers will be out and about on our routes giving free tips and advice and on occasions, if you're lucky, free coffee. They may even have some goodies in those panniers – so don't be shy, if there's something you want to know more about, how to ride better and improve your skills, just ASK! (Around the Corner, 2009a)

All three elements are based on experienced police motorcyclists giving other riders the benefit of that experience, via police comments on the website, riding routes that the police enjoy using themselves coupled with safety tips contained within the booklet, on the website, or via a discussion with a police rider. The idea of establishing this 'hearts and minds' campaign has the aim of communicating with riders in a way that is accepted by riders, allowing messages that can have a positive affect on rider safety to be communicated effectively.

Intervention evaluation

In 2008, Lothian and Borders Police Force launched its second year of the AtC campaign. The evaluation of the 2008 AtC intervention is based on 642 responses from riders gathered through face-to-face interviews (N = 546) and an Internet survey (N = 96). Further data were collected from a subset of the original 642 respondents (N = 83) in an end of season survey (EOSS). The data from EOSS is used to explore any long-term effects from AtC.

Demographics

For the questionnaires that were completed during the season (excluding the EOSS) the gender split was 85 per cent male (N = 499) and 15 per cent female (N = 87), this split is generally in line with the profile of UK bikers found in other studies (Broughton 2007). Therefore the respondents are quite representative of the general biker population in terms of gender. However, female riders did seem to be more willing to answer the EOSS questionnaire than male riders, a split of 28 per cent (N = 21) and 72 per cent (N = 54) respectively.

The mean respondent age was 44.3 years (SD = 10.66; 95 per cent CI 43.4–45.37) with a standard deviation of 10.7 years. The median age was 45 years. Table 15.1 is a comparison between the survey respondents' age and the age of riders according to the Department for Transport (DfT, 2006). This suggests that the respondents are over-represented in the 40–60-year-old age group, which may be due to the sample being leisure riders, with the DfT sample also containing those who ride for functional reasons. It also shows that the campaign is targeting riders within the 40–60-year-old age group, a group that is involved in 39 per cent of all PTW KSI crashes (DfT, 2008).

Survey respondents were asked for the make and model of their bikes. This information was used to categorise bikes by performance based on the power index system that uses bike data to produce a measure of bike performance (Broughton, 2008). This Power Index (PI) was used to classify respondents' bikes into high- and low-performance categories, the boundary for the low/high split was an index of 75, which equates to bikes like a Ducati 748 or Honda CB600 Hornet. Approximately two-thirds of respondents (64 per cent) rode the lower performance bikes with 36 per cent riding the higher grouped machines. The younger (under 40) age group were more likely to ride the higher performance machines (42 per cent), compared with 30 per cent of older riders (χ^2 (1 df, N = 582) = 4.259, P = 0.039). There was also a gender split, with only 23 per cent of female respondents riding the higher category bikes, compared with 38 per cent of males (χ^2 (1 df, n = 583) = 7.720, P = 0.005).

Table 15.1   Comparison of respondents' age profile against the Department for Transport data on PTW users

Age	Survey %	DfT %	KSI %
Under 19	2.7	9	6
20–29	7.9	11	24
30–39	16.6	31	26
40–49	44.4	28	27
50–59	21.2	14	12
60+	7.2	8	4

Source: Survey data and DfT (2008).

Overall the review of demographics shows that the campaign tends to interact with those in the problem age group, and riders of a cross-section of bikes by performance.

Acceptance of Advice from Police Riders

For this type of intervention to have a chance of being successful, motorcyclists must be willing to accept that the police are trying to help them; there must be a distinction between an educational intervention and one that is based on enforcement. With this help coming from a motorcyclist it is less likely that it will be rejected because of a belief that the person is not an expert; however if the advice were given by someone who was not perceived as an expert, such as a police car driver, then this might not be the case. For example, one biker when commenting on their beliefs on the difference between police motorcyclists and drivers said: 'They aren't as biased towards bikers as standard police seem to come across as'.

When motorcyclists were asked if they would be happy to have a conversation with a police rider in a non-enforcement situation, 95 per cent either agreed or strongly agreed that they would (N = 529).

The evidence shows that motorcyclists are generally happy for the police to discuss road safety with them and these discussions may have a direct affect on some riders' behaviour while for others the positive effects may be of a more secondary nature, but still positive. This conversation has to be balanced with the risk of prosecution, and riders tended to think that a conversation with the police might have better road safety outcomes that a prosecution would:

Sometimes a talk can be better than a charge.

Coaching not prosecution.

More talk. I think there is a fear that the police only want to charge you with something. Not true.

The other elements of the campaign, namely the booklet and the website, also heavily rely on content from the police riders. This should reinforce the idea of the campaign as based on communicators generally respected by the biking community.

Efficacy of the Campaign Elements

The AtC campaign consists of three elements: the booklet, police contact and the website. All three elements were designed to complement each other and to reinforce the overall message of safe enjoyment of the Lothian and Scottish Borders motorcycle routes. Each of these elements was assessed by asking riders to give their opinions on each.

The booklet contains maps of recommended biking routes along with useful information, such as places to stop for a cup of tea, and descriptions of some of the potential hazards. It is in a small fold-out format to make it easy to store and use. Of those questioned 476 had previously seen the booklet, with the majority of these feeling that the booklet was good; only 4 per cent disagreed with this. Respondents reported that the booklet had a positive affect on their riding with 85 per cent stating that the booklet made them think about how they rode (N = 469) and 69 per cent stating that the booklet helped them ride in a safer manner (N = 465). The booklet also had the affect of directing riders to the website, with only 7 per cent saying that they were either unsure or would not visit the website (N = 469). Of the self-reported changes made by riders investigated with the EOSS, 39 per cent said that the booklet had helped them ride in a safer manner (N = 83). Some of the positive effects can be assessed by reviewing some of the rider comments:

The booklet raised my awareness about how dangerous the roads around my area were, I did not realise how many accidents had taken place involving motorcyclists in Lothian.

To be much more aware of the unknown, i.e. what is around the next bend, cows, someone who has broken down, a tractor pulling out?

The danger hotspots and good rides. Very clear small succinct. Nicely done.

It was good to refresh your mind with what could happen if you don't concentrate.

The most novel element of this campaign is the direct non-enforcement intervention by the police motorcyclists. During the 2009 launch of the campaign, an AtC representative commented:

> Around the Corner aims to engage and empathise with bikers in the Borders using our best asset, Police bikers. It's hoped that through this interaction we can encourage safe riding and signpost individuals towards any necessary further training. (Lothian and Borders Safety Camera Partnership, 2009)

As discussed earlier, riders are generally happy to have a conversation with police motorcyclists. Riders were questioned about what effect these conversations had on their riding behaviour in the EOSS with 26 per cent stating that they rode safer on the day that they were spoken to, and 74 per cent saying that it had no effect on their riding on the day (N = 81). Similar percentages are reported when riders were asked if they have ridden in a safer manner since their conversation (25 and 75 per cent respectively, N = 79).

While it is encouraging that 25 per cent of riders are self-reporting and that they now ride in a safer manner because of their contact with the police, reported secondary benefits adds to the overall efficacy of the intervention. Comments made by riders suggest it has had a role in building awareness of other road safety initiatives and improving the overall relationship between the police and the bikers:

> Learnt that L&B run a 'BikeSafe' programme and there was an information evening planned.

> They had a friendly attitude and talked about accidents in the local area, asking us to calm down and take things a little easier.

> Not much that I didn't already know, but they did reinforce their humanity and their shared love of biking. I'll be more comfortable approaching police from here on.

The third strand of the campaign is the website. The website is in an interactive format giving information, weather reports, biking news and details of upcoming events. It also contains riding tips and maps of routes.

Riders who had visited the website were asked what their overall opinion of the website was. All the responses were between OK and very good, with no one rating it as poor or very poor (Table 15.2).

Of those who completed a questionnaire while visiting the website, 82 per cent (N = 79) of those said they would probably or definitely recommend the site to others, only one respondent said that it was unlikely that they would recommend the site to another person. When asked on the EOSS if they had visited the website, 52 per cent stated they had (N = 82).

Table 15.2   Overall opinion of the website

	N	%
OK	18	22.8
Good	47	59.5
Very good	14	17.7
Total	**79**	**100.0**

The website was useful in getting riders to consider how they rode, with 91 per cent agreeing that it did (N = 77). However, this high figure was not reflected in those who said it would change the way they rode, with only 43 per cent saying it probably or definitely would; a further 43 per cent said that it may effect a behaviour change (N = 76). The EOSS asked riders if the website had actually had an effect on their riding style, 80 per cent of riders stated that it had no effect at all, but 18 per cent did say that they now rode in a safer manner (N = 81).

The comments given indicated that riders were using the website for a variety of purposes and had gained a range of different benefits from it. This ranged from improving their skills, to sharing information and knowledge, to highlighting new aspects of riding safety.

> Learned about limit points and positioning approaching corners. Learned to use brakes before bends, not during.

> Relay road info to the bikers through the internet and speaking to bikers about road safety.

Some comments suggested that a similar approach might be useful for police forces in other areas of the country to consider:

> The website is quite well organised and informative to bikers in the border region shame that the region I live in has not taken this idea on board for bikers in the north east region of England as we have similar roads and accident black-spots where biker injury and deaths occur each season.

The website also served an additional purpose of promoting the police force in a positive light and promoting other initiatives and events happening in the area.

> That the police do care for the safety and well being of motorcyclists, and that they are willing to share their knowledge/road-craft skills to others.

> There were a number of new items which I have never thought of before.

About the different things that the Around the Corner campaign is doing to help motorcyclists in the Lothian and Borders area.

Effectiveness of Reaching Target Audience

Road safety intervention campaigns such as AtC not only have to have an appropriate message, they must reach those who need to hear the message and affect change within this target audience. Therefore the point of the campaign is to influence those whose safety behaviours and attitudes need to be improved; so for evaluating AtC campaign components some questions were asked to assess rider attitudes and behaviours; for example, braking hard is often a sign that a rider is out of, or nearly out of, control (Stradling et al., 2008). This can therefore be used as an indicator of the level of control that a rider has. Within a 60mph speed limit area about 35 per cent of riders admitted to braking hard at least sometimes (Table 15.3), while in a 30mph speed limit within an urban area this reduces to 18 per cent (Table 15.4). This suggests that riders are less likely to push close to their limits within a town compared to riding in a non-urban situation.

Table 15.3   How often do you brake hard out of town (60 mph limit) on a typical journey?

	N	%
Never	111	19.1
Rarely	270	46.4
Sometimes	168	28.9
Often	26	4.5
Always	7	1.2
Total	**582**	**100.0**

Younger riders (under 40) are more likely to brake hard out of town compared to older (over 40) riders, 43 per cent compared to 30 per cent (χ^2 (1 df, N = 567) = 9.513, P = 0.002). Those who ride the higher performance machines are more likely to brake hard out of town, 44 per cent compared to 29 per cent of those on the less powerful bikes (χ^2 (1 df, N = 579) = 12.706, P < 0.001).

Those braking more frequently out of town were more likely to believe that speaking to the police had a positive effect on their riding that day (mean answer of 3.33 compared to 2.31, t(106) = 4.754, P < 0.001, r = 0.419). This effect is also evident, although less strong, when considering longer-term positive effects (3.37 compared to 2.46, t(106) = 3.738, P < 0.001, r = 0.341). This group also indicated that they were more likely to be positively influenced by the booklet

Table 15.4   How often do you brake hard within town (30 mph limit) on a typical journey?

	N	%
Never	142	28.4
Rarely	269	53.8
Sometimes	74	14.8
Often	9	1.8
Always	6	1.2
Total	**500**	**100.0**

(3.57 compared to 2.38, t(106) = 4.779, $P < 0.001$, r = 0.421) and the website (3.18 compared to 2.62, t(106) = 2.747, $P = 0.007$, r = 0.258). The advice given within the AtC intervention also influenced those who are more likely to brake hard to consider undertaking training (1.61 compared to 1.15, t(101) = 2.872, $P = 0.005$, r = 0.275). Thus the group of riders who tend to be close to being out of control are the most positively affected by this intervention – suggesting that the intervention is reaching and influencing those most at risk.

Changes in Attitudes and Behaviour

At the end of the riding season, riders were asked what impact the AtC campaign had had on their riding style. Several positive responses were made in relation to speed, positioning and use of controls:

> I have enjoyed myself more this summer probably due to having slowed down a bit, there have been less scary moments than the previous summer.

> I believe it made me think more about how I ride.

> Probably made me pay more attention to smooth throttle operation and positioning and speed entering bends.

Some respondents indicated a more general benefit in terms of awareness, both of their own skills and what others may do:

> Makes me think that you are always still learning every day and to expect the unexpected, whilst out on the road.

Nice to know there are police bikers looking out for us and who understand. Also nice to know that folk pull out and run out in front of you guys even with all your reflective gear.

A few comments while seeming to say that AtC had no effect on them, actually suggest that it has, for example:

It had no direct effect.

Being brutally honest none at all. However it was a useful reminder of just how vulnerable motorcyclists are, especially when they forget to turn off the stupid switch.

These comments suggest that while they may not believe that there has been an impact on their riding behaviour, it has an indirect impact reminding them of the need to ride defensively.

There would seem to be a generally positive view on the impact of AtC on rider safety. Most of the respondents strongly agreed (38 per cent) or agreed (53 per cent) with 'I think "Around the Corner" can make motorcycling safer'; with the remaining nine per cent being neutral (N = 76). Over half of the respondents (53 per cent) strongly agreed with 'I think "Around the Corner" is a good use of police motorcyclists' time', with a further 40 per cent agreeing with this statement, the remaining six riders were neutral on the subject (N = 76). When asked if AtC made them think about how they rode nearly 70 per cent agreed or strongly agreed (Table 15.5), however about 7 per cent disagreed or strongly disagreed. Around 50 per cent agreed that they are safer riders because of AtC (Table 15.6).

Table 15.5   'Around the Corner' has made me think about how I ride

	N	%
Strongly agree	14	18.4
Agree	38	50.0
Neither	19	25.0
Disagree	3	3.9
Strongly disagree	2	2.6
Total	**76**	**100.0**

Some researchers have identified knowledge as a key aspect in safer road use (Wallen Warner and Aberg, 2008), therefore respondents to the end of season questionnaire were asked if they agreed with the statement '"Around the Corner" increased my knowledge on motorcycle safety'. Around 60 per cent (N = 76) either agreed or strongly agreed with this statement, and only 7 per cent did not agree with it (Table 15.7)

Table 15.6   'Around the Corner' has made me a safer rider

	N	%
Strongly agree	7	9.2
Agree	32	42.1
Neither	31	40.8
Disagree	4	5.3
Strongly disagree	2	2.6
Total	**76**	**100.0**

Table 15.7   'Around the Corner' increased my knowledge on motorcycle safety

	N	%
Strongly agree	13	17.1
Agree	32	42.1
Neither	26	34.2
Disagree	4	5.3
Strongly disagree	1	1.3
Total	**76**	**100.0**

Discussion

Motorcyclists are vulnerable road users with their risk of being killed or seriously injured being considerably higher than that of car drivers (RoSPA, 2001; DfT, 2008; Institute of Highway Incorporated Engineers 2005). There are many safety interventions currently being used to try to reduce the rider KSI rate. However, for any non-enforcement based campaign to succeed there must be some 'buy in' from the end user, in this case motorcycle riders. Motorcyclists are in the minority, making up about 3 per cent of traffic (DfT, 2008). They are often viewed as belonging to a subgroup where membership is restricted to other riders with distain for those who have excluded themselves from 'this club' by deciding not to ride (Broughton and Walker, 2009; Gutkind, 2008). This leaves road safety personnel with two potential problems; how to understand the motivations, goals, beliefs and behaviour of riders and how to successfully penetrate the sometimes seemingly closed world of biking to put across safety messages.

The idea of the AtC road safety intervention was to use motorcyclists, who also happened to be police riders, to communicate the safety message to riders. The data presented in the preceding sections indicate that riders saw the police motorcyclists as mostly bikers and part of the biker fraternity. This, coupled with respect for the riding skills of the police motorcyclists, provided a solid platform from which a safety message could be communicated.

The communication used was designed to give riders the information they felt they would want, such as which roads are good for riding on, together with potential hazards, such as particularly bad patches of loose gravel. The use of the booklet containing two popular riding routes is a good example of this, where the information that the rider would want is provided, but it is supplemented with specific road safety information.

This policy is about communicating road safety advice with the aim of changing attitudes and behaviour; therefore the resultant intervention has been designed to avoid sending out mixed messages. When police riders are involved in engagement in a particular area this message will not be discouraged by a parallel enforcement initiative. However, such a strategy does not negate obvious enforcement issues that may develop. This strategy has been endorsed by riders who are supportive of AtC.

The website is also used to promote the safety message from police riders, with large sections of content coming from the police motorcyclists. Certainly the website was well received, and riders generally would like to see its content increased to include such content as links/contact details for other organisations that could provide help for motorcyclists; the inclusion of videos on safety training and statistics and information on accidents involving motorcyclists.

Although very little numerical evidence could be found on behaviour change when statistical tests were run on riders' behaviour during their interaction with the intervention and the EOSS, there was considerable evidence of positive effects when riders were considered at the individual level; for example, riders reporting that they are riding slower or taking the wearing of protective equipment more seriously.

One comment often heard is that those who need extra training are those who are least likely to take it and that often training is taken up by those who tend to ride safer and tend towards being risk-averse (Broughton, 2007). However, in this case the use of respected communicators seems to encourage the more extreme riders to consider training. While skills-based training itself is not the full answer (Broughton, 2008), what is encouraging is an attitude where these riders are admitting to themselves that their skills may not be up to a level that they would like them to be.

Conclusion

This survey suggests that the implementation of a 'hearts and minds' policy can, and does, increase rider road safety knowledge and influence attitudes. Although there is no statistical evidence of behaviour change, the prerequisites for this change are being put into place (Rothengatter, 2002; Parker, 2002).

The three elements of this particular intervention form a consistent, but welcoming, road safety message of 'have fun – but do it safely' rather than the idea of trying to reduce PTW KSI incidents by moving the problem to a different area by sending out a 'bikers not welcome' message (Broughton and Walker, 2009). This policy of welcoming riders to the area may also have a positive knock-on effect for the local tourism industry; encouraging this relatively affluent segment of the population to visit the area.

It can be difficult to quantify the impact of safety policies and interventions, especially if the relatively low number of KSI events is used to measure intervention effectiveness. The policy decision to use a precious resource, such as a police officers' time, to go out and speak to a few riders may seem difficult to justify, even more so in the results-driven world that we reside in. However, the evaluation research carried out for this particular intervention demonstrates that taking a long-term policy view of a real problem can generate road safety results, not just in the short term but also into the future.

References

Around the Corner (2009a). *Around the Corner*, http://www.aroundthecorner.org.uk.

Around the Corner (2009b). *Lothian and Borders Re-launch Motorbike Campaign*, http://www.aroundthecorner.org.uk.

Broughton, P.S. (2007). *Risk and enjoyment in powered two-wheeler use.* Unpublished PhD thesis, Transport Research Institute, Napier University.

Broughton, P.S. (2008). Flow, task capability and powered-two-wheeler (PTW) rider training. In L. Dorn, ed., *Driver Behaviour and Training, Vol III*. Aldershot, Ashgate, pp. 415–424.

Broughton, P.S. and Walker, L. (2009). *Motorcycling and Leisure: Understanding the Recreational PTW Rider*. Aldershot, Ashgate.

Delaney, A. et al. (2004). *A Review of Mass Media Campaigns in Road Safety*. Victoria, Australia, Monash University, Accident Research Centre.

DfT. (2006). *Compendium of Motorcycling Statistics 2006*. London, DfT.

DfT. (2008). *Compendium of Motorcycling Statistics 2008*. London, DfT.

Goldberg, M.E. et al. (1997). *Social Marketing: Theoretical and Practical Perspectives*. Philadelphia, PA, Lawrence Erlbaum Associates.

Gutkind, L. (2008). *Bike Fever*. River Grove, AZ, Follett.

Institute of Highway Incorporated Engineers (2005). *IHIE Guidelines for Motorcycling. 2005*. London, IHIE.

Kotler, P. and Lee, N.R. (2008). *Social Marketing: Influencing Behaviours for Good*. London, Sage.

Lothian and Borders Safety Camera Partnership (2009). *Around the Corner Biker Campaign 2009.* Statement made by Sandy Allan, Road Safety Coordinator at the Scottish Motorcycle Show, 14 March 2009.

Ludwig, T.D. et al. (2005). Using social marketing to increase the use of helmets among bicyclists. *Journal of American College Health,* 54(1), 51–58.

Parker, D (2002). *Changing Drivers' Attitudes to Speeding: Using the Theory of Planned Behaviour.* Buckingham, Open University Press.

RoSPA. (2001). *Motorcycling Safety Position Paper.* Birmingham, The Royal Society for the Prevention of Accidents.

Rothengatter, T. (2002). Drivers' illusions – no more risk. *Transportation Research Part F: Traffic Psychology and Behaviour,* 5(4), 249–258.

Stradling, S. et al. (2008). *Understanding Inappropriate High Speed: A Quantitative Analysis.* Road Safety Research Report 93, London, Department for Transport.

Wallen Warner, H. and Aberg, L. (2008). Drivers' beliefs about exceeding the speed limits. *Transportation Research Part F: Traffic Psychology and Behaviour,* 11(5), 376–389.

Chapter 16

The Motorcycle Rider Behaviour Questionnaire (MRBQ) and Commercial Motorcycle Riders in Nigeria

Oluwadiya Kehinde Sunday and Ladoke Akintola

Introduction

All over the world, motorcycle riding is becoming increasingly popular: in the USA, more than 1.1 million units were sold in 2008; and in Nigeria, where comparable data are not available, 52 per cent of all licence plates issued in 2004 were for motorcycles (Federal Road Safety Commission, 2006; Morris, 2009). The motorcyclist is also inherently more vulnerable to injury than the car occupant: motorcyclists were about 35 times more likely to die than passenger car occupants and eight times more likely to be injured (National Highway Traffic Safety Administration 2008). The increasing popularity of the motorbike in Nigeria has led to a frightening increase in fatality rates. The casualty problem is compounded by an upsurge in the use of motorcycles for commercial use (Oluwadiya et al., 2004, 2009).

In the last two decades, it has become increasingly recognised that epidemiological principles will be important in understanding and controlling the factors contributing to injuries (Peden et al., 2004). Thus, in the case of motorcycle trauma, the factors contributing to road crashes can be divided into the motorcycle, the environment in which rider rides (e.g. traffic, road type and weather conditions) and the rider. Two of the most important characteristics of the rider are rider behaviour and attitude, about which relatively little has been written (Elliott et al., 2007). Aspects of riding behaviours that have been found to correlate with crash risks include speeding, drink-driving, poor observation and signalling at junctions (Elliott et al., 2007). Such findings are useful, but what would probably be more beneficial is to have a tool that will combine all these factors to reliably measure all aspects of riding behaviour. The Driver Behaviour Questionnaire (DBQ) was developed by Reason and collegues to measure driving behaviour among motor vehicle drivers (Elliott et al., 2007). It has proved to be a valuable research tool for investigating driver behaviour and its tripartite typology has been successfully reproduced in many other studies from across the world (Aberg and Rimmo, 1998; Parker et al., 1995)

Because of the success of the DBQ, a similar tool, the MRBQ was developed and validated in the UK for use among motorcycle riders (Elliott et al., 2007). The study, which was based on a large sample of motorcycle riders in the UK, revealed that motorcycle rider behaviours could be explained by a five-factor structured model: traffic errors, control errors, speed violations, performance of stunts and use of safety equipment. Traffic errors were mostly associated with hazards perception or observational skills; control errors concerned with speed or errors in speed selection. Traffic errors, followed by control errors, were found to be the most consistent predictors of crash involvement. Stunt and speeding behaviours were less consistent in predicting accident liability. The authors identified riding styles, getting pleasure from riding and a liking for speed as predictors of behavioural errors, which were themselves found to be predictors of crashes. They argued that such a relationship shows that an important aspect of motorcycle safety stems directly from the motivation that prompts people to ride motorcycles in the first place. This presents an important challenge to the problem of motorcycle safety.

Rider behaviour questionnaires need to be benchmarked for different countries and cultures (Ozkan et al., 2006) due to variation in social norms and traffic systems. This effect has been reported in the use of the DBQ across different countries in Europe, the Middle East, Asia and Oceania. For example in Australia, even though the tripartite typology of factors was retained, they were found to be different: the three factors were general errors, dangerous errors and violations (Blockey and Hartley, 1995). It was also found that in certain subsets of drivers some factors loaded separately, for example, in a set of company drivers, some violations and errors loaded onto the same factors (Dimmer and Parker, 1999). Finally, some studies have reported a different numbers of factors from the original three-factor structure (Dimmer and Parker, 1999; Parker et al., 2000).

In Nigeria, as in many low income countries, riding or driving for pleasure is rare (Oluwadiya et al. 2009). Riding motorcycles is a livelihood and a necessary part of daily commuting to and from work. Thus, because the motivation to ride motorcycles in Nigeria may be quite different from the motivations of riders in higher income countries (HIC's), the six-factor typology of the MRBQ might not be perfectly replicated. The aim of the present study was to investigate the factor structure of the MRBQ amongst Nigerian motorcyclists, compared to its performance in the original study (errors), and to examine the relationship between the MRBQ factors and self-reported crash involvement of the riders. This chapter will discuss possible ways in which findings from this study can help in formulating policies directed at reducing motorcycle crash rates and injuries.

Method

Participants

The study participants consisted of 500 motorcycle riders in Osogbo, the capital of Osun State in south-west Nigeria. Two trained research assistants administered the questionnaires to participants at petrol stations, roadsides and commercial motorcyclists' parks. The participants were assured of confidentiality and anonymity. Most motorcycles in Nigeria have small engines and are mostly ridden for either commercial commuting or private business purposes.

The Study Instrument

The study instrument was the MRBQ, which is a 43-item questionnaire. The original version includes 5 factors which are: traffic errors (13 items), control errors (7 items), speed violations (10 items), performance of stunts (6 items) and use of safety equipment (7 items). A Nigerian version was devised by rewording and making necessary modifications to adapt the questionnaire items to a Nigerian sociocultural context. Overall, 7 items were removed, 11 reworded, 6 new items added, 22 unchanged, leaving a total of 40 items in the new version (Table 16.1). Finally, the six-point rating scale (1 – never, 2 – hardly ever, 3 – occasionally, 4 – quite often, 5 – frequently and 6 – nearly all the time), was reduced to a four-point rating scale (1 – never, 2 – occasionally, 3 – frequently and 4 – nearly all the time), after the initial pilot study, which showed that respondents had difficulty distinguishing between the pairs of hardly ever/occasionally and quite often/frequently.

Table 16.1   Original MRBQ and the modified MRBQ used for the study

Original MRBQ	Modifications in present study
Fail to notice that pedestrians are crossing when turning into a side street from a main road	Unchanged
Not notice someone stepping out from behind a parked vehicle until it is nearly too late	Unchanged
Not notice a pedestrian waiting to cross at a zebra crossing, or a pelican crossing that has just turned red	Not notice a group of pedestrians waiting to cross the road
Pull out on to a main road in front of a vehicle that you had not noticed, or whose speed you have misjudged	Unchanged
Miss 'GiveWay' signs and narrowly avoid colliding with traffic having the right of way	Removed
Fail to notice or anticipate that another vehicle might pull out in front of you and have difficulty stopping	Unchanged
Queuing to turn left on a main road, you pay such close attention to the main traffic that you nearly hit the vehicle in front	Unchanged
Distracted or preoccupied, you belatedly realize that the vehicle in front has slowed and you have to brake hard to avoid a collision	Unchanged

Table 16.1   Original MRBQ and the modified MRBQ used for the study
         continued

Original MRBQ	Modifications in present study
Attempt to overtake someone that you had not noticed to be signalling a right turn	Attempt to overtake someone that you had not noticed to be signaling a left turn
When riding at the same speed as other traffic, you find it difficult to stop in time when a traffic light has turned against you	When riding at the same speed as other traffic, you find it difficult to stop in time a traffic warden has signalled you to stop
Ride so close to the vehicle in front that it would be difficult to stop in an emergency	Ride so close to the vehicle in front that it would be difficult to stop in an emergency
Run wide when going round a corner (verge to the middle of the road)	Run wide when going round a corner (verge to the middle of the road)
Ride so fast into a corner that you feel like you might lose control	Ride so fast into a corner that you feel like you might lose control
Exceed the speed limit on a country/rural road	Removed
Disregard the speed limit late at night or in the early hours of the morning	Unchanged
Exceed the speed limit on a motorway	Exceed the speed limit on a highway
Exceed the speed limit on a residential road	Unchanged
Race away from traffic lights with the intention of beating the driver/rider next to you	Race away from bumps and traffic stops with the intention of beating the driver/rider next to you
Open up the throttle and just 'go for it' on country roads	Removed
Ride between two lanes of fast-moving traffic	Unchanged
Get involved in unofficial 'races' with other riders or drivers	Unchanged
Ride so fast into a corner that you scare yourself	Unchanged
Attempt to do, or actually do, a wheelie	Attempt to do, or actually ride with the front wheels off the ground
Pull away too quickly and your front wheel comes off the road	Unchanged
Intentionally do a wheel spin	Unchanged
Unintentionally do a wheel spin	Unchanged
Wear riding boots?	Removed

Table 16.1   Original MRBQ and the modified MRBQ used for the study
concluded

Original MRBQ	Modifications in present study
Wear a protective jacket (leather or non-leather)?	Unchanged
Wear body armour (elbow pads, shoulder pads, knee pads, etc.)	Removed
Wear no protective clothing?	Unchanged
Wear gloves?	Unchanged
Wear bright/fluorescent strips/patches on your clothing	Unchanged
Brake or slow down when going round a corner or bend	Unchanged
Change gear when going round a corner or bend	Unchanged
Find that you have difficulty controlling the bike when riding at speed (e.g. steering wobble)	Find that you have difficulty controlling the bike when riding at speed
Skid on a wet road or manhole cover	Skid on a wet road
Have trouble with your visor or goggles fogging up	Unchanged
Driver deliberately annoys you or puts you at risk	Unchanged
Ride when you suspect you might be over the legal limit for alcohol	Ride when you suspect you might drunk
Wear a leather one-piece suit?	Removed
Wear bright/fluorescent clothing?	Removed
Not present	Wear agbada or other flowing gowns
Not present	Ride in tandem with another bike with your leg on the other bike or vice versa
Not present	Overtake or attempt to overtake a vehicle on the right (passenger side)
Not present	Ride or attempt to ride against the traffic
Not present	Carry more than one passenger on the bike
Not present	Wear helmets

Demographic and Riding Variables

Participants were asked to indicate their age, gender, educational level, place of residence, religion, riding experience, formal training, possession of licence, whether the licence was with the rider at the moment, alcohol intake, whether the motorcycle had a speedometer, the usual purpose of riding, history of traffic violation and history of crash involvement.

Statistical Analyses

Statistical analyses were performed using SPSS version 15. Principal component analysis (PCA) with Varimax rotation was run to examine the factor structure of the MRBQ. Cronbach's alpha reliability coefficients were also calculated for assessing the internal consistency of the MRBQ scale scores. Generalized linear modelling (GLM) was performed with crash involvement as dependent variable, and MRBQ factors, age, experience, motorcycle usage and alcohol use as independent variables.

Results

All respondents were males. The mean age was 27 years (range = 17–70 years) and the mean riding experience was 4.72 years (range = 1–19 years). The correlation between age and riding experience was 0.571 (significant at 0.01 levels). Other demographics are shown in Table 16.2. Just 12 (2.4 per cent) had formal training, 36 (7.4 per cent) admitted to having a riding licence; but only 14 (2.8 per cent) had it with them at the time of the interview. One hundred and ten (22 per cent) had been involved in a crash previously while 124 (24.8 per cent) had been involved in one to six traffic violations in the past.

Factor Structure

Analysis using PCA with Varimax rotation was used to analyse the 40-item data (Table 16.3). The criteria used to determine the number of factors that would best fit the data were the Kaiser criterion of eigenvalues over 1.0, the Scree plot and parallel analysis. Initially, 15 factors had eigenvalues above 1.0, but the Scree plot and parallel analysis showed that the data were best fitted by a four-factor structure, indicating that there were four basic types of behaviours underpinning in the MRBQ among Nigerian motorcycle riders. The rotated factors accounted for 32.5 per cent of the total variance, which was not as robust as the 41.2 per cent MRBQ performance among British motorcycle riders in the original MRBQ (errors), but the robustness is comparable to that of the DBQ (33 per cent) among British drivers (Reason 1990).

Table 16.2   Demographic and riding data of participants

Variables	Number	Percentage
Education (N = 484)		
Primary	25	5.0
Secondary	367	76.0
Post-secondary	74	15.3
None	18	3.7
Religion (N = 497)		
Christian	248	49.9
Muslim	215	43.3
Others	34	6.8
Primary use of motorcycle (N = 491)		
Commercial	156	31.8
Private	70	14.3
Both	213	43.4
Business	48	9.8
Government	4	0.8
Alcohol intake (N = 468)		
Yes	170	36.3
No	298	63.7
Have functioning speedometer (N = 425)		
Yes	304	71.7
No	121	28.3

The first factor included eleven items which dealt with issues of safety and control and was thus labelled *control/safety* (Table 16.3). It accounted for 10.5 per cent of the total variance in the data. The mean of the items was used to produce the control/safety composite scale for further data analysis.

The second factor accounted for 8.6 per cent of the total variance and it contains nine items which described activities related to *stunt*. However, two of the items ('Wear gloves' and 'Distracted or pre-occupied, you belatedly realise that the vehicle in front has slowed and you have to brake hard to avoid a collision') could not be considered to belong to this factor, and were therefore excluded from the composite scale that was used for further analysis. This left seven items remaining in the scale.

The third factor consisted mainly of items that related to error and it accounted for 8 per cent of the total variance. Four items loaded under this factor, and all were included in the composite scale called error and used for further analysis.

There were seven items loading on to the fourth factor. All seven items are related not only to speeding but also to hurrying and were labelled *speeding/ impatience*. They were all used for the composite scale for this factor. Five items did not load under any of the components.

Table 16.3   Factor structure of the MRBQ items

	Factors			
	Control and safety	Stunt	Error	Speed – hurrying
Brake or slow down when going round a corner or bend	−0.773			
Wear helmets	−0.578			
Driver deliberately annoys you or puts you at risk	−0.547	0.464		
Ride when you suspect you might be drunk	0.536			
Wear no protective clothing	0.502			
Unintentionally do a wheel spin	0.501			
Skid on a wet road	0.496			
Pull away too quickly	0.404			
Wear a protective jacket	−0.367			
Ride or attempt to ride against the traffic	0.350			0.313
Attempt to, or actually do ride with the front wheels off the road	0.327			
Pull out on to a main road in front of a vehicle				
Ride between two lanes of fast-moving traffic				
Wear bright/fluorescent strips				
Run wide when going round a corner		0.530		
Find that you have difficulty controlling the bike		0.509		
Race away from bumps and traffic stops		0.495		
Ride in tandem with another bike		0.455		
Wear gloves		0.441		
Carry more than one passenger on the bike		0.393		
Ride so fast into a corner that you feel like you might lose control		0.387		
Distracted or preoccupied		0.349		
Ride so fast into a corner that you scare yourself		0.303		
Wear agbada or other flowing gowns				
Standard deviation	0.21	0.30	0.44	0.28

Table 16.3   Factor structure of the MRBQ items *concluded*

	Factors			
	Control and safety	**Stunt**	**Error**	**Speed – hurrying**
Fail to notice that pedestrians are crossing			0.794	
Not notice someone stepping out from behind parked vehicle			0.753	0.339
Not notice a group of pedestrians waiting to cross the road			0.743	
Change gear when going round a corner or bend			0.517	
When riding at the same speed as other traffic, fail to notice or anticipate that another vehicle might pull out				0.512
Disregard the speed limit late at night				0.488
Overtake or attempt to overtake a vehicle on the right				0.479
Get involved in unofficial 'races' with other riders				0.469
Ride too close to the vehicle in front				0.437
Intentionally do a wheel spin	−0.365			0.368
Queuing to turn left on a main road				0.361
Mean*	1.78 (2.67)	1.55(2.33)	1.37(2.06)	1.60(2.4)

* The figures in brackets were the scores in the present study multiplied by 1.5 to obtain the scores that would have been obtained if we had used the six–item Likert scale employed by Sexton et al.

There were many differences between the factors under which the MRBQ items loaded in British compared with Nigerian motorcycle riders (Table 16.4). Three items ('Fail to notice or anticipate that another vehicle might pull out in front of you and have difficulty stopping', 'Queuing to turn left on a main road, you pay such close attention to the main traffic that you nearly hit the vehicle in front' and 'Ride so close to the vehicle in front that it would be difficult to stop in an emergency') which loaded under the error factor for British participants, and loaded under speed/impatience in Nigerian riders. Similarly, three of the items that loaded under stunt in British riders loaded under safety/control in Nigerian riders.

Table 16.1 also showed that the most common behaviour reported by the participants was control/safety and the least reported behaviour was error.

Table 16.4   Factor structure of the original MRBQ compared to the present study

		Original MRBQ	Present MRBQ
1	Fail to notice that pedestrians are crossing when turning into a side street from a main road	Error	Error
2	Not notice someone stepping out from behind a parked vehicle until it is nearly too late	Error	Error
3	Not notice a group of pedestrians waiting to cross the road	Error	Error
4	Pull out on to a main road in front of a vehicle that you had not noticed, or whose speed you have misjudged	Error	Unloaded
5	Fail to notice or anticipate that another vehicle might pull out in front of you and have difficulty stopping	Error	Speed/impatience
6	Queuing to turn left on a main road, you pay such close attention to the main traffic that you nearly hit the vehicle in front	Error	Speed/impatience
7	Distracted or preoccupied, you belatedly realize that the vehicle in front has slowed and you have to brake hard to avoid a collision	Error	Loaded under stunt, but does not belong
8	Attempt to overtake someone that you had not noticed to be signalling a left turn	Error	Unloaded
9	When riding at the same speed as other traffic, you find it difficult to stop in time a traffic warden has signalled you to stop	Error	Unloaded
10	Ride so close to the vehicle in front that it would be difficult to stop in an emergency	Error	Speed/impatience
11	Run wide when going round a corner (verge to the middle of the road)	Error	Stunt
12	Ride so fast into a corner that you feel like you might lose control	Error	Stunt
13	Disregard the speed limit late at night or in the early hours of the morning	Speed	Speed/Impatience
14	Exceed the speed limit on a highway	Speed	Excluded. Too many missing data*

Table 16.4   **Factor structure of the original MRBQ compared to the present study *concluded***

		Original MRBQ	Present MRBQ
15	Exceed the speed limit on a residential road	Speed	Excluded. Too many missing data*
16	Race away from bumps and traffic stops with the intention of beating the driver/rider next to you	Speed	Stunt
17	Wear helmets	Safety	Control/safety
18	Ride between two lanes of fast-moving traffic	Stunt	
19	Get involved in unofficial 'races' with other riders or drivers	Stunt	Speed/ impatience
20	Ride so fast into a corner that you scare yourself	Stunt	Stunt
21	Attempt to, or actually do ride with the front wheels off the ground	Stunt	Control/safety
22	Pull away too quickly and your front wheel comes off the road	Stunt	Control/safety
23	Intentionally do a wheel spin	Stunt	Speed/ impatience
24	Unintentionally do a wheel spin	Stunt	Control/safety
25	Wear a protective jacket (leather or non-leather)?	Safety	Control/safety

Reliability analysis

The Cronbach alpha for all the items was 0.526; also, the Cronbach alpha for control/safety, stunts, error and speed/impatience was 0.514, 0.493, 0.668 and 0.435 which, compared to the original MRBQ among British subjects, was generally low.

The relationship between MRBQ factors and accident involvement

The GLM was used to identify which MRBQ factors were related to crash risk. Table 16.5 shows that only speed/impatience, among the MRBQ, was a significant predictor of motorcyclists' crash involvement. Also, Alcohol use, experience

Driver Behaviour and Training

and usual purpose of riding (commercial commuting was the reference) were all significant predictors. Perhaps surprisingly, age and previous traffic violations did not significantly predict previous crash involvement.

Table 16.5   General linear modelling of the MRBQ and other factors on crash liability

Variables	B	Standard error	Z	95% CI	P
MRBQ factors					
Speed/impatience	0.092	0.0402	2.29	0.014–0.171	0.021
Error	0.040	0.0282	1.42	–0.015–0.096	0.153
Stunt	–0.064	0.0406	–1.58	–0.144–0.015	0.113
Control/safety	–0.006	0.0554	–0.11	–0.114–0.103	0.918
Demographic characteristics					
Experience	–0.021	0.0047	–4.47	–0.030– – 0.012	0.000
Age	–0.002	0.0019	0.01	–0.005–0.002	0.405
Previous traffic violation	0.018	0.0310	0.03	–0.043–0.078	0.569
Alcohol use = Yes	0.110	0.0240	0.02	0.063–0.157	0.000
Motorcycle usage					
For business	0.198	0.0433	0.04	0.113–0.283	0.000
Government	–0.423	0.1429	0.14	–0.703– – 0.143	0.003
Both commercial and private	0.014	0.0254	0.03	–0.035–0.64	0.568
Private	0.073	0.0355	0.04	0.004–0.143	0.039
Commercial	0				

Discussion

The main objective of this study was to investigate the performance of the MRBQ among Nigerian motorcycle riders, and compare the findings from this study with the findings from the original MRBQ study among British motorcyclists. The differences will be discussed in the following sections.

The MRBQ Factor Structure

There are many differences between the classification of behaviours among the motorcyclists for the present study and that provided by the MRBQ among samples of motorcyclists in Great Britain. First, in this study, factor analysis showed that MRBQ items loaded under four factors: control/safety, stunt behaviours, error and speed/impatience; which is distinctly different from the five-factor loading of the MRBQ items among British motorcyclists. The second difference is the way in which items loaded under factors that were different from those factors under which they loaded in the original study. For example, three factors which loaded under stunt among motorcyclists in Britain; 'Attempt to do, or actually ride with the front wheels off the ground', 'Pull away too quickly and your front wheel comes off the road', and 'Unintentionally do a wheel spin', loaded under control/safety in the present study. Thirdly, there is a blurring of the distinction between control and safety which was apparent in the original MRBQ study, so that the two are perceived as one among Nigerian motorcyclists. A possible reason for this may be that motorcyclists in Nigeria perceived many of the items classified as stunts by their fellows in Britain as control and safety behaviours; and this may reflect a fundamental difference in what, in the first place, motivated riders to start riding motorcycles in the two countries. In Britain and most high income countries, motorcycles are often used for recreation and sport. In this context, persons with a liking for speed and performance of stunts are primarily attracted to the sport; they therefore see and classify some of the behavioural items in the context of what attracted them to the sport in the first place. However, in Nigeria, motorcycles are hardly ever ridden for pleasure; for most, it is a means of making a living (Oluwadiya et al., 2009; Odeleye, 2003). Therefore items that were perceived as stunt among British riders were classified under control/safety among Nigerian riders. Stunt was defined as a difficult or unusual or dangerous feat; usually done to gain attention (Word Web Pro 5.2, 2007). It is therefore an intentional action which is dangerous because it can lead to a loss of control. Such intentional acts of dangerous feats are not likely to be employed by the kind of persons attracted to riding in Nigeria. In Nigeria, riders perceived such acts in the context of losing control and safety. Nothing illustrates these differences in perception more than the way two items, 'Intentionally do a wheel spin' and 'Unintentionally do a wheel spin', were classified by the motorcyclists in the two countries. In Britain, the two were classified together as stunts, whereas in Nigeria, the unintentional act was classified as control/safety while the intentional act was classified as speed/ impatience. Thus what was seen in Britain as sensation-seeking (stunt) was either classified as control/safety issue when unintentional, or speed/impatience (get to where you are going faster and thus make more money), when intentional.

Another major difference in the classification of behaviours among the two motorcyclist populations is the way in which items which could be classified as pure speed behaviours among British riders has become blurred to include items that imply impatience in Nigerian motorcyclists. Hence, items such as 'Fail to notice or anticipate that another vehicle might pull out in front of you and have difficulty stopping', 'Queuing to turn left on a main road, you pay such close attention to the main traffic that you nearly hit the vehicle in front', 'Ride so close to the vehicle in front that it would be difficult to stop in an emergency', and 'Ride or attempt to ride against the traffic' may be perceived by the Nigerian motorcyclist as a measure of someone in a hurry, who would not mind giving all it takes to arrive, as soon as possible, at his destination. Previous studies from Nigeria have alluded to the congested and chaotic traffic situation on Nigerian roads and have implied that this might be responsible for some peculiar crash characteristics in Nigeria (Oluwadiya et al., 2009; Odeleye, 2003). It is therefore possible that the traffic context of Nigerian roads is responsible for shaping the factor structure of MRBQ in Nigeria. This will be in keeping with the suggestion by Ozkan and colleagues (2006) who posited that external factors, such as lack of traffic enforcement and traffic congestion, could sometimes be more important than such internal factors as cognitive mechanisms, attention etc., for shaping the factor structure of the DBQ.

Consistently with the original MRBQ, most of the study participants scored each item of the questionnaire at the lower end of the scale (i.e. never/occasionally). Nevertheless, the composite scores of the factors was higher among the Nigerian motorcyclists than their British counterparts, meaning that the frequency of such behaviours is higher among the former. This finding is consistent with findings from studies of the DBQ among different cultures which showed that the scores on the factors were higher in southern European countries than in their northern counterparts (Ozkan, 2006), and were even higher among the Arabs (Bener et al.. 2008). The only factor that was higher among British motorcyclists than their Nigerian counterparts was for the safety factor.

Finally, both the general reliability scores were lower in this study than in the British study. This was probably due to the scaling method (four-item Likert scale) employed in this study compared with the six-item scaling used in the original study. On the other hand, it might be due to actual cultural differences in the way the two populations of motorcyclists viewed the questionnaire items. Several studies have shown that language has an impact on the way respondents answer questions relating to cultural values and it has been suggested that it would be extremely difficult to design reliable scales in a multi-country setting (Corless et al., 2001; Harzing, 2004). Further studies could be directed at how the performance and reliability of the questionnaire could be improved by addressing theoretical, methodological and practical issues that may have been responsible for low reliability.

Prediction of Motorcyclists' Crash Risks

General linear modelling raised some important differences between the present study and the original MRBQ data among British motorcyclists. The first important difference was that even though only one of the MRBQ factors was predictive of crash involvement in both sets of data, the significantly predictive factor was different in the two settings. In the British MRBQ study, traffic error was the significant predictive factor, whereas the predictive factor in the present study was speed/impatience. In their discussion, Elliot and colleagues (2007) gave the following explanation for their finding:

> motorcycles are inherently more demanding to control than are cars and, given the dynamics of motorcycling, the commission of an error when riding is likely to have more severe consequences than making an error when driving a car. For example, it is often possible for a car driver to recover from making an error without losing control of the vehicle. However, the recovery from an error when riding a motorcycle is potentially more difficult due to the relative instability of a two-wheeled vehicle compared with a four-wheeled vehicle. This is likely both to strengthen the causal link between errors and crash risk and to make riders more aware of their errors.

So why is our finding different? One reason might be that three of the items (one of them had the highest loading value) that loaded under the error factor in the original study, actually loaded under speed/impatience in the present study. The other reason might be that in the Nigerian chaotic driving culture and poor road conditions, errors may not be an important contributor to crashes because opportunities to speed are not always present; but when such opportunities occur, pent-up impatience may encourage them to speed and take unnecessary risks.

The study also showed that crash liability increased with self-admitted alcohol intake, and decreased with experience. Surprisingly, and unlike in the original MRBQ data, age was not predictive of crash liability. This finding may be due to the fact that the mean age of the respondents in our study is 27.03 years which is much lower than the mean age (43 years) of the respondents in the original MRBQ data.

The usual reason for riding was also a significant predictor of crash liability. For this analysis, commercial motorcycle usage was the reference category. The GLM showed that unemployed motorcyclists were negative predictors of crash liability while motorcyclists whose purposes were for business or private use were positive predictors compared with commercial motorcyclists. This later finding was rather surprising in view of the popular perception of commercial motorcyclists as being more reckless than all other road users.

Implication for Road Safety Intervention

The fact that speed/impatience violations were significant predictors of crash liabilities has important implications for formulating policies directed at reducing motorcycle crash rates and injuries. It suggests that interventions targeted at improving road behaviours will be important in reducing motorcycle crashes. Such interventions will aim at developing and improving insights into dimensions of crash risks and how it relates to self-appraisal. It will have the added advantage that such programmes can also simultaneously reduce the incidence of all other MRBQ factors. Policy makers should also target certain groups of motorcyclists for safety intervention campaigns, including private motorcyclists and company vehicles. Hitherto, commercial commuter motorcyclists, popularly called okada, had been the focus of public outcry, but this study has shown that private and company motorcycles had higher risks of crash involvement. Experience is the most important predictor of crash risk ($z = -4.47$, Table 16.2). One way of ensuring that motorcyclists gain some experience before riding is by enforcing training, and making it a condition for obtaining riding licences. The period of training would serve as a means of attaining some experience. Finally, ways must be found to ensure that only licensed riders use Nigerian roads.

Conclusion

This study showed that in Nigeria, the MRBQ performed differently from British motorcyclists. The factor structure has decreased to four, and it is not as reliable. However, one of the factors, speed/impatience, as well as other sociodemographic variables including experience, alcohol use and motorcycle usage are important predictors of crash liability among Nigerian motorcycle riders. Policy makers can achieve reduction in motorcycle crashes by focusing on these factors.

Limitations

Because this study is based on self-reported data, it may suffer from limitations that have been highlighted in previous studies (Bener et al., 2008). Also, licensing status could not be included in the GLM procedure, because the licensing rate was low; therefore, the effect of licence possession could not be determined.

References

Aberg, L. and Rimmo, P.A. (1998). Dimensions of aberrant driver behaviour. *Ergonomics*, 41(1), 39–56.

Bener, A. et al. (2008). The Driver Behaviour Questionnaire in Arab Gulf countries: Qatar and United Arab Emirates. *Accident Analysis and Prevention*, 40(4), 1411–1417.

Blockey, P.N. and Hartley, L.R. (1995). Aberrant driving behaviour: Errors and violations. *Ergonomics*, 38(9), 1759–1771.

Corless, I.B. et al. (2001). Issues in cross-cultural quality-of-life research. *Journal of Nursing Scholarship*, 33(1), 15–20.

Dimmer, A.R. and Parker, D. (1999). The accidents, attitudes and behaviour of company car drivers. In G.B. Grayson, ed., *Behavioural Research in Road Safety IX*. Crowthorne, Transport Research Laboratory, pp. 78–85.

Elliott, M.A. et al. (2007). Errors and violations in relation to motorcyclists' crash risk. *Accident Analysis and Prevention*, 39(3), 491–499.

Federal Road Safety Commission (2006). *2005 Annual Report*. Abuja, Federal Road Safety Commission.

Harzing, A.-W. (2004). Does language influence response styles? A test of the cultural accommodation hypothesis in fourteen countries. In B.N. Setiadi et al., eds, *Ongoing Themes in Psychology and Culture*, online edition. Melbourne, FL, International Association for Cross-Cultural Psychology. Retrieved from http://www.iaccp.org 15 April 2009.

Morris, C.C. (2009). *Special Report, Motorcycle Trends in the United States*. Washington, DC, US Department of Transportation, Bureau of Transporation Statistics.

National Highway Traffic Safety Administration (NHTSA) (2008). Traffic safety facts: Motorcycles, 2007 data. *Annual of Emergency Medicine*, 52(4), 453–454.

Odeleye, J.A. Improved road traffic environment for better child safety in Nigeria. 14th ICTCT Workshop Proceedings.

Oluwadiya, K.S. et al. (2004). Epidemiology of motorcycle injuries in a developing country. *West Africa Journal of Medicine*, 23(1).

Oluwadiya, K.S. et al. (2009). Motorcycle crash characteristics in Nigeria: Implication for control. *Accident Analysis and Prevention*, 41(2), 294–298.

Ozkan, T. (2006). *The Regional Differences between Countries in Traffic Safety: A Cross-Cultural Study and Turkish Case*. Accessed 30 April 2008 from http://ethesis.helsinki.fi/julkaisut/kay/psyko/vk/ozkan/theregio.pdf

Ozkan, T. et al. (2006). Cross-cultural differences in driving skills: A comparison of six countries. *Accident Analysis and Prevention*, 38(5), 1011–1018.

Parker, D. et al. (2000). Elderly drivers and their accidents: The Aging Driver Questionnaire. *Accident Analysis and Prevention*, 32(6), 751–759.

Parker, D. et al. (1995). Behavioural characteristics and involvement in different types of traffic accident. *Accident Analysis and Prevention*, 27(4), 571–581.

Peden, M. et al. (2004). *World Report on Road Traffic Injury Prevention*. Geneva, World Health Organization.

Reason, J., Manstead, A., Stradling, S., Baxter, J. and Campbell, K. (1990). Errors and viloations on the roads: A real distinction? *Ergonomics*, 33, 1315–1332.

Word Web Pro 5.2 (Ed.) (2007) Word Web Pro 5.2.

PART 4
At Work Road Safety

Chapter 17

Contemporary Behavioural Influences in an Organisational Setting and Implications for Intervention Development

Bevan Rowland, Jeremy Davey, James Freeman and Darren Wishart

Introduction

Professional drivers and safety

Within the industrialised world, work-related crashes are the most common cause of work-related death, injury and reduced productivity (Charbotel et al., 2001; Toscano and Windau, 1994). Likewise in Australia, road crashes are the most common cause of work-related fatalities, injuries and absence from work (Haworth et al., 2000), with the average time lost being greater than any other workplace claim (Stewart-Bogle, 1999). There are obvious costs related to work crashes such as vehicle and property repair costs. There are also many hidden expenses including third-party costs, workers compensation, medical costs, rehabilitation, customer-related costs, increased insurance premiums, administrative costs, legal fees and loss of productivity (Collingwood, 1997; Haworth et al., 2000).

Drivers who drive for work purposes may face many risks that may directly, or indirectly, influence their driving behaviour and ability to drive safely (Broughton et al., 2003). Direct influences may include secondary tasks while driving, such as aspects of multi-tasking, as many drivers are required to work away from their home office or depot. For example, use of mobile phones while driving may be required or encouraged (Broughton et al., 2003; Salminen and Lähdeniemi, 2002) or eating a meal while driving (Broughton et al., 2003). Indirect influences may include time pressures or unrealistic work schedules placed on professional drivers by organisations to achieve job quotas or attend meetings on time (Boufous and Williamson, 2009; Broughton et al., 2003; Newnam et al., 2002; Rowland et al., 2008). In addition, issues of time pressure can have negative consequences, such as speeding, aggressive driving and increased fatigue or stress. For example, time pressure may require or encourage drivers to engage in aberrant driving behaviours to make up time, such as speeding (Newnam et al., 2002), close following and risky overtaking (Broughton et al., 2003). Furthermore, a number of research studies have highlighted the influence of fatigue on professional driver behaviour and crash risk (Broughton et al., 2003; Freeman et al., 2008; Mitchell et al., 2004;

Salminen and Lähdeniemi, 2002; Taylor and Dorn, 2006). Research suggests that increased risk of fatigue is associated with an elevated exposure (including distances to be travelled and travel time), driving during hours most associated with sleepiness (e.g., shift work) and road monotony and travel after a hard day's work (Broughton et al., 2003).

Measurement of driver behaviour

The Manchester Driver Behaviour Questionnaire (DBQ) has become one of the most prominent measurement scales to examine self-reported driving behaviours (Lajunen and Summala, 2003) as well as investigate the relationship between self-reported driving behaviour and crash/offence involvement (Davey et al., 2006; Freeman et al., 2008; Gras et al., 2006; Rowland et al., 2008; Wishart et al., 2006).

Initially, the questionnaire was developed to distinguish between two empirically different classes of behaviour, errors and violations (Reason et al., 1990), and further modified to include 'lapses' (Lajunen and Summala, 2003). However, recent versions of the DBQ have four different subscales: errors, lapses, violations and aggressive violations (Gras et al., 2006; Lajunen et al., 2003; Sullman et al., 2002) or even five factors (Parker et al., 2000). In addition to the differing number of factors identified, research has generally reported differences in factor structure, as specific items often load on different factors depending on the driving context (Davey et al., 2006), which ultimately influences the naming and interpretation of each factor (Freeman et al., 2008). Furthermore, previous applied research has demonstrated that the DBQ is robust to minor changes to some items that have been made to suit specific organisational, cultural and environmental contexts (Blockey and Hartley, 1995; Davey et al., 2007; Özkan and Lajunen, 2005; Parker et al., 2000).

Professional drivers and the DBQ

There is an increasing quantity of research that has utilised the DBQ to investigate the driving behaviours of professional drivers (Davey et al., 2006; Freeman et al., 2008; Newnam et al., 2002, 2004; Rowland et al., 2008; Sullman et al., 2002; Wishart et al., 2006; Xie and Parker, 2002). Within a professional driver setting, research utilising the DBQ have also reported a substantial level of factor structure variability. For example, Sullman et al. (2002) and Xie and Parker (2002) examined the driving behaviours of professional drivers and identified four factors. In contrast, Dimmer and Parker (1999) focused on company car drivers and reported a six-factor DBQ structure. In addition, an Australian study by Davey and colleagues (2006) utilised the DBQ to examine the behaviours of a group of work-related drivers and reported a traditional three factor solution of errors, aggressive and speeding violations. However, within this study a greater number of traditional items considered to be speeding violations actually loaded on the aggressive violation factor. More recently, Freeman et al. (2008) included an additional 15 items to the traditional 20-item DBQ, and reported a three-factor solution (e.g. speeding/aggression, fatigue and errors). A substantial proportion of the additional items, which related to fatigue, time pressure and multi-tasking behaviours, loaded on what was named the fatigue factor. This factor also proved predictive of self-reported offences (i.e., demerit point loss), identifying the potential value of the contemporary items in assessing aberrant behaviours within a professional driver setting. Furthermore, this research also highlighted the need to include additional items addressing non-traditional driver behaviours.

Research aims

The current study aims to extend the above mentioned research (Freeman et al., 2008) and utilise the expanded DBQ to examine a group of professional drivers. Additionally, the study aims to explore the potential outcomes with regard to intervention strategy development. To meet this criterion, further analyses will be conducted to determine the behavioural predictors of crash and offence involvement.

Method

Participants

A total of 444 individuals volunteered to participate in the study and were all employees of state government departments within Queensland (Australia) who drove company-owned vehicles as part of their daily work tasks. Recruitment of participants was facilitated at industry fleet safety workshops held within the various city and regional areas. There were 307 (69 per cent) males and 136 (31 per cent) females (gender not stated for one respondent). The average age of the sample was 42 years (range 17–68 years).

Study questionnaire

A modified version of the DBQ was used in the current study that consisted of 40 items. In addition to the traditional 20 items incorporated with the DBQ, the authors of the current research also included the 15 additional items utilised in previous research (Davey et al., 2006; Freeman et al., 2008), as well as a further five items. The extra five items included: 'Exceed speed limit on residential road without realising', 'Exceed speed limit on highway without realising', 'Hit/bump/ scrape something while manoeuvring', 'Intentionally disobey a "Stop" or "Give way" sign', and 'Lose concentration while driving'. One item (e.g., Hit/bump/ scrape something while manoeuvring) was originally included within the DBQ slips and lapses subscale. However, this item was included as a separate item primarily because of the high rate of low speed manoeuvring incidents (including parking and reversing) by professional drivers in Australia (Rowland et al., 2005). Unintentional speeding items were included due to speed variability reported by many drivers as a resultant factor of inattention and/or fatigue. For similar reasons the item *Lose concentration while driving* was included. Dissimilar to the original DBQ error factor item *Miss 'Stop' or 'Give way' signs,* the new item *Intentionally disobey a 'Stop' or 'Give way' sign* was included to gauge the propensity of the sample in relation to a violation (potentially an aggressive act or as a result of time pressure, etc.) and not an error.

The additional 20 items added to the traditional DBQ focused primarily on contemporary work-related driving safety issues such as fatigue, time pressure, distraction and multi-tasking. Similar to previous research, these items were derived from the implementation of focus groups and interviews with work-related drivers from a number of large Australian industry organisations which facilitated the identification of key themes proposed to influence driving performance such as fatigue/tiredness, time/work pressures, multi-tasking and general distraction. Some of the additional items used previously (Freeman et al., 2008) included *'Drive while under time pressure', 'eat a meal while driving for work'* and *'drive while using a mobile phone'.* Respondents were required to indicate on a seven-point scale (0 = never to 7 = always) how often they commit each of the errors (eight items), highway-code violations (eight items) aggressive violation items (four items), as well as the additional 20 items. In addition, minor word changes were performed on the original DBQ items to ensure they were more reflective of Australian driving conditions.

Sociodemographic questions were included in the questionnaire to determine participants' age, gender, driving history (e.g., years experience, number of traffic offences and crashes) and their driving exposure (e.g., driving hours per week, number of kilometres per year).

Results

Factor analysis of the Modified Driver Behaviour Questionnaire (MDBQ)

Exploratory factor analysis was conducted with the complete 40-item questionnaire. The MDBQ items were subjected to principal components analysis (PCA) using SPSS Version 17. Prior to performing PCA, the suitability of data for factor analysis was assessed. Inspection of the correlation matrix revealed the presence of many coefficients of 0.5 and above. The Kaiser–Meyer–Oklin value was 0.94, exceeding the recommended value of 0.6 (Kaiser, 1970) and Bartlett's Test of Sphericity (Bartlett, 1954) reached statistical significance, supporting the factorability of the correlation matrix. All items and factors for the modified DBQ for professional drivers are reported in Table 17.1.

Based on previous research utilising similar items (Freeman et al., 2008) and examining Catell's (1966) scree test, it was decided to use three factors for further investigation. The three-factor solution explained a total of 49 per cent of the variance, with factor one contributing 36 per cent. The first factor contained the eight DBQ error items and a combination of two DBQ aggressive violations, one DBQ Highway Code violation and four additional items. In relation to the additional items, two focused on non-seatbelt use and one each on alcohol use before driving and low speed manoeuvring, all of which could be considered as errors of observation and judgement. Therefore, this factor was labelled 'Errors' as the predominant theme to collectively emerge from the items focused on driver errors.

The second factor accounted for eight per cent of the total variance and contained 13 of the 20 additional items that focused on a combination of fatigue, time pressures, distraction and multi-tasking. However, one of the new items, (e.g., *exceed speed limit on residential road without realising*) which specifically relates to speed also loaded on the factor. Unintentional speeding on a residential road could be identified as a result of fatigue and inattention; as the driver becomes more inattentive due to fatigue (etc.) they driver may not be active in checking their speed. The factor items highlight a number of reported contemporary work-related driving risks and due to the variability of the items, the factor was labelled 'Contemporary risks'.

Finally, the third factor accounted for approximately 6 per cent of the overall variance and comprised of ten items. Seven of the eight traditional Highway Code violations loaded on this factor. However, it is also noted that two traditional aggressive violation items also loaded on the factor (e.g., *become angered by another driver and indicate hostility and sound horn to indicate annoyance*) and one new item relating to speeding behaviour (e.g. *exceed speed limit on highway unintentionally*). Although many of the items are associated with speeding behaviour (e.g. *intentionally exceed speed limit on highway, exceed speed limit on highway without realising, race away from traffic lights to beat car beside you,* etc.) the authors decided to label the factor 'Highway Code violations', due primarily to a majority of Highway Code violations that loaded on the items.

Reliability and intercorrelations of the MDBQ

The Cronbach alpha coefficients for the three MDBQ subscales were errors = 0.91, contemporary risks = 0.92 and Highway Code violations = 0.87, indicating a robust internal consistency. In addition, bivariate analysis indicated that the strongest relationship existed between errors and Highway Code violations ($r = 0.66**$), followed by errors and contemporary risks ($r = 0.58**$) and then contemporary risks and Highway Code violations ($r = 0.55**$). Similar to previous research (Freeman et al., 2008), there was only a moderate bivariate relationship between contemporary risks and hours driven per week ($r = 0.26**$) or number of kilometres driven per year ($r = 0.21**$). Likewise, there were moderate relationships between contemporary risks and both age ($r = -0.27**$) and years driving experience ($r = -0.25**$), indicating older drivers and those with more driving experience are less likely to engage in aberrant driving behaviour.

Table 17.1   Factor structure of the MDBQ

Become angered by another driver and give chase (AV)	0.77		
Whilst turning nearly hit cyclist (E)	0.70		
Remove your seatbelt for some reason while driving	0.69		
Drive while over the blood alcohol limit (HV)	0.69		
Pull out of junction and disrupt traffic flow (AV)	0.68		
Not wear your seatbelt	0.64		
Fail to notice pedestrians crossing (E)	0.63		
Skid while braking or cornering on slippery road (E)	0.61		0.55
Hit/bump/scrape something while manoeuvring	0.61		
Attempt overtake of someone turning in front (E)	0.61		
Non attention and nearly hitting vehicle in front (E)	0.61		
Have one or two alcoholic drinks before driving	0.60		
Underestimate speed of oncoming vehicle while overtaking (E)	0.60		0.53
Fail to check rear mirror before pulling out (E)	0.59		
Miss stop or give way signs (E)	0.58		
Have difficulty driving because of tiredness or fatigue		0.81	
Save time by driving quicker between jobs		0.79	
Drive while tired		0.79	
Find yourself nodding off while driving	0.58	0.76	
Find your attention being distracted from road		0.74	
Drive while under time pressure		0.72	
Eat a meal while driving for work		0.72	
Do paperwork or admin. while driving	0.51	0.72	
Lose concentration while driving		0.70	
Exceed speed limit on residential road without realising		0.69	
Find yourself driving on autopilot		0.65	
Drive while using hand-held mobile phone		0.55	
Drive home from work after long day		0.54	
Intentionally exceed speed limit on highway (HV)			0.73
Become angered by another driver and indicate hostility (AV)			0.72
Exceed speed limit on highway without realising			0.71
Race away from traffic lights to beat car beside you (HV)			0.69
Cross junction even though traffic lights have already changed (HV)			0.68
Drive especially close to the car in front to signal faster driving (HV)			0.66
Sound horn to indicate annoyance (AV)			0.65
Impatient with slow driver and overtake on inside (HV)			0.63
Intentionally disregard speed limit on residential road (HV)	0.54	0.58	0.60
Stay in a closing lane and force your way into another (HV)			0.53

(E) = Errors; (HV) = Highway Code violations; (AV) = Aggressive violations.

Two questions did not load: (a) Intentionally disobey a 'Stop' or 'Give Way' sign, and (b) Drive while using a 'hands-free' mobile phone.

Self-reported frequent driving behaviours

Table 17.2 reports the overall mean scores for the three factors, revealing that participants reported a higher frequency for both the contemporary risks and Highway Code violation factors compared to the errors factor. The means are higher than previous research that has focused on Australian professional drivers (Freeman et al., 2008), indicating that the current sample engaged in, or at least reported, a higher level of aberrant driving behaviours. However, it is noted that the MDBQ questionnaire utilised in the current study has an additional five items compared to previous DBQ research involving Australian professional drivers (Freeman et al., 2008), and this should be borne in mind when making comparisons with previous research. In addition, Table 17.2 shows the mean and standard deviation scores for the six highest-ranked items from the complete 40-item questionnaire. The results indicate that while speeding remains one of the most common forms of aberrant behaviour reported by the fleet drivers (Davey et al., 2006; Freeman et al., 2008; Newnam et al., 2004; Sullman et al., 2002), drivers are also at risk of driving while under time pressure, fatigued/tired or while distracted. Interestingly, one of the additional items, *Exceed the speed limit on a highway unintentionally*, was revealed as the second-highest ranked item. Speeding behaviour, both intentional and unintentional, is a significant risk to professional driver safety (Davey et al., 2006; Freeman et al., 2008). Notably, apart from the additional item, the highest ranked items are comparable to previous research by Freeman and colleagues (2008).

Table 17.2   Mean and standard deviations for the MDBQ factors

MDBQ	M	SD
Errors (15 items)	1.59	0.57
Contemporary risks (13 items)	2.40	0.86
Highway Code violations (10 items)	2.20	0.77
Highest ranked items		
1. Drive while under time pressure	3.15	1.34
2. Exceed the speed limit on a highway unintentionally	2.97	1.20
3. Intentionally exceed the speed limit on a highway	2.96	1.32
4. Find yourself driving on autopilot	2.91	1.46
5. Find your attention being distracted from the road	2.86	1.06
6. Drive while tired	2.86	1.07

Prediction of work-related crashes and offences

The final part of the study aimed to examine the utility of the MDBQ to predict self-reported work crashes (N = 81) as well as demerit point loss (N = 77). Logistic regression analyses were performed to examine the contributions of the three factors (e.g., errors, contemporary risks and Highway Code violations) as well as driving exposure (e.g., kilometres driven each year and hours driving per week) to the prediction of self-reported crashes and offences (demerit point loss) in the past 12 months. The number of kilometres driven each year and hours of driving per week were entered in the first step of each regression model to investigate, as well as control for, the influence of driving exposure before the inclusion of the MDBQ factors.

For both crashes and offences, several additional regression models were examined to determine the sensitivity of the results. A test of the full model with all five predictors entered together, as well as the two models entered separately, confirmed the same significant predictors (e.g., exposure, errors, contemporary risks). Forward and backward stepwise regression identified the same predictors. Inclusion of gender, age and years driving experience did not increase the predictive value of the models.

Prediction of crashes

With regard to crashes, and interestingly, hours per week ($P = 0.001$) was the only exposure variable that predicted work-related crashes. This result indicates that time driving rather than distance is predictive in the current sample. The three MDBQ factors were then entered into step two of the model to assess whether any of the factors are predictive of demerit crashes. Collectively, the additional variables were significant, with a chi-square statistic of X^2 (3, N = 444) = 35.97, $P = 0.000$. More specifically, the model indicates that as participants become more rushed, tired and/or lose concentration (contemporary risks factor), they are more likely to be involved in a work-related crash ($P = 0.010$). Furthermore, those participants that make mistakes/errors while driving are twice as likely to be involved in a work-related crash.

Prediction of offences

A similar analysis was conducted to identify the factors associated with incurring demerit point loss. In relation to controlling for driving exposure, unlike previous research which identified only the number of kilometres driven per year as being predictive of incurring demerit point loss (Freeman et al., 2008), both hours per week ($P = 0.016$) and kilometres per year ($P = 0.016$) were predictive in the current model.

Subsequently, the three MDBQ factors were entered into step two of the model to assess whether any of the factors were predictive of demerit point loss, over and above the exposure variables. Collectively, the additional variables were significant, with a chi-square statistic of X^2 (3, N = 444) = 52.15, $P = 0.000$. Similar to crashes and

Table 17.3   Logistic regression for crashes

Variables	B	SE	Wald	p	Odds ratio Exp (B)	95% CI	
						Lower	Upper
Step 1							
Hours per week	0.56	0.17	11.18	0.001	1.75	1.26	2.44
Kms per year	0.14	0.12	1.39	0.238	1.15	0.91	1.48
Model chi-square 30.20** (df = 2)							
Step 2							
Hours per week	0.43	0.17	5.84	0.016	1.53	1.08	2.17
Kms per year	0.16	0.13	1.67	0.196	1.18	0.91	1.53
Errors	0.71	0.27	6.67	0.010	2.05	1.19	3.54
Contemporary risks	0.43	0.17	6.56	0.010	1.55	1.10	2.16
Highway Code violations	0.01	0.21	0.006	0.940	1.01	0.66	1.55
Model chi-square 66.18** (df = 5)							
Block chi-square 35.97** (df = 3)							

after controlling for exposure, the model indicates that as participants' become more rushed, tired and/or lose concentration, the corresponding likelihood of engaging in infringements that results in demerit point loss increases ($P = 0.001$). Additionally, those participants that make mistakes/errors ($P = 0.001$) while driving are two-and-a-half times more likely to commit offences resulting in demerit point loss.

Table 17.4   Logistic regression for offences

Variables	B	SE	Wald	P	Odds ratio Exp (B)	95% CI	
						Lower	Upper
Step 1							
Hours per week	0.41	0.17	5.81	0.016	1.50	1.08	2.10
Kms per year	0.30	0.12	5.83	0.016	1.35	1.05	1.72
Model chi-square 31.82** (df = 2)							
Step 2							
Hours per week	0.23	0.18	1.49	0.222	1.26	0.87	1.81
Kms per year	0.36	0.14	7.11	0.008	1.43	1.10	1.87
Errors	0.98	0.29	10.99	0.001	2.66	1.49	4.74
Contemporary risks	0.60	0.17	11.46	0.001	1.82	1.28	2.57
Highway Code violations	−0.16	0.23	0.46	0.497	0.85	0.54	1.34
Model chi-square 83.98** (df = 5)							
Block chi-square 52.15** (df = 3)							

Discussion

The considerable evolution of the DBQ since its inception reflects the popularity of the scale to assess current driving performance (Freeman et al., 2008; Lajunen et al., 2003). Furthermore, the DBQ has demonstrated that it is adaptable and is robust to changes to items that have been made to suit specific organisational, cultural and environmental contexts (Blockey and Hartley, 1995; Davey et al., 2007; Özkan and Lajunen, 2005; Parker et al., 2000). To further determine the adaptability of the DBQ, to examine and predict aberrant driving behaviours within professional driving settings, the present study aimed to utilise a modified version of the DBQ to explore a group of fleet drivers' behaviours. More specifically, the present research aimed to further explore the contemporary changes to the DBQ and build upon research initiated by Freeman and colleagues (2008) by including an additional five items highlighted as important issues in the work-related driving context.

Factor analytic techniques were implemented to assist with the interpretation of the scale scores and in line with previous research utilised a three-factor solution. The first factor was the clearest factor to interpret and contain all the traditional DBQ error factor items, as well as two traditional aggressive and one Highway Code violations. All except one of the items appeared to be associated with errors, including failures of observation and judgement. However, similar to results in Freeman and colleagues (2008), the item *'become angered by another driver and give chase'*, seems out of place as it is considered purely an aggressive act, and further research is required to clarify this issue.

The second factor consisted purely of the additional items that focused on themes associated with tiredness, fatigue, time pressures, loss of concentration and distraction. These themes are considered quite broad and thus identifying a clearly definable title for the factor again proved difficult. Nonetheless, the items highlight a number of additional contemporary risky driving behaviours within the professional driving context, especially within Australia. The results are consistent with other recent research indicating factors such as fatigue and multi-tasking affect driving performance (Freeman et al., 2008). The authors believe that these additional items are of importance for future intervention development and for overall improvement of professional driver safety within industry organisational settings. Previously, issues relating to fatigue, time pressure, distraction, inattention and multi-tasking were not considered for professional driver interventions. Rather, organisations have tended to focus on traditional skills-based driver training in an attempt to address increases in crash or offence involvement.

A combination of seven traditional Highway Code violations, two aggressive violations and one new item loaded on the third factor and were labelled Highway Code violations. Consistent with previous research (Freeman et al., 2008) the factor was characterised by items that were a combination of traditional speeding behaviours as well as some aggressive acts. It is noteworthy that some of the highway violations that loaded on this factor may be interpreted as aggressive violations, especially for professional drivers (Freeman et al., 2008). For example, *while driving especially close to a car in front of you to indicate for the driver to drive faster* and *crossing a junction knowing that the lights have already turned against you* have traditionally been considered to be Highway Code violations. However, the two items may also signify an aggressive behaviour or at least indicate some level of frustration. Furthermore, frustration and even stress (e.g., work/personal pressures, traffic congestion, etc.) may result in aggressive violations and may subsequently lead to an increase in speeding violations (e.g., Highway Code violations).

Two additional questions did not load on any of the factors, and of particular interest was that of 'hands-free' mobile phone use. Unlike previous research (Freeman et al., 2008), the item relating to 'hand-held' phone did load on the contemporary risks factor and the item is consistent with issues of distraction and inattention (Gordon, 2005). Within Australia, hands-free mobile phone use is legal and even though there are deemed no considerable safety benefits of using either hands-free or hand-held (McEvoy et al., 2005), the former practice may not be considered an aberrant driving behaviour by professional drivers. In addition, in relation to the item *intentionally disobey a 'Stop' or 'Give way' sign*, wording of the item may be confusing for professional drivers, as the propensity for intentionally disobeying either pieces of traffic legislation may vary.

Calculation of Cronbach's alpha reliability coefficients for the new factors (errors, contemporary risks and Highway Code violations) revealed higher reliability scores compared to previous research, indicating that the items are highly consistent in their measurement ability to assess aberrant driver behaviour for this sample. In addition, once the additional items were analysed, the mean scores for the individual items revealed that participants in the current sample were most likely to report driving while under time pressure as well as exceeding the speed limit on the highway (both intentionally and unintentionally), followed by driving while tired and inattention. Similar to Freeman and colleagues (2008), the results indicate that although speeding remains one of the most common forms of aberrant behaviour reported by the professional drivers (Newnam et al., 2004; Sullman et al., 2002), drivers are also at risk of driving while fatigued, tired, distracted or under time pressure.

With regard to the prediction of self-reported driving offences and crashes, logistic regression analyses revealed that exposure to the road, making mistakes (e.g., errors) and reporting risk characteristics (e.g., fatigue, time pressures, distraction/inattention) were predictive of both crashes and offences (demerit point loss). As per previous research (Davey et al., 2006, 2007; Freeman et al., 2008; Wishart et al., 2006), exposure to the road was expected to be a significant predictor given that increasing driving distances is likely to increase the probability of making driving errors which may lead to crashes and/or offences. In addition, the regression models indicate that participants who reported a higher number of driving errors were most likely to be involved in a work-related crash or incur demerit point loss. This result is similar to previous Australian research utilising the DBQ, which identified errors as the primary predictor of work-related crashes and offences (Rowland et al., 2008, 2008a). Interestingly, the contemporary risks factor was also identified as a predictor of both crashes and demerit point loss. This result is of particular importance because not only were contemporary risks factor items some of the most commonly reported aberrant behaviours by the sample, but this factor also predicted both crashes and offences. Not surprisingly, feeling tired, perceptions of time pressure, and distraction/inattention may be considered some of the most likely reasons for crash and offence involvement. These results have implications for future intervention development, highlighting a need to address contemporary risky behaviours and not only driver skills and abilities. Practically, this may include a revision of driver training and education programmes incorporating factors addressing contemporary risks and developing new intervention strategies targeting specific risky behaviours (e.g., speeding, fatigue, time pressure, distraction and multi-tasking).

When interpreting the results of this study a number of limitations should be taken into account. The response rate of participants was relatively low, but consistent with previous research utilising a version of the DBQ scale in Australia (Dobson et al., 1999; Freeman et al., 2008; Wishart et al., 2006). Concerns also remain regarding the reliability of self-report questionnaires, such as the propensity of professional drivers to provide socially desirable responses. However, previous research indicated that bias caused by socially desirable responding is relatively small in DBQ responses (Lajunen and Summala, 2003). In addition, the participant sample was mainly government fleet drivers and results may not be reflective of other professional driving populations. Unlike previous research which utilised insurance company employees, the participants in this study may prove to be more similar to the greatest proportion of the Australian work-related driver population. For example, the present study's participants are primarily field-type employees who drive in similar road and environmental contexts (city, urban, rural and off-road) and to a lesser extent perform similar types of work. Therefore, although this sample may not be a true representation of all professional driving populations in Australia and especially overseas, it does represent a substantial cohort.

Conclusions

The present study has provided some additional preliminary evidence that modifying the DBQ to suit applied settings can produce favourable results with regard to identifying the factors that influence the driving task. Overall, factors relating to fatigue, time pressure, distraction, inattention, multi-tasking and also speed are shown to be considerable risky behaviours within a professional driving context and can result in negative outcomes. This research has identified a need to address these additional factors and include these aspects within intervention development and content. However, further research is required to assess the predictive utility of the MDBQ within other contexts and cultures.

References

Bartlett, M.S. (1954). A note on the multiplying factors for various chi square approximations. *Journal of the Royal Statistical Society*, 16(Series B), 296–298.

Blockey, P.N. and Hartley, L.R. (1995). Aberrant driving behaviour: errors and violations. *Ergonomics*, 38, 1759–1771.

Boufous, S. and Williamson, A. (2009). Factors affecting the severity of work related crashes in drivers receiving a worker's compensation claim. *Accident Analysis and Prevention*, 41, 467–473.

Broughton, J. et al. (2003). *Work-related Road Accidents*. TRL Report TRL582. Crowthorne, Transport Research Laboratory.

Catell, R.B. (1966). The scree test for number of factors. *Multivariate Behavioural Research*, 1, 245–276.

Charbotel, B. et al. (2001). Work-related road accidents in France. *European Journal of Epidemiology*, 17, 773–778.

Collingwood, V. (1997). *Promoting the Safe Driving Policy in NSW Fleets of Twenty or more Vehicles. Staysafe 36: Drivers as Workers, Vehicles as Workplaces: Issues in Fleet Management.* Report No. 9/51. Ninth report of the Joint Standing Committee on Road Safety of the 51st Parliament. Sydney, Parliament of New South Wales.

Davey, J. et al. (2006). A study predicting crashes among a sample of fleet drivers. In *Proceedings of the Road Safety Research, Policing and Education Conference*, Gold Coast, Australia, CD-ROM.

Davey, J. et al. (2007). An application of the Driver Behaviour Questionnaire in an Australian organizational fleet setting. *Transportation Research, Part F*, 10(1), 11–21.

Dimmer, A.R. and Parker, D. (1999). The accidents, attitudes and behaviour of company car drivers. In G.B. Grayson, ed., *Behavioural Research in Road Safety IX*. Crowthorne, Transport Research Laboratory.

Dobson, A. et al. (1999). Women drivers' behaviour, socio-demographic characteristics and accidents, *Accident Analysis and Prevention*, 31, 525–535.

Freeman, J. et al. (2008). A study of contemporary modifications to the Manchester Driver Behaviour Questionnaire for organizational fleet settings. In L. Dorn, ed., *Driver Behaviour and Training, vol. 3, Human Factors in Road and Rail Safety*. Aldershot, Ashgate, pp. 201–214.

Gordon, C. (2005). Driver distraction related crashes in New Zealand. Paper presented at the Driver Distraction conference, 2–3 June, Sydney, Australia.

Gras, M.E. et al. (2006). Spanish drivers and their aberrant driving behaviours. *Transportation Research Part F*, 9, 129–137.

Haworth, N. et al. (2000). *Review of Best Practice Fleet Safety Initiatives in the Corporate and/or Business Environment*. Report No. 166. Melbourne, Monash University Accident Research Centre.

Kaiser, H. (1970). A second generation Little Jiffy. *Psychometrika*, 35, 401–415.

Lajunen, T. et al. (2004). The Manchester Driver Behaviour Questionnaire: A cross-cultural study. *Accident Analysis and Prevention*, 36, 231–238.

Lajunen, T. and Summala, H. (2003). Can we trust self-reports of driving? Effects of impression management on driver behaviour questionnaire responses. *Transportation Research, Part F*, 6, 97–107.

McEvoy, S. et al. (2005). Role of mobile phones in motor vehicle crashes resulting in hospital attendance: A case-crossover study. *BMJ*, doi:10.1136/bmj.38537.397512.55.

Mitchell, R. et al. (2004). Work-related road fatalities in Australia. *Accident Analysis and Prevention*, 36, 851–860.

Newnam, S. et al. (2002). A comparison of the factors influencing the safety of work-related drivers in work and personal vehicles. In *Proceedings of the Road Safety Research, Policing and Education Conference, Adelaide*, CD-ROM. Melbourne, Tulip Meetings Management.

Newnam, S. et al. (2004). Factors predicting intentions to speed in a work and personal vehicle. *Transportation Research Part F*, 7, 287–300.

Özkan, T. and Lajunen, T. (2005). A new addition to DBQ: Positive driver behaviours scale. *Transportation Research Part F*, 8, 355–368.

Parker, D. et al. (2000). Elderly drivers and their accidents: The aging driver questionnaire. *Accident Analysis and Prevention*, 32, 751–759.

Reason, J. et al. (1990). Errors and violations: a real distinction? *Ergonomics*, 33, 1315–1332.

Rowland, B. et al. (2008b). The influence of driver pressure on road safety attitudes and behaviours: A profile of taxi drivers. Paper presented at 18th Canadian Multidisciplinary Road Safety Conference, Whistler, Canada.

Rowland, B. et al. (2008a). Work-related road safety risk assessment: Utilisation of self-report surveys to predict organizational risk. In Australasian Road Safety Research, Policing and Education Conference 2008, 9–12 November 2008, Adelaide, SA. Melbourne, Tulip Meetings Management.

Rowland, B. et al. (2005). Occupational fleet safety research: A case study approach. In *Proceedings Occupational Health and Safety Visions Conference 2005*, 27–30 September, Cairns, submitted papers CD-ROM. Queensland, Deprartment of Workplace Health and Safety.

Salminen, S. and Lähdeniemi, E. (2002). Risk factors in work-related traffic. *Transportation Research, Part F*, 5, 77–86.

Stewart-Bogle, J.C. (1999). Road safety in the workplace. The likely savings of a more extensive road safety training campaign for employees. Paper presented at the Insurance Commission of Western Australia Conference on Road Safety, 'Green Light for the Future'.

Sullman, M.J. et al. (2002). Aberrant driving behaviours amongst New Zealand truck drivers. *Transportation Research Part F*, 5, 217–232.

Taylor, A.H. and Dorn, L. (2006). Stress, fatigue, health, and risk of road traffic accidents among professional drivers: The contribution of physical inactivity. *Annual Review Public Health*, 27, 371–391.

Toscano, G. and Windau, J. (1994). The changing character of fatal work injuries. *Monthly Labour Review*, 17–28.

Wishart, D. et al. (2006). Utilising the Driver Behaviour Questionnaire in an organizational fleet setting: Are modifications required? *Journal of the Australasian College of Road Safety*, 17(2), 31–38.

Xie, C. and Parker, D. (2002). A social psychological approach to driving violations in two Chinese cities. *Transportation Research Part F*, 5, 293–308.

Chapter 18

A Review of the Effectiveness of Occupational Road Safety Initiatives

Tamara Banks, Jeremy Davey, Herbert Biggs and Mark King

Introduction

Several reports have recently been published that offer reviews of current industry practice and risk management guidelines for organisations striving to achieve best practice in managing occupational road risks (Anderson et al., 1998; Haworth et al., 2000, 2008; Murray and Pratt, 2007). Practices and processes recommended in these reports include: having a fleet safety policy that defines safe driving responsibilities and communicates to employees the organisation's commitment to safe driving; recruiting and selecting drivers based on safe driving records and awareness of safety issues; including safe driving components in employee inductions; conducting safety training needs analyses and providing and evaluating any required road safety education; recognising good and poor driving behaviours through an official incentive and disincentive scheme; eliminating or minimising exposure to road hazards when planning road journeys; selecting vehicles based on safety features and documenting maintenance procedures; managing access to vehicles with regards to job needs and ensuring that drivers are physically/medically fit to drive safely; monitoring employee driving activities, for example distances driven and public feedback regarding their driving performance; and recording and monitoring individual driver, individual vehicle and overall fleet incident involvement and managing identified high risks.

These reports provide some guidance to practitioners. However, as many of the recommended initiatives have not yet been scientifically evaluated, enthusiastic endorsement of these guidelines is cautioned. To assist practitioners in making informed decisions about how they manage occupational road risks, this review identifies what outcomes have been observed in previously investigated occupational road safety initiatives.

Method

An extensive document review of published articles was conducted to investigate the effectiveness of occupational road safety interventions. To minimise the chance of overlooking relevant articles, both key road safety databases and multidisciplinary databases were searched. The six electronic bibliographic databases searched included: ATRI, the Australian Transport Index (via Informit Search); Business Source Elite (via EBSCOhost); ISI Web of Science; PsycINFO (via EBSCOhost); ScienceDirect; TRIS Online (Transportation Research Information Services).

Furthermore, searches were not limited to road safety research journals because several key articles on the subject appear in health, psychology and workplace health and safety journals. A master list of search terms was generated and used with each electronic database. Search terms included: work road safety, organisation road safety, occupation road safety, fleet safety, company driver, fleet driver, initiative, strategy, programme and intervention. These terms were selected based on the terminology used within relevant published research papers and industry reports. To keep the review current and manageable, the search was limited to articles published between January 1990 and September 2008.

The search identified 181 non-duplicated titles. These titles were reviewed using predetermined inclusion/exclusion criteria. Included articles had to appear in English in a peer-reviewed journal, conference proceeding or book. As the purpose of the document review was to explore scientific evaluations of various initiatives, non peer-reviewed literature was excluded as the scientific quality of these documents had not previously been established. In addition, included articles had to focus on occupational road safety initiatives. This criterion eliminated articles that focused on more general community road safety initiatives and general health and safety initiatives. Using these criteria, 122 articles were excluded at the title-review stage. It was observed that the ATRI and TRIS databases generated many non-peer reviewed articles. Some of these articles were government funded reports, while other articles were from less credible sources such as industry magazines.

The abstracts of the remaining 59 articles were then reviewed using the same criteria and process. This review resulted in the exclusion of 33 additional articles. The search terms selected in this review were designed to have a broad scope to maximise the chance of finding all relevant articles. Unfortunately this process also generated many 'hits' in bibliographic databases that were outside the scope of this review. For example the search term 'fleet safety' yielded several articles pertaining to aviation and sailing vessels that were unrelated to the subject of occupational road safety.

The full text of the remaining 26 articles was then reviewed. Seventeen articles that mentioned occupational road safety but did not focus on initiatives were also excluded as they were considered to be extraneous. This review left nine peer-reviewed relevant articles.

The researcher also used ISI Web of Science and Google Scholar to obtain key articles cited in the reviewed papers. This process identified an additional 11 articles that survived the title-review, abstract-review, and full-text-review process. This review process generated a total of 20 peer-reviewed relevant articles. A structured data abstraction form was then used to extract key information from each of the 20 articles that were retained after the full-text review.

Results

For ease of interpretation, the initiatives reviewed have been grouped according to whether they target safety at an organisational level, employee-level, or through implementing protective equipment.

Organisational-level initiatives

Initiatives reviewed targeting safety at the organisational-level comprised: policy development; driver selection criteria; a web-based risk management tool; and raised wages.

Ludwig and Geller (1999a) investigated the effectiveness of store managers creating a *policy* mandating turn-signal use and attaching it to drivers' pay cheques in America. They observed a slight increase in the targeted safe driving behaviour during the policy implementation phase. The authors noted that the safest drivers were the first to comply with the safe driving policy. The article did not indicate if these outcomes were significant. It was also unclear whether the effects would be maintained as no post-intervention data were collected. Additional research by White and Murray (2007) explored the influence of policy development. Insurance data, analysed one year pre-intervention and one year post-intervention, in an Australian case study organisation indicated a reduction in all major crash types and an improved loss ratio from 69 to 48 per cent. The article did not indicate if these outcomes were significant. Also, as policies were implemented in conjunction with other initiatives, it is unclear whether policy development had a unique contribution to safety enhancements.

A questionnaire, administered to a sample of the best safety performers in the American trucking industry, indicated that the best safety performing trucking firms utilised *screening criteria in driver hiring situations*. Top screening criteria when hiring included a lack of alcohol or drug-related violations, speeding tickets, traffic violations, and chargeable crashes, together with a preference for honesty, self-discipline, self-motivation, and patience (Mejza et al., 2003). Given that no comparative research was conducted to explore the prevalence of this practice among less safe carriers it is unclear how effective this initiative is.

Research has investigated the effectiveness of a commercial *web-based risk management tool* designed to carry out risk assessments and monitor employee safety. In an Australian case study organisation, it was found that the use of the tool was associated with a reduction in all major crash types and an improved loss ratio (White and Murray, 2007). However, the unique contribution of the web-based risk management tool to safety enhancements is unclear as it was implemented in conjunction with other initiatives.

The relationship between *pay rises* and crash risk has been investigated. Using a driver crash involvement model, it has been found that as pay rises, crash probability becomes lower. Using a sample of 2,368 truck drivers from an American trucking and logistics firm, Rodriguez and colleagues (2006) calculated that a 1 percentage increase in pay corresponded to a 1.3 percentage decrease in crash risk. The authors did note that other minor policy changes occurred simultaneously with the pay rise, including a safety bonus for safety performance and greater effort to return drivers to their homes after shifts. The researchers reported that they were not able to control for any potential effects of these other initiatives.

Employee-level initiatives

Initiatives reviewed targeting safety at the employee level and comprised: driver training; group discussions; awareness and information campaigns; goal setting; performance feedback; enlisting employees as community road safety change agents; self-monitoring forms; signing safety pledge cards; safety reminders; and rewards.

The effectiveness of *driver training* as an occupational road safety initiative has been explored by several researchers. Practical driver training has been found to be associated with a significant decrease in accident risk based on accident and mileage data collected two years pre-intervention and two years post-intervention in a Swedish company (Gregersen et al., 1996). Similarly, a combination of practical and classroom based training has been found to be associated with significant improvements in audit ratings of occupational traffic risk management in Japan (Salminen, 2008). Furthermore, participation in an American training programme on visual search and scanning patterns was found to be associated with significantly enhanced overall driving ability, performance during curve negotiation, visual search monitoring and detection of brake malfunction in a simulated driving exercise (Llaneras et al., 1998). In addition to these empirical studies, questionnaire research has also been conducted. A sample of the best safety performers in the American trucking industry indicated that the best safety performing trucking firms required pre-service and in-service driver training to build competence in regulatory compliance, driving ability, vehicle condition assessment, operational and safety procedures and disciplinary policies. These firms utilised vehicle-based and classroom-based training and evaluated whether learning was applied (Mejza et al., 2003).

Several researchers have investigated the effectiveness of *group discussions* focusing on occupational road safety problems and solutions. Interactive group discussions have been found to be associated with decreased traffic-related work incidents in Japan (Salminen, 2008) and increases in safe driving behaviours in America (Ludwig and Geller, 1991). Ludwig and Geller found that the targeted behaviour of safety-belt use increased during the intervention and remained high for at least three months post-intervention. As group discussions were implemented in conjunction with other initiatives in this study, the unique effects of discussions are unclear. A more thorough investigation of the effects of group discussions was achieved by Gregersen and colleagues (1996). This study found that group discussions were associated with a significant decrease in both accident risk and accident cost based on data collected two years pre-intervention and two years post-intervention in a Swedish company. It was found that accident cost reduction was greatest in the group discussion condition, as compared to the four other study conditions including driver training, campaigns, rewards and control.

Research suggests that *safety awareness and information campaigns* may have limited utility as an occupational road safety initiative. American research investigating the effects of an intensive two-week safety campaign (including posters, fact sheets, email, promotion booth at annual health fair, thought provoking survey, pledge card, letter from CEO, messages on pay cheque stubs and promotion on internal website) found that the initial improvements immediately after the campaign were not maintained. The researchers observed that safety behaviours measured one month post-intervention and three months post-intervention, had returned to almost the pre-intervention levels (Scheltema et al., 2002). Furthermore, research investigating a less intensive safety campaign (including the presentation of videos and pamphlets) found no effect of the initiative on accident risk (Gregersen et al., 1996).

The effectiveness of *group goal setting* has been investigated in several American studies. It was found that both participative group goal setting and assigned group goal setting were associated with increased safe target behaviours including performing a complete stop at stop signs (Ludwig and Geller, 1997) and turn-signal use (Ludwig et al., 1999). In relation to non-targeted safety behaviour, results are mixed. Ludwig et al. (1999) observed that regardless of whether goal setting was participatory or assigned, no overall differences in safety belt use occurred and an increase in complete intersection stops occurred in response to a turn-signal intervention. In comparison, Ludwig and Geller (1997) found that improvements in non-targeted safety behaviour, including turn-signal use and safety-belt use, were only observed for employees in the participative goal setting condition. Furthermore no changes were observed in safety-belt use and a slight decrease in turn signal use was observed among employees in the assigned goal setting condition. It is important to note that Ludwig and Geller did not indicate whether the observed increases in safety behaviours were significant.

Performance feedback has been found in several American studies to be an effective occupational road safety initiative. More specifically, feedback presented publicly on individual driving behaviours has been found to be associated with increases in the targeted safe driving behaviours of turn-signal use and complete stopping at intersections (Ludwig et al., 1999, 2001). Similarly, the use of graphs displayed on worksite noticeboards presenting individual and group performance feedback has been found to be associated with an increase in overall safe driving performance (Olson and Austin, 2001). Overall safe driving performance was assessed in relation to the loading and unloading of passengers, cornering safely and allowing adequate following distances while vehicles were in motion and the positioning of vehicles when coming to a complete stop. It is important to note that none of the above three studies indicated whether the observed increases in safety behaviours were significant.

The effectiveness of enlisting delivery drivers to serve as *change agents of a community road safety campaign* has been investigated in two American pizza stores. This initiative was found to be associated with increases in both the targeted behaviour of safety-belt use and the non-targeted behaviour of turn-signal use (Ludwig and Geller, 1999b). It is important to note that Ludwig and Geller did not indicate whether the observed increases in safety behaviours were significant. Cognitive dissonance theory (Festinger, 1957) suggests that individuals are motivated to achieve consistency between their beliefs and actions. Based on this theory, Ludwig and Geller suggested that employees would be motivated to achieve consistency between their own driving behaviours and the safe driving behaviours that they were advocating.

Self-monitoring forms have been investigated as an occupational road safety initiative. The use of self-monitoring forms recording safe behaviour estimates has been found to be associated with a 12 per cent increase in overall safe driving performance over baseline performance within a sample of American bus drivers (Olson and Austin, 2001). As previously indicated, overall safe driving performance pertained to passenger loading and unloading and the operation of a vehicle both while in motion and stopped. It is important to note that Olson and Austin did not indicate whether the observed increase in safe driving behaviours was significant.

The *signing of safety pledge cards* has been researched in America. Ludwig and Geller (1991) found that the signing of pledge cards promising personal commitment to buckle-up was associated with increases in both the observed targeted behaviour of safety-belt use and the non-targeted behaviour of turn-signal use. Although Ludwig and Geller observed safety behaviours to remain high for at least three months post-intervention, research by Scheltema and colleagues (2002) suggests that the initial improvements observed in safety-belt use associated with the signing of pledge cards may not be maintained after one month. Furthermore Ludwig and Geller did not indicate whether the increases in safety-belt use and turn-signal use were significant.

The presence of driver *safety reminders* has been found to be associated with safer driving behaviours. More specifically, employee designed seatbelt buckle-up reminder signs displayed in the workplace and co-workers reminding delivery drivers to buckle-up their seatbelts when driving was found to be associated with increases in both seatbelt use and the non-targeted safe driving behaviour of turn-signal use in American employees (Ludwig and Geller, 1991). Ludwig and Geller did not indicate whether these increases were significant.

Several researchers have explored the relationship between *rewards* and road safety outcomes. In relation to group rewards, it has been found that rewarding a Swedish work group with money for incident-free driving was associated with a significant reduction in accident risk (Gregersen et al., 1996). In relation to individual rewards, in America it has been found that running a competition and rewarding only the safest driver with a vehicle maintenance prize was associated with increases in turn-signal use and complete intersection stopping (Ludwig et al.,

2001). Further support for the use of rewards in managing occupational road risks comes from a questionnaire administered to a sample of the best safety performers in the American trucking industry. This study found that the best safety performing trucking firms utilised a range of driver reinforcement methods to encourage safe driving. The most popular rewards included verbal praise, public recognition, congratulatory letters, safety decorations and cash (Mejza et al., 2003). Given that Mejza and colleagues did not investigate the prevalence of these practices among less safe carriers it is unclear from their research whether the provision of rewards was associated with safer driving performing. Furthermore, in contrast to the above three studies, Newnam and colleagues (2006) found no support for the utility of financial incentives, in the form of tailored insurance premiums for organisations based on incident rates, in positively changing Australian fleet managers' safety attitudes.

Protective equipment initiatives

Safety initiatives reviewed pertaining to implementing protective equipment comprised: alcolock devices; fatigue management technologies devices; in-vehicle compensatory devices to target ability deficiencies of older commercial drivers; in-car data recorders; and gasoline vapour recovery devices.

Alcolock devices installed in Swedish commercial transport company vehicles have successfully prevented vehicles from starting in cases where breath tests indicated that drivers' blood alcohol content levels exceeded the legal limit (Bjerre, 2005; Bjerre and Kostela, 2008). Questionnaire data indicated that although some employees had initial suspicion and concerns of increased workloads when the devices were installed, the alcolocks were very well accepted by most employers, employees and passengers after the installation (Bjerre, 2005; Bjerre and Kostela, 2008).

A *fatigue management technologies device* that provided information on driver sleep need, driver drowsiness and lane tracking performance, and also reduced driver work involved in controlling vehicle stability, has been researched using a sample of Canadian and American truck drivers. It was found that feedback from the device reduced driver drowsiness and lane tracking variability during night driving. The authors did caution that the benefits of this device may be reduced due to increased attention and compensatory behaviours needed to respond to the device (Dinges et al., 2005).

The effectiveness of *in-vehicle compensatory devices* (comprising an auditory navigational system, an automatic transmission and an advanced auditory warning system) to target ability deficiencies of older commercial drivers has been researched using a sample of American truck drivers. It was found that drivers with the device, as compared to drivers with no device, demonstrated significantly enhanced overall driving ability, performance during curve negotiations, visual search monitoring and detection of brake malfunction (Llaneras et al., 1998). It is important to note that these effects were obtained in simulated driving experiences.

Therefore it is unclear whether the effects would be replicated in real traffic environments.

Research has investigated the effectiveness of *in-car data recorders* in managing occupational road safety in seven European fleets. It was found that data recorders and the feedback they generated and displayed for drivers, was associated with a significant reduction in accident rate (Wouters and Bos, 2000). An average 20 per cent reduction in accident rate was observed between the data collected one year pre-intervention and one year post-intervention.

As solvents exposure is known to impair psychomotor performance, *gasoline vapour recovery devices* were installed in attempt to reduce traffic injuries among Taiwanese gasoline workers when commuting home after shifts. Chiang and colleagues (2005) compared occupational injury registry data in relation to traffic commuting incidents between employees exposed and not exposed to vapour. Their analysis of pre-installation data and post-installation data revealed that the installation of gasoline vapour recovery devices was associated with a significant reduction in the cumulative injury rate for exposed employees only.

Discussion

All of the initiatives reviewed, except for driver selection criteria, appeared to be effective during the intervention period. However, only six continued to be effective in the post-intervention period. Initiatives found to be positively associated with occupational road safety both during and after the intervention period included: a pay rise; driver training; group discussions; enlisting employees as community road safety change agents; safety reminders; and group and individual rewards. It is interesting to note that of these, five were employee-level interventions. Although it needs to be taken into account that half the interventions reviewed were at the employee level, the finding appears to call into question the role of organisational-level interventions. However, it is also clear that none of the employee-level interventions would be feasible without active implementation and support by the organisation. Similarly, it is interesting that none of the protective equipment interventions showed a continued effect after the intervention period.

The authors offer three suggestions to future researchers to further explore occupational road safety initiative effectiveness. Firstly, although the structured procedure used in this review was beneficial in restricting the review to documents previously assessed by academic experts to be of scientific merit, this process may have overlooked other relevant documents. Future research may expand upon the current review by including non peer-reviewed documents or by adopting a broader scope. The authors recognise that some non peer-reviewed documents may potentially be of an equally high standard to peer-reviewed documents. Furthermore the inclusion of research on community-based initiatives may have provided a different view on the effectiveness of some initiatives. For example, community-based research on the effectiveness of driver training has found little evidence for the effectiveness of driver training in reducing crashes for experienced drivers and that it may even have a negative impact on novice drivers (Christie, 2001).

Secondly, research should be conducted to identify which initiatives offer the greatest opportunity for advancing road safety. As many of the studies reviewed investigated the effects of initiatives when implemented in combination with other initiatives, it was not possible to distinguish the unique impact of some initiatives. Therefore, observed positive safety outcomes in relation to some initiatives may have actually resulted from other initiatives implemented at the same time. Only one of the reviewed studies that investigated multiple initiatives allowed for the unique effect of each initiative to be observed. This study by Gregersen and colleagues (1996) adopted a between groups, pre-post design implementing one initiative per experimental group. Based on the findings of Gregersen et al., group discussions appeared to achieve better safety outcomes than driver training, campaigns or rewards. Future research should follow this initiative and strive to eliminate confounds when researching the effectiveness of occupational road safety initiatives.

Thirdly, researchers should explore whether initiatives with scientific merit are perceived to be effective by employees. Watson (1997) identified that community perception of road safety countermeasures did not align with evidence, but rather appeared to suffer from a misunderstanding of behaviour change principles and crash causation. Therefore although several empirical studies have found interactive group discussions to be related to increased occupational road safety outcomes, employees may not perceive this to be a very effective safety initiative. Future studies should explore employee perceptions, as employees' are more likely to embrace initiatives that they believe will assist them in achieving a goal and to resist initiatives that they believe will have little utility in achieving goals.

In conclusion, this chapter has provided a current review of 20 empirical evaluations of 19 occupational road safety initiatives. The findings from this chapter may assist health and safety practitioners by allowing them to make informed decisions when developing occupational road risk management strategies. The authors suggest that practitioners consider initiative effectiveness, cost and involvement level. Some low-cost initiatives that require minimal involvement, such as group safety discussions, have been found to be effective with a large majority of employees. However, other low-cost initiatives that require minimal involvement, such as implementing a policy, were found to be effective for safer employees but had minimal effect on higher-risk employees. It is suggested that higher cost and higher involvement initiatives, such as driver training focusing on issues such as fatigue, driver stress, distractions, journey planning etc., may be required to manage higher-risk employees effectively.

References

Anderson, W. et al. (1998). *Workplace Fleet Safety System*. Brisbane, Australia, Queensland Transport.

Bjerre, B. (2005). Primary and secondary prevention of drink driving by the use of alcolock device and program: Swedish experiences. *Accident Analysis and Prevention*, 37(6), 1145–1152.

Bjerre, B. and Kostela, J. (2008). Primary prevention of drink driving by the large-scale use of alcolocks in commercial vehicles. *Accident Analysis and Prevention*, 40(4), 1294–1299.

Chiang, W. et al. (2005). Reduction of post-shift traffic injuries among gasoline station workers: Are they related to the reduction of occupational gasoline vapour exposure. *Accident Analysis and Prevention*, 37, 956–961.

Christie, R. (2001). *The Effectiveness of Driver Training as a Road Safety Measure: A Review of the Literature*. RACV Literature Report No. 01/03. Nobel Park, Victoria, Royal Automobile Club of Victoria Ltd.

Dinges, D.F. et al. (2005). Pilot test of fatigue management technologies. Paper presented at the Transportation Research Record: Journal of the Transportation Research Board.

Festinger, L. (1957). *A Theory of Cognitive Dissonance*. New York, Harper and Row.

Gregersen, N. et al. (1996). Road safety improvement in large companies. An experimental comparison of different measures. *Accident Analysis and Prevention*, 28(3), 297–306.

Haworth, N. et al. (2000). *Review of Best Practice Road Safety Initiatives in the Corporate and/or Business Environment*. Clayton, Australia, Monash University Accident Research Centre.

Haworth, N. et al. (2008). *Improving Fleet Safety – Current Approaches and Best Practice Guidelines*. Sydney, Australia, Austroads.

Llaneras, R. et al. (1998). Enhancing the safe driving performance of older commercial vehicle drivers. *International Journal of Industrial Ergonomics*, 22, 217–245.

Ludwig, T.D. (2000). Intervening to improve the safety of delivery drivers: A systematic behavioral approach. *Journal of Organizational Behavior Management*, 19(4), 1–124.

Ludwig, T. and Geller, E. (1991). Improving the driving practices of pizza deliverers: Response generalization and moderating effects of driving history. *Journal of Applied Behavior Analysis*, 24, 31–34.

Ludwig, T. and Geller, E. (1997). Managing injury control among professional pizza deliverers: Effects of goal setting and response generalization. *Journal of Applied Psychology*, 82, 253–261.

Ludwig, T. and Geller, E. (1999a). Behavioral impact of a corporate driving policy: Undesirable side-effects reflect countercontrol. *Journal of Organizational Behavior Management*, 19(2), 25–34.

Ludwig, T. and Geller, E. (1999b). Behavior change among agents of a community safety program: Pizza deliverers advocate community safety belt use. *Journal of Organizational Behavior Management*, 19(2), 3–24.

Ludwig, T. et al. (1999). Using publicly-displayed feedback to increase turn-signal use: Examining spread of effect to safe stops and safety belt use. *Journal of Organizational Behavior Management*, 19(2), 3–24.

Ludwig, T. et al. (2001). Using public feedback and competitive rewards to increase the safe driving of pizza deliverers. *Journal of Organizational Behavior Management*, 21, 75–104.

Mejza, M.C. et al. (2003). Driver management practices of motor carriers with high compliance and safety performance. *Transportation Journal*, 42(4), 16–29.

Murray, W. and Pratt, S. (2007). *Worldwide Occupational Road Safety (WORS) Review Project*. Prepared for the Department of Health and Human Services, Centers for Disease Control and Prevention, National institute for Occuptaional Safety and Health.

Newnam, S. et al. (2006). Using psychological frameworks to inform the evaluation of fleet safety initiatives. *Safety Science*, 44(9), 809–820.

Olson, R. and Austin, J. (2001). Behavior-based safety and working alone: The effects of a self-monitoring package on the safe performance of bus operators. *Journal of Organizational Behavior Management*, 21(3), 5–43.

Rodriguez, D. et al. (2006). Pay incentives and truck driver safety: A case study. *Industrial and Labor Relations Review*, 59(2), 205–225.

Salminen, S. (2008). Two interventions for the prevention of work-related road accidents. *Safety Science*, 46, 545–550.

Scheltema, K. et al. (2002). Seat-belt use by trauma center employees before and after a safety campaign. *American Journal of Health Behavior*, 26(4), 278–283.

Watson, B.C. (1997) When common sense just won't do: Misconceptions about changing the behaviour of road users. In F. Bullen and R Troutbeck, eds, *The Second International Conference on Accident Investigation, Reconstruction, Interpretation and the Law*, 20–23 October 1997, Brisbane, Queensland. Brisbane, School of Civil Engineering, Queensland University of Technology, pp. 347–359.

White, J. and Murray, W. (2007). Occupational Road Safety Case Study: Rocke Australia cuts risks, collisions and costs. *Journal of the Australasian College of Road Safety*, 18(3), 28–29.

Wouters, P. I. and Bos, J. M. (2000). Traffic accident reduction by monitoring driver behaviour with in-car data recorders. *Accident Analysis and Prevention*, 32(5), 643–650.

Chapter 19

Developing Risk-assessment Tools for Fleet Settings: Where to From Here?

James Freeman, Darren Wishart, Jeremy Davey and Bevan Rowland

Introduction

The costs of work-related crashes

In Australia and overseas, fleet safety or work-related road safety is an issue gaining increased attention from researchers, organisations, road safety practitioners and the general community. This attention is primarily in response to the substantial physical, emotional and economic costs associated with work-related road crashes. The increased risk factors and subsequent costs of work-related driving are also now well documented in the literature. For example, it is noteworthy that research has demonstrated that work-related drivers on average report a higher level of crash involvement compared to personal car drivers (Downs et al., 1999; Kweon and Kockelman, 2003) and in particular within Australia, road crashes are the most common form of work-related fatalities (Haworth et al., 2000).

Earlier estimations indicated that work-related road crash injuries are also approximately twice as likely to result in death or permanent disability as other workplace injuries (Wheatley, 1997) and the average time lost due to injury is greater than any other workplace claim (Stewart-Bogle, 1999). However, it is also noted that the proposed reasons for such elevated risk remain varied and include factors ranging from greater road exposure to fatigue and increased mobile phone use. Despite such uncertainty, research has also found that driving fatalities account for up to 23 per cent of work-related fatalities in Australia, and 13 per cent of the national road toll (Murray et al., 2003). The magnitude of the work-related road safety issue within Australia is also demonstrated with the Australian Safety and Compensation Council's (2007) data that relates to workers compensation claims. The organisation indicated that in the financial year 2004–2005, a total of 405 fatalities were recorded that resulted in claims for workers' compensation. However, 287, or at least 70 per cent, were associated with work-related road safety. Furthermore, vehicle crashes involving a fatality were almost three times more common than the next most common form of injury which involved chemicals or substance. Similar statistics continue to be reported within a number of other motorised companies, and in fact, road accidents are the most common cause of injuries within some organisations (Fletcher, 2005). In regards to the costs, research has suggested that to calculate true financial costs of work-related crashes, organisations should utilise a multiplier of between three and five times the damage only expenses (Mooren and Sochon, 2004). As a result, companies are increasingly implementing safety interventions and training programmes in order to enhance safety culture (Darby et al., 2009).

The collection of data: An asset management approach

Historically, the fleet industry has examined the above outcomes (particularly crashes) through an asset management perspective, and thus focused on the frequency and extent of vehicle fleet crashes. As a consequence, data reporting and recording mechanisms for work-related vehicle incidents are often asset-focused and do not provide comprehensive information relating to the powerful underlying attitudinal or behavioural factors influencing such negative outcomes. Many organisations thus often rely on insurance claims data which primarily records data of interest relating to the asset that the insurance provider is insuring. Unfortunately for fleet safety risk management, the information obtained within these records all too often lacks comprehensive information that assists in informing intervention strategies. Furthermore, across many organisations this type of data also contains a plethora of missing data. The information contained in the claims forms is incomplete, with drivers or persons involved failing to complete the information in substantial detail which could provide some insight into underlying contributing factors. For example, the authors have showcased industry claims data that have been recorded and utilised as the primary data recording mechanism to inform intervention strategies, with the data being supplied missing up to 50 per cent of the data across various data recording fields (Wishart et al., 2008a). It is also noted that some researchers have experienced success in utilising claims data to identify at-risk drivers (Murray et al., 2005; Wåhlberg, 2005). However, it should be further acknowledged that utilising this type of information is still somewhat reliant on the reactive nature of crashes. Furthermore, it may be suggested that organisations primarily concerned with asset management have historically taken on a reactive approach. This perspective often includes a 'silver bullet' frame of reference and involves a single countermeasure implemented in an ad hoc manner that is expected to address all fleet safety-related issues (Davey et al., 2007c). One shortcoming of this process is that the interventions usually only provide short-term relief from the problem, and thus fail to address the underlying behavioural, organisational and situational factors that influence work-related driving outcomes. Thus, it may be argued that comprehensive assessments of either employee driving performance or proactive risk management analyses have traditionally been rather one-dimensional and often reactive ('after the event') rather than proactive. However, there has been considerable shift in recent times in regards to the recognition of utilising a range of data sources to develop an accurate understanding of the extent of the problem as well as inform intervention development. Nevertheless, it is commonly accepted that a considerable level of uncertainty remains regarding identifying the most effective way to undertake driver assessments (Darby et al., 2009).

Self-report measures of driving behaviour

Given the above-mentioned need to look beyond simplistic approaches of collecting and analysing fleet safety problems, researchers and practitioners are beginning to direct an increasing level of focus towards developing and utilising self-report measurement scales to determine whether practical relationships exist between self-reported attitudes and behaviours as well as subsequent crash involvement. One of the most prominent reasons for this focus is that work-related vehicle accidents have been found to be attributable to employees' attitudes and behaviours (Chapman et al., 2001). However, it is noted that the strength of such relationships has yet to be defined. Additionally, it is noted that the vast majority of research (to date) that has focused on self-reported data still refers to general motorists rather than professional drivers or work-related driving. Nonetheless, preliminary efforts have begun to provide a promising insight into the factors associated with crash involvement, and have used a variety of the following scales: the Driving Skill Inventory (DSI) (Lajunen and Summala, 1997), Driver Anger Scale (Deffenbacher et al., 1994), the Driver Behaviour Questionnaire (DBQ) (Reason et al., 1990), the Driver Attitude Questionnaire (DAQ) (Parker et al., 1995a) and various safety climate questionnaires. While a complete review of all assessment tools related to this topic is beyond the scope of this chapter for reasons of parsimony, a brief summary of some of the more popular assessment tools to date is provided below.

Driver behaviour questionnaire

The Manchester Driver Behaviour Questionnaire (DBQ) is arguably the most widely used measurement scales to examine self-reported driving behaviours (Lajunen and Summala, 2003). In fact, the DBQ (which measures highway and aggressive violations as well as errors in the last six months), has been extensively utilised in a range of driver safety research areas, such as: age differences in driving behaviour (Dobson et al., 1999), older drivers (Owsley et al., 2003), the genetics of driving behaviour (Bianchi and Summala, 2004), driver aggression and traffic congestion (Lajunen et al., 1999), trait anxiety (Shahar, 2009), trait aggression (King and Parker, 2008), cross-cultural studies (Lajunen et al., 2003), four-wheel driving (Bener et al., 2008a), driving exposure (Ozkan et al., 2006), and impression management (Lajunen and Summala, 2003), as well as factors contributing to accident involvement (Gras et al., 2006; Parker et al., 1995b), different types of crashes (Parker et al., 1995a), and demerit point loss (Davey et al., 2007a). Furthermore, the versatility of the DBQ has also been demonstrated via the utilisation of the instrument in a number of countries, including Great Britain (Lajunen et al., 1999), China (Xie and Parker, 2002), France (King and Parker,

2008), Australia (Davey et al., 2007b; Newnam et al., 2004), Spain (Gras et al., 2006), Netherlands (Lajunen et al., 2004), Finland (Lajunen et al., 2004; Bianchi and Summala, 2004), United Arab Emirates (Bener et al., 2008b), New Zealand (Sullman et al., 2002) and Greece (Chliaoutakis et al., 2005).

Unsurprisingly, the extent of the application in varying situations has led to considerable variation in regards to the number of factors identified from using the DBQ. Initial research confirmed the original three factors of *errors, violations* and *lapses* (Aberg and Rimmo, 1998; Parker et al., 1995b), while more recent efforts have identified either four factors (Gras et al., 2006; Lajunen et al., 2003; Mesken et al., 2002; Sullman et al. 2002) or five factors (Parker et al., 2000). In addition to the different number of factors identified, research has generally reported differences in factor structure, as specific items often load on different factors depending on the driving context (Davey et al., 2007a) which, not surprisingly, ultimately influences the naming and interpretation of each factor. For example, recent research has also suggested that within an Australian setting there appears to be some overlap between acts of aggression and highway violations, indicating that instances of highway violations may also contain certain aspects of aggression (Davey et al., 2007b; Freeman et al., 2009). These researchers have also proposed that other factors such as fatigue and multi-tasking (e.g., distraction) should also be considered in work-related driving assessment tools as they appear relevant to Australian fleet settings (Freeman et al., 2008), which is also discussed in a following section of this paper.

However, a central issue is that the DBQ has predominantly been utilised with general motoring samples (as compared to work-related settings), which is a common theme within the driving assessment literature. Nevertheless, the comparatively small body of research that has utilised the DBQ in work settings has indicated that errors are predictive of crashes even after controlling for exposure (Davey et al., 2007b; Freeman et al., 2008) as well as aggressive violations (Xie and Parker, 2002), and that the DBQ is in fact associated with safety culture (Oz and Lajunen, 2007). However, it is also noted that such preliminary work has also indicated that the DBQ is not extremely efficient at predicting real crashes but is associated with self-reported ones (Freeman et al., 2008; Wåhlberg et al., 2009), and some of the reasons for this are also presented in a following section.

Driver attitude questionnaire

Another measurement tool that has been utilised within a range of driving settings is the Driver Attitude Questionnaire (Parker et al., 1996). The DAQ was developed by Parker et al. (1996), and measures respondents' attitudes towards four major driving issues: (a) drink driving, (b) following closely to other vehicles, (c) risky overtaking and (d) speeding. Research has begun to utilise the DAQ within a number of different applied settings such as: speed awareness training (Meadows, 2002), speeding behaviours (Fildes et al., 1991), general driver training programs (Burgess and Webley, 2000), general driver improvement schemes (Conner and Lai, 2005), bicycle interventions (Anderson and Summala, 2004), as well as more recently work-related settings such as fleet drivers (Davey et al., 2007a; Wishart et al., 2006) and taxi drivers (Rowland et al., 2008). Initial evidence suggests that the DAQ can assist in investigating motorists' attitudes towards key road safety behaviours, with motorists generally reporting the most lenient attitudes towards speeding violations (Davey et al., 2007b; Meadows, 2002; Rowland et al., 2008).

There has also been some preliminary research conducted within work settings. For example, Wishart and colleagues (2006) combined the DAQ with a number of other self-reported driving assessment questionnaires (e.g., DBQ and climate safety questionnaire) to investigate the driving behaviours of 443 fleet motorists in a sample of Australian fleet drivers. Reliability of each of the subscales ranged between 0.51 and 0.65 with results indicating that participants were more likely to report that drink-driving was unacceptable, followed by attitudes toward risky overtaking. In contrast, the least high factor was attitude towards speeding, which suggested that participants were more likely to report an acceptance that it was ok to speed. Taken together, these results suggest that within a sample of Australian fleet drivers, it is less acceptable to drink and drive but more acceptable to speed while driving for work. Rowland et al. (2008) also utilised the DAQ to provide a profile of a sample of taxi drivers' (N = 182) attitudes towards road safety. Similar to above, participants were again most likely to believe that speeding was more acceptable than drink-driving, following too closely or engaging in risky overtaking manoeuvres. Additionally, more lenient attitudes about risky overtaking practices were also predictive of incurring demerit point loss while driving for work in the last 12 months. Furthermore, Wills and colleagues (2004) also utilised the DAQ along with a variety of other tools to examine the driving behaviours of a group of fleet drivers in Queensland and also reported a correlation (albeit weak) between fleet safety climate, driving attitudes and driving behaviour. Finally, Davey et al. (2007) also utilised the DAQ when examining a group of Australian fleet drivers (N = 4195), but found that driving exposure and the DBQ sub-factors were predictive of crashes rather than the DAQ constructs.

Given the above results, it may be suggested that further research is warranted to enhance the development and application of attitude-type questions to identify underlying assumptions associated with work-related driving which in turn can inform intervention initiatives. However, it is also noted that, similar to DBQ results, researchers have yet to demonstrate that the DAQ is an efficient predictor of either real or self-reported crash involvement (Fildes et al., 1991; Rowland et al., 2008; Wills et al., 2004). In fact, researchers have suggested that considerable care needs to be taken when interpreting such results considering the relatively small amount of variance (and thus predictive power) that is explained by such measurement tools (Fildes et al., 1991).

Organisational safety culture assessment tools

The aspect of a company's safety climate is also receiving increasing attention within the fleet arena, which is now being reflected in an array of measurement tools being developed to assess an organisation's attitudes towards fleet and road safety issues. According to Robbins et al. (1994), organisational culture is a concept that refers to a system of shared meaning that employees or members of an organisation have that distinguishes their organisation from other organisations. An organisation's culture will often consist of certain characteristics based around a set of values that the organisation or its members uphold. Furthermore, Robbins and colleagues (1994) suggest that within an organisation's culture, there exists a series of subcultures which reflect common problems, experiences or situations that employees come across. One such subculture of interest in organisational road safety is that of safety culture, and safety culture has been identified as a key predictor of safety performance (Ostram et al., 1993). The concept of the safety climate (or culture) is beginning to gain increasing attention within the fleet safety literature (Wills et al., 2004). In fact, evidence is starting to mount that indicates that creating a positive safety culture within organisations has a positive effect on employee's driving performance (Haworth et al., 2000; Moser, 2001; Newnam et al., 2004).

For safety climate, a growing body of research is demonstrating a link between safety culture and a variety of outcomes, ranging from vehicle crash rates (Stradling 2000; Sullman et al., 2002), to injury severity (Dimmer and Parker, 1999). Currently, a range of different assessment tools that focus on safety climate have been developed, including: The Safety Climate Questionnaire (Glendon and Litherland, 2001), Organisational Safety Climate Questionnaire (Griffin and Neal, 2000) and the Organisational Safety Culture Scale (Oz and Lajunen, 2007). In regards to fleet settings, company car drivers who perceive stronger management values relating to safety have been reported to adopt a higher level of safety-conscious driving behaviours (D'Silva, 2004). As highlighted above, Wills and colleagues (2004) also utilised the Safety Climate Questionnaire along with a variety of other tools to examine the driving behaviours of a group of fleet drivers in Queensland and also reported a correlation (albeit weak) between fleet safety climate, driving attitudes and driving behaviour. Similarly, Machin (2005) used another Organisational Safety Climate Questionnaire by Griffin and Neal (2000) and reported that the climate predicted safety behaviour at work among a group of coach drivers. Additionally, Oz and Lajunen (2007) utilised the Organisational Safety Culture Scale and found that an organisational safety culture predicted the frequency of errors, violations and accidents among a group of taxi and cargo company drivers. In contrast, Oz and Lajunen (2007) developed and utilised an Organisational Safety Culture Scale to collect drivers' perceptions of the safety culture of the organisation in which they were working. The three subscales were conceptualised as 'traffic safety' incorporating seven items, 'general safety' containing only three items, and 'work safety' consisting of five items. Although it is noted that none of the three organisational safety factors of traffic safety, work safety or general safety were predictive of self-report positive driver behaviour. Results suggested a positive relationship existed between high levels on the work safety dimension and subsequent errors and violations. The authors suggested this finding may have occurred as a result of the organisation (and subsequently the employees) having a high priority for safety and thus encouraging employees to report incidents and hazardous situations. Taken together, research is beginning to suggest that perceptions regarding the safety policies and practices of organisations may have a direct knock-on effect for driving outcomes, and not surprisingly, this is an area that is receiving an increasing amount of research attention.

Additional driving measurement tools and future directions

It is noted that a number of additional measurement scales have been utilised to assess self-reported driving performance, although such tools have again traditionally focused on the general driving population rather than on fleet settings. Additionally for reasons of parsimony, all the measurement scales cannot be reviewed in the following section. However, some of the scales that have proven useful include: the Driver Skill Inventory (Lajunen and Summala, 1995), Driver Anger Scale (Deffenbacher et al., 1994), The Road Behaviour Questionnaire (Antonio et al., 2005), Driver Stress Inventory (Matthews et al., 1997), Police Driver Risk Index (Gandolfi and Dorn, 2005), Propensity for Angry Driving Scale (DePasquale et al., 2001), and Driver Coping Questionnaire (Matthews et al., 1997).

Additionally and more recently, research attention is being directed towards online assessment possibilities. This is to be expected given the possibilities of increasing the efficiency of assessment processes. Currently, the effectiveness of online approaches has been well established in other risk-assessment domains such as health screening (Menard and Boatwright, 2001). Furthermore, from a scientific perspective, researchers are increasingly embracing computer technology in regards to utilising driving simulators to investigate a variety of issues that has produced valuable results in a range of areas, in particular fatigue (Karrer et al., 2005) driver simulation (Allen et al., 2007), and driver distraction (Berthelon et al., 2007). Additionally, researchers are now incorporating online data collection methods to increase the efficiency and sample size of projects. Such online methods have been successfully utilised within a range of road safety projects, including: drug-driving (Freeman and Davey, 2008), younger drivers (Wundersitz and Burns, 2005) and the implementation of driving diaries (de Craen and Twisk, 2005).

The advent and widespread nature of the computer within working lives has also provided an opportunity to utilise online assessments of employees' attitudes and driving behaviours in order to identify at-risk drivers. While only preliminary, some initial results suggest that this approach has considerable potential. For example, a preliminary study by Darby et al. (2009) of 16,004 employees of a United Kingdom firm who completed an online driver risk-assessment tool has produced promising results. This study has previously been reported and is ongoing (Murray et al., 2005) and involves the assessment of participants' attitudes, behaviour and knowledge of the road rules and perceptions of hazards. Driver attitudes, behaviours, knowledge and hazard perception were associated with self-reported crashes. Furthermore, detection of poor attitudes or behaviours resulted in the implementation of a variety of interventions that could include one-on-one intervention, in-vehicle, classroom or computer-based training, monthly communications and performance reviews.

Additionally, researchers from Cranfield University have developed similar online assessment tools based on the Driver Stress Inventory such as the Fleet Driver Risk Index (FDRI), and are now tailoring the instrument to meet the needs of different driver groups e.g., fleet, bus, police, etc. For example, the FDRI is currently being used by over 50 organisations nationally and internationally, with the aim being to highlight attitudes, feelings and behaviours for driver coaching purposes rather than predicting risk of crash involvement. However, preliminary results suggest that the scale can be utilised to predict self-report crashes (Dorn and Gandolfi, 2007). Overall, the FDRI has a strong empirically based foundation that originates from the stress research.

Given the array of available driving scales/questionnaires, the challenge for researchers remains identifying the most salient driver-assessment tools (or essential components within the above tools), to not only identify 'at-risk' individuals but also diagnose fleet-safety problems at an organisational level. In regards to this pursuit, one of the primary aims should be to improve the predictive power of such scales, as research has yet to be published that demonstrates the above tools are extremely efficient at achieving one of their primary purposes, even among databases that include thousands of employees (Murray et al., 2005). More specifically, given that the motoring group in question not only has an increased exposure to the road but also may be prone to engage in habitual and/ or patterned behaviour, it may be considered surprising that greater models of predictive power have yet to be ascertained. For example, Freeman and colleagues (2008) examined the driving behaviour and crash involvement of a large group of telecommunication employees by utilising the DBQ and found the overall model (that included demographic characteristics and traditional self-report measurement scales) was not very efficient at predicting those most likely to be involved in work-related crashes. Wåhlberg and colleagues (2009) reported a similar result utilising data from samples in the United States, United Kingdom, Sweden and Canada. Additionally, research that has focused on other professional drivers has produced similar results (Sullman et al., 2002), as has research that has examined general motorists (Parker et al., 1995b). Finally, an interesting study reported by Murray and colleagues (2005) that compared self-reported versus claims data indicated that the latter were more efficient at predicting employees' risk, although both forms of data did not account for a large proportion of model variance e.g., claims data = 13.6 per cent vs self-reported data = 7 per cent.

While to some extent the above results may be expected as there is arguably a plethora of factors that may influence both driving performance and the likelihood of crash involvement, from a research perspective the results suggest that further scientific effort is required to determine the efficacy of current assessment tools to identify at-risk drivers. Given elements of exposure and possibly patterned or habitual behaviour, actuarial models and other attempts to identify at-risk drivers within fleet settings should at least be greater than for general motorist samples, as there is arguably a greater probability that the former group's behaviour will result in negative outcomes compared to motorists who drive less often and whose behaviour may be more likely to change with the driving environment. Some of the reasons for this failure are outlined below.

Limitations of current measurement tools

Despite the increasing popularity of the above-mentioned measurement tools, a number of limitations (or restrictions) should be borne in mind when implementing and interpreting the outcomes of the collected data. While not intending to be exhaustive, one of the primary limitations is that such tools rely heavily on self-report data. This form of data can be influenced by many factors including self-report bias, either through impression management or self-deception. A recent study by Wåhlberg and colleagues (2010) found that self-reported crash involvement is influenced by social desirability, and thus the authors suggest that any research using such data sources should also incorporate a lie scale. It may be argued that this propensity may be increased within the current context as employees will naturally be reluctant to reveal negative attitudes or practices that may jeopardise their employment status. However, it is also noted that socially desirable responding in regards to the DBQ has been found to be quite small in some preliminary studies (Lajunen and Summala, 2003). On a different level, disparity will often exist between stated intentions and subsequent behaviours, which naturally increases the difficulty associated with identifying individuals who are most likely to engage in aberrant behaviours in the proceeding months. From a practical point of view, many of the current assessment tools are not necessarily conducive for administration to large-scale commercial driving environments due to their length. The present authors continue to encounter resistance from fleet managers to utilise such measure scales as both managers and fleet drivers are not willing or not able to devote the appropriate period of time necessary to accurately complete these driving assessment tools. Rather, the length of some scales may promote a 'tick and flick' effect, which would naturally reduce the effectiveness of such scales. Additionally, it has been suggested that some of the traditional scales are becoming increasingly antiquated as contemporary issues that influence fleet drivers' performance have not been included in assessment scales such as fatigue and time pressure (Freeman et al., 2008). However, it is noted that new contemporary scales (especially online versions such as the FDRI) are increasingly incorporating additional factors proposed to influence driving outcomes. In order for these tools to be fully embraced by the industry, they will need to be easily administered and user-friendly, as well as provide diagnostic, evaluative and appraisal outcomes that can directly inform the implementation of corresponding interventions.

Conclusion

Taken together, an increasing amount of research attention is being directed towards developing and implementing practical and effective driver risk-assessment tools within the fleet arena to reduce the burden of crashes. While the field has made substantial inroads in regards to developing the plethora of assessment tools currently available to practitioners, questions still remain about which scales (or combination of scales) should be utilised to improve the capacity to predict drivers most at risk of damaging company vehicles and/or injuring themselves as well as other motorists. It is also noted that the majority of current research into driver risk-assessment tools has been cross-sectional rather than longitudinal, and thus there is little published research that involves follow-up components to determine whether identified at-risk drivers actually become involved in crashes. Despite this, it is also accepted that it is difficult to accurately measure infrequent phenomena such as deviant driving by methods other than self-report data (Lajunen and Summala, 2003). Perhaps researchers and practitioners need to look beyond the individual's attitudes and behaviours and also the corresponding risk of the company, which may ultimately prove to increase the predictive efficiency of any actuarial model. For example, although various organisations may contain a proportion of higher-risk drivers, some organisations are likely to place drivers at a higher risk simply due to organisational practices and procedures. For instance, if two organisations have similar levels of at-risk drivers but organisation A lacks comprehensive work-related policy procedures and initiatives in comparison to organisation B, the driver from organisation A is likely to be at an elevated risk. This in turn may exacerbate the level of risk to both the individual and the company, and preliminary work by the authors into this area has confirmed this principle (Wishart et al., 2008b). Additionally, further advances may also be found by combining self-report data with claims data, as initial attempts have produced some positive results in this area (Murray et al., 2003). No doubt further technological advances such as the advent and uptake of data tracking systems will also assist in improving our understanding of work-related driving problems (Isler et al., 2007). As a result, it is likely that future challenges will continue to revolve around sifting through countless data possibilities to identify the most relevant information to improve fleet safety. Regardless of the above possibilities and similar to what is known about novice drivers about moving from reactive to anticipatory responding, researchers and practitioners in the work-related driving arena need to continue to develop risk-assessment tools and interventions that increase and promote proactive practices rather than post-event approaches that have historically proven to be relatively inefficient and expensive.

References

Aberg, L. and Rimmo, P. (1998). Dimensions of aberrant driver behaviour. *Ergonomics*, 41, 39–56.

Allen, R.W. et al. (2007). The effect of simulation training on novice driver accident rates. In L. Dorn, ed., *Driver Behaviour and Training, Volume III*. Aldershot, Ashgate, pp. 265–276.

Anderson, A. and Summala, H. (2004). Commuter bicyclists' self image, attitudes, behaviour and accidents. Paper presented at the Third International Conference on Traffic and Transport Psychology, University of Helsinki.

Antonio, P. et al. (2005). Driving at fifteen: Assessment of moped rider training among teens. In L. Dorn, ed., *Driver Behaviour and Training, Volume II*. Aldershot, Ashgate, pp. 253–260.

Australian Safety and Compensation Council (2007). *Compendium of Workers Compensation Statistics Australia 2003–04*. Canberra, Commonwealth of Australia.

Bener, A. et al. (2008a). The impact of four-wheel drive on risky driver behaviours and road traffic accidents. *Transportation Research Part F*, 11, 324–333.

Bener, A. et al. (2008b). The driver behaviour questionnaire in Arab Gulf countries: Qatar and United Arab Emirates. *Accident Analysis and Prevention*, 40, 1411–1417.

Berthelon, C. et al. (2007). Driving experience and simulation of accident scenarios. In L. Dorn (ed.) *Driver Behaviour and Training, Volume III*. Aldershot, Ashgate, pp. 277–289.

Bianchi, A. and Summala, H. (2004). The 'genetics' of driving behaviour: Parents' driving style predicts their children's driving style. *Accident Analysis and Prevention*, 36, 655–569.

Burgess, C. and Webley, P. (2000). *Evaluating the Effectiveness of the United Kingdom's National Driver Improvement Scheme*. Exeter, School of Psychology, University of Exeter.

Chapman, P. et al. (2001). A study of the accidents and behaviours of company car drivers. In *Behavioural Research in Road Safety: Tenth Seminar Proceedings*. London, Department of Transport.

Chliaoutakis, J. et al. (2005). Lifestyle traits as predictors of driving behaviour in urban areas of Greece. *Transportation Research Part F*, 8, 413–428.

Conner, M. and Lai, F. (2005). *Evaluation of the Effectiveness of the National Driver Improvement Scheme*. London, Department of Transport.

Darby, P. et al. (2009). Applying online fleet driver assessment to help identify, target and reduce occupational road safety risks. *Safety Science*, 47(3), 436–442.

Davey, J. et al. (2007a). Predicting high risk behaviours in a fleet setting: Implications and difficulties utilising behaviour measurement tools. In L. Dorn, ed., *Driver Behaviour and Training, Vol. III. Human Factors in Road and Rail Safety.* Aldershot, Ashgate, pp. 175–187.

Davey, J. et al. (2007b). An application of the driver behaviour questionnaire in an Australian organisational fleet setting. *Transportation Research Part F: Traffic Psychology and Behaviour*, 10, 11–21.

Davey, J., Freeman, J. and Wishart, D. (2007). Predicting high risk behaviours in a fleet setting: Implications and difficulties utilising behaviour measurement tools. In L. Dorn, ed., *Driver Behaviour and Training, vol. 3, Human Factors in Road and Rail Safety.* Aldershot, Ashgate, pp. 175–187.

De Craen, S. and Twisk, D. (2005). Assessment of a diary to study development of higher-order-skills during driving experience. In L. Dorn, ed., *Driver Behaviour and Training, Volume II.* Aldershot, Ashgate, pp.179–192.

Deffenbacher, J.L. et al. (1994). Development of a driving anger scale. *Psychological Reports*, 74, 83–91.

DePasquale, J.P. et al. (2001). Measuring road rage: Development of the Propensity for Angry Driving Scale. *Journal of Safety Research*, 32, 1–16.

Dimmer, A.R. and Parker, D. (1999). The accident, attitudes and behaviour of company car drivers. In G Grayson, ed., *Behavioural Research in Road Safety XI.* Crowthorne, Transport Research Laboratory, pp. 78–85.

D'Silva, C. (2004). An investigation of driver behaviour and its association with organisational safety climate in company car drivers. MSc Thesis, Cranfield University.

Dobson, A., Brown, W., Ball, J., Powers, J. and McFadden, M. (1999).Women drivers' behaviour, socio-demographic characteristics and accidents. *Accident Analysis and Prevention*, 31, 525–535.

Dorn, L. and Gandolfi, J. (2007). Designing a psychometrically-based self assessment to address fleet driver risk. In L. Dorn, ed., *Driver Behaviour and Training, Volume III.* Aldershot, Ashgate, pp. 235–247.

Downs, C. et al. (1999). *The Safety of Fleet Car Drivers: A Review.* TRL Report 390. Crowthorne, Transport Research Laboratory.

Fildes, B. et al. (1991). *Speed Behaviour and Drivers' Attitudes to Speeding.* Melbourne, Monash University Accident Research Centre.

Fletcher, S. (2005). A qualitative analysis of company car driver road safety. In L. Dorn, ed., *Driver Behaviour and Training, Volume II.* Aldershot, Ashgate, pp. 327–336.

Freeman, J. and Davey, J. (2008). The impact of new oral fluid drug driving detection methods in Queensland: are motorists deterred? In *Proceedings of the High Risk Road Users Conference*, Parliament House, Brisbane, pp. 86–98.

Freeman, J. et al. (2008). Risk assessment in work-related fleet driving settings: Can self-report questionnaires be used to predict crash involvement? In *Proceedings of the International Symposium on Safety Science and Technology*, Beijing, pp. 1948–1957. Beijing, Beijing Institute of Technology.

Freeman, J., Wishart, D., Davey, J., Rowland, B. and Williams, R. (2009). Utilising the driver behaviour questionnaire in an Australian organisational fleet setting: Can it identify risky drivers? *Australasian College of Road Safety Journal*, 20(2), 38–45.

Gandolfi, J. and Dorn, L. (2005). Development of the Police Driver Risk Index. In L. Dorn, ed., *Driver Behaviour and Training, Volume II*. Aldershot, Ashgate, pp. 337–347.

Glendon, A.I. and Litherland, D.K. (2001). Safety climate factors, group differences, and safety behaviour in road construction. *Safety Science*, 39, 157–188.

Gras, E. et al. (2006). Spanish drivers and their aberrant driving behaviours. *Transportation Research Part F*, 9, 129–137.

Griffin, M. and Neal, A. (2000). Perceptions of safety at work: A framework for linking climate to safety performance, knowledge, and motivation. *Journal of Occupational Health Psychology*, 5, 347–358.

Haworth, H. et al. (2000). *Review of Best Practice Road Safety Initiatives in the Corporate and/or Business Environment*, No. 166. Clayton, Monash University Accident Analysis Research Centre.

Isler, R. et al. (2007). Piloting a telemetric data tracking system to assess post-training real driving performance of young novice drivers. Development of the Police Driver Risk Index. In L. Dorn, ed., *Driver Behaviour and Training, Volume III*. Aldershot, Ashgate, pp.17–29.

Karrer, K. et al. (2005). Fatigue-related driver behaviour in untrained and professional drivers. In L. Dorn, ed., *Driver Behaviour and Training, Volume II*. Aldershot, Ashgate, pp. 349–358.

King, Y. and Parker, D. (2008). Driving violations, aggression and perceived consensus. *Revue europeenne de psychologie appliqué*, 58, 43–49.

Kweon, Y. and Kockelman, K. (2003). Overall injury risk to different drivers: Combining exposure, frequency, and severity models. *Accident Analysis and Prevention*, 35, 441–450.

Lajunen, T. et al. (1999). Does traffic congestion increase driver aggression? *Transportation Research Part F*, 225–236.

Lajunen, T. et al. (2003). The Manchester Driver Behaviour Questionnaire: A cross-cultural study. *Accident Analysis and Prevention*, 36(2), 231–238.

Lajunen, T. et al. (2004). The Manchester Driver Behaviour Questionnaire: A cross-cultural study. *Accident Analysis and Prevention*, 36, 231–238.

Lajunen, T. and Summala, H. (1995). Driving experience, personality, and skill and safety-motive dimensions in drivers' self-assessments. *Personality and Individual Differences*, 19, 307–318.

Lajunen, T. and Summala, H. (1997). Effects of driving experience, personality, and driver's skill and safety orientation on speed regulation and accidents. In J. Rothengatter and E. Carbonell Vaya, eds, *Traffic and Transport Psychology: Theory and Application*. Amsterdam, Pergamon, pp. 283–294.

Lajunen, T. and Summala, H. (2003). Can we trust self-reports of driving? Effects of impression management on driver behaviour questionnaire responses. *Transportation Research Part F*, 6, 97–107.

Machin, M. (2005). Predictors of coach drivers' safety behaviour and health status. In L. Dorn, ed., *Driver Behaviour and Training. Volume II*. Aldershot, Ashgate, pp. 359–371.

Matthews, G. et al. (1997). A comprehensive questionnaire measure of driver stress and affect. In R. Carbonell Vaya and J. Rothengatter, eds., *Traffic and Transport Psychology: Theory and Application*. Pergamon, Amsterdam, pp. 317–324.

Meadows, M. (2002). Speed Awareness Training. Paper presented at the 67th Road Safety Congress, 4–6 March, Stratford, UK.

Menard, M. and Boatwright, M. (2001). Preconception health promotion: evaluation of an online interactive risk assessment tool. *American Journal of Obstetrics and Gynaecology*, 185(6), 161.

Mesken, J., Lajunen, T. and Summala, H. (2002). Interpersonal violations, speeding violations and their relation to accident involvement in Finland. *Ergonomics*, 7, 469–483.

Mooren, L. and Sochon, P. (2004). *Road Safety Towards 2010: 2004 Year Book of the Australasian College of Road Safety*. Canberra, Australian College of Road Safety.

Moser, P. (2001). Rewards for creating a fleet safety culture. *Professional Safety*, 46, 39–41.

Murray, W. et al. (2003). *Evaluating and Improving Fleet Safety in Australia. Report for the Australian Transport Safety Bureau*. Canberra, Australian College of Road Safety.

Murray, W. et al. (2005). Comparing IT-based driver assessment results against self-reported and actual crash outcomes in a large motor vehicle fleet. In L. Dorn, ed., *Driver Behaviour and Training, Volume II*. Aldershot, Ashgate, pp. 373–382.

Newnam, S. et al. (2004). Factors predicting intentions to speed in a work and personal vehicle. *Transportation Research Part F: Traffic Psychology and Behaviour*, 7, 287–300.

Ostram, L. et al. (1993). Assessing safety culture. *Nuclear Safety*, 34, 163–172.

Owsley, C. et al. (2003). Impact of impulsiveness, venturesomeness, and empathy on driving by older drivers. *Journal of Safety Research*, 34, 353–359.

Oz, B. and Lajunen, T. (2007). Effects of organisational safety culture on driver behaviours and accident involvement amongst professional drivers. In L. Dorn, ed., *Driver Behaviour and Training, Volume III*. Aldershot, Ashgate, pp. 143–153.

Ozkan, T. et al. (2006). Driver behaviour questionnaire: a follow-up study. *Accident Analysis and Prevention*, 38, 386–395.

Parker, D. et al. (1995a). Driving errors, driving violations, and accident involvement. *Ergonomics*, 38, 1036–1048.

Parker, D. et al. (1995b). Behavioural characteristics and involvement in different types of traffic accidents. *Accident Analysis and Prevention*, 27(4), 571–581.

Parker, D., McDonald, L., Rabbitt, P. and Sutcliffe, P. (2000). Elderly drivers and their accidents: the aging driver questionnaire. *Accident Analysis and Prevention*, 32, 751–759.

Parker, D. et al. (1996). Modifying beliefs and attitudes to exceeding the speed limit: An intervention study based on the theory of planned behaviour. *Journal of Applied Social Psychology*, 26, 1–19.

Reason, J. et al. (1990). Errors and violations: A real distinction? *Ergonomics*, 33, 1315–1332.

Robbins, S. et al. (1994). *Organisational Behaviour Concepts Controversies and Applications Australia and New Zealand*. Sydney: Prentice Hall.

Rowland, B. et al. (2008). The influence of driver pressure on road safety attitudes and behaviours: A profile of taxi drivers. *Proceedings of the Canadian Multidisciplinary Road Safety Conference*, CD-ROM, Whistler, Canada. Canadian Association of Road Safety Professionals.

Shahar, A. (2009). Self–reported driving behaviors as a function of trait anxiety. *Accident Analysis and Prevention*, 41, 241–245.

Stewart-Bogle, J.C. (1999). Road safety in the workplace. The likely savings of a more extensive road safety training campaign for employees. Paper presented at the Insurance Commission of Western Australia Conference in Road Safety 'Green Light for the Future'. Available at www.transport.wa.gov.au/roadsafety/Facts/papers/contents.html

Stradling, S.G. (2000). Driving as part of your work may damage your health. In G.B. Grayson, ed., *Behavioural Research in Road Safety IX*. Crowthorne, Transport Research Laboratory, pp. 0–9.

Sullman, M.J. et al. (2002). Aberrant driving behaviours amongst New Zealand truck drivers. *Transportation Research Part F*, 5, 217–232.

Wåhlberg, A. (2005). Differential accident involvement of bus drivers. In L. Dorn, ed., *Driver Behaviour and Training, Volume II*. Aldershot, Ashgate, pp. 383–393.

Wåhlberg, A.E. et al. (2010). The effect of social desirability on self-reported and recorded road traffic accidents: implications for self reports of driver behaviour. *Transportation Research Part F*, 13(2), 106–114.

Wåhlberg, A.E. et al. (2009). The Manchester Driver Behaviour Questionnaire as a predictor of road traffic accidents. *Theoretical Issues in Ergonomics Science*, published online and available at http://dx.doi.org/10.1080/14639220903023776.

Wheatley K. (1997). An overview of issues in work-related driving. In *Staysafe 36: Drivers as Workers, Vehicles as Workplaces: Issues in Fleet Management. Report No. 9/51. Ninth Report of the Joint Standing Committee on Road Safety of the 51st Parliament*. Sydney, Parliament of New South Wales.

Wills, A. et al. (2004). The relative influence of fleet safety climate on work-related driver safety. *Proceedings of the Road Safety Research, Policing and Education Conference*, Perth, CD-ROM. Melbourne, Tulips Meeting Management.

Wishart, D. et al. (2006). An application of the driver attitude questionnaire to examine driving behaviours within an Australian organisational fleet setting. *Proceedings of the Road Safety Research, Policing and Education Conference*, Gold Coast, Australia, CD-ROM. Melbourne, Tulip Meetings Management.

Wishart, D. et al. (2008a). Paper presented at the CARRS-Q Fleet Safety Seminar, Victoria Park Golf, April.

Wishart, D., Davey, J., Rowland, B., Freeman, J. and Banks, T. (2008b). Situational analysis: Logan City Council. Unpublished report. Carrs-Q, Queensland University of Technology.

Wundersitz, L. and Burns, N. (2005). Identifying young driver subtypes: Relationship to risky driving and crash involvement. In L. Dorn, ed., *Driver Behaviour and Training, Volume II.* Aldershot, Ashgate, pp. 155–168.

Xie, C. and Parker, D. (2002). A social psychological approach to driving violations in two Chinese cities. *Transportation Research Part F*, 5, 293–308.

Chapter 20

From Research to Commercial Fuel Efficiency Training for Truck Drivers using TruckSim

Nick Reed, Stephanie Cynk and Andrew M. Parkes

TruckSim: A Brief History

In 2002, the UK road haulage was considered to be undergoing a crisis of recruitment of new drivers and the retention of existing ones. European Union (EU) directives on working time and training were expected to put more pressure on transport operators to recruit more drivers and to increase their fleet sizes.

The European Commission Directive on Training for Professional Drivers (EU Commission, 2003) that was adopted in April 2003 (Directive 2003/59/EC) stipulates obligatory basic and continuous training for drivers of goods and passenger vehicles from September 2009. All drivers wishing to drive large goods vehicles (LGVs) in excess of 7.5 tonnes in a professional capacity will have to undergo training for, and obtain, a vocational certificate of professional competence (CPC), further to the LGV licence. Professional drivers will be required to undergo continuous training of 35 hours every five years to refresh their knowledge and skills. The directive offers scope for a proportion of this training to be conducted on a 'top of the range' driving simulator.

The UK Department for Transport (DfT) established a research programme to determine the potential role of the synthetic training in both *ab initio* licence acquisition and skills development in experienced drivers on behalf of the Road Haulage Modernisation Fund (RHMF). TRL commissioned EADS to produce an advanced full-motion-base truck simulator and developed bespoke UK road databases and courseware. This system was christened 'TruckSim'. The main objectives of the initial research programme were to expose a large number of students and freight companies to the potential of synthetic training and to inform the DfT and the RHMF of how the synthetic training could be best integrated into training and testing programmes. Subsequent research focused on fuel efficiency training.

The Simulator

TruckSim at TRL is a dedicated facility to provide training for drivers of commercial vehicles. The Full Mission Simulator (FMS) was built by EADS and consists of a Mercedes Actros cabin mounted within a pod and surrounded by a curved screen.

Seven projectors in the pod provide the driver with a 270° field of view along with the use of rear-view mirrors as normal. A flat screen monitor is mounted on the nearside of the cabin to supply the equivalent view of a kerb mirror. The simulator display has a refresh rate of 60Hz, a resolution of 1280 × 1024 pixels per channel and approximately 2.9 arc minutes per pixel.

The pod is mounted on a combination of hydraulic and electric actuators to give

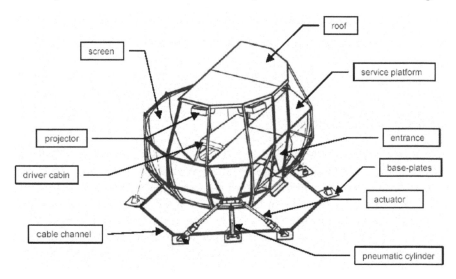

Figure 20.1   Schematic of TruckSim

Figure 20.2   Interior and exterior of TruckSim

full motion with six degrees of freedom: pitch, roll, heave, yaw, surge and sway. An eight-speed manual gearbox (four over four with range change) is provided in the cab. The motion parameters are further described in Table 20.1.

Table 20.1   Summary of full motion of truck simulator

	Pitch	Roll	Yaw	Heave	Surge	Sway
Displacement	10°	12°	8°	±0.5m	±0.65m	±0.6m
Acceleration	>100°s^{-2}	100°s^{-2}	100°s^{-2}	0.5g	0.75g	0.75g

A simulated road network was created containing generic motorway, rural, urban and suburban areas with correct UK-specification junction layouts and signage, and a distribution centre for parking manoeuvres. Various features of the driving experience, including weather, ambient lighting, road friction and truck load, can be adjusted from the simulator control room. Exercises were created to examine drivers' responses to the different elements of the simulated environment.

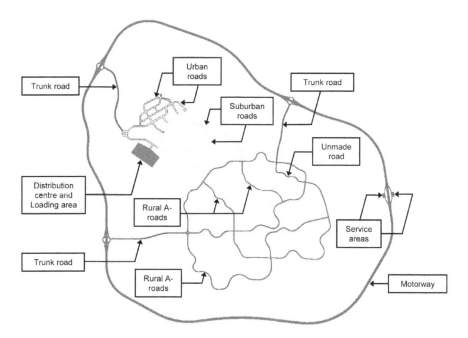

Figure 20.3   TruckSim road database

The truck simulator provides a 'real' experience; it duplicates the operation of a vehicle and reproduces a world outside. Scenes are delivered with sufficient reality to ensure that the driver becomes truly immersed in the experience. Up to 40 other vehicles with intelligent behaviour can be displayed in the scene at any one time.

Figure 20.4   Screenshots from TruckSim

Both internal and external information is accurately reproduced ensuring drivers feel as if they are sitting in the cab of their own vehicle. Quality audio (3D sound) and visual systems contribute to making the experience as real as possible.

Evaluation Study

Having implemented the simulator, the next step was to evaluate it in terms of its effectiveness as a training tool. The truck simulator training was evaluated using qualitative methods investigating issues such as simulator sickness, similarity to driving a real truck, quality of simulation, handling realism and general acceptance.

The evaluation study used 616 professional truck drivers. They participated in two computer-based training (CBT) tasks similar to theory tests that drivers must pass before licence acquisition. The first task asked 35 multiple-choice questions and the second was a hazard perception task of a computer demonstration of truck driving. Next, trainees completed an exercise on a part-task simulator (a static version of the full-motion TruckSim) with guidance from a professional driver trainer to introduce the trainees to the appearance and controls of a simulated environment.

Under the supervision of a professional trainer, each trainee completed three exercises in the TruckSim. The first drive was a familiarisation drive on a motorway. The second and third exercises were from two selected from a poor weather drive, an urban drive, a route including high-speed bends, and a reversing exercise. Questionnaires were distributed to obtain opinions of the exercises and a further questionnaire assessed levels of simulation sickness, subjective evaluations of realism and acceptance of the simulator as a training tool.

The truck simulator received a positive response from the truck drivers involved. Although the overall scores for similarity to driving, quality of simulation and handling realism were generally high, the simulator was reported as being particularly similar to real truck driving in terms of overall cab design, headlights and the mirrors. The aspects of simulation quality that received the most praise were the road layout, brightness, clarity of the display and the overall driving environment. Handling realism received the most positive scores for signal turns and seeing the road and other road users.

The research programme also identified areas that needed to be addressed to improve the TruckSim further as a training tool. The similarity of the steering and deceleration to driving a real truck was rated as low. In relation to the simulation quality, trainees regarded the traffic and realism of cyclists as poor. Finally, the realism of controlling in a turn and service braking received lower scores.

Simulator sickness was generally low. However, there was a high variation in the experience of simulator sickness with some participants experiencing high levels of sickness. High levels of simulator sickness correlated with lower ratings of simulator similarity, quality and handling realism. A further finding was that participants older than 30 years were more likely to experience higher levels of simulator sickness. The simulator may therefore be better suited to training younger drivers.

The exercises that were conducted affected drivers' perceptions of the TruckSim. Those who experienced the poor weather exercise viewed it favourably and had more positive perceptions of the similarity to driving, quality and realism of the simulator. The reversing exercise was less favoured and had the opposite effect on perceptions of the similarity to driving, quality and realism of the simulator.

Of particular importance to the research, the majority of drivers did not feel that training in the simulated environment affected their learning in their normal cab. The trainer was also deemed influential in their learning experience.

Fuel Efficiency

The TruckSim evaluation phase demonstrated that drivers had a subjective preference for exercises involving continuous driving rather than slow speed manoeuvres. Fuel efficiency was therefore an obvious target for simulator training scenarios, particularly since fuel consumption is a directly measurable property of every drive completed. It is therefore easy to compare performance between drives/drivers; unlike accident risk, the level of which is harder to capture in a short, naturalistic drive. The DfT commissioned TRL to undertake a further study using TruckSim to develop the simulator training capability and to gain further feedback from an additional group of drivers.

Performance in the simulator scenarios was assessed on the critical measures used in the safe and fuel efficient driving (SAFED) standard (DfT, 2003); fuel used, gear changes and time taken. The SAFED training programme uses a real vehicle and requires drivers to complete a route lasting one hour. Drivers are then given some classroom training on safety and fuel efficiency before drivers complete the same one-hour route again. Results demonstrated that drivers typically reduce fuel consumption by 2–12 per cent and make fewer gear changes but do not take any longer to complete their journey (DfT, 2005).

After a familiarisation drive in order to become acquainted with the controls of the simulator and the feel of driving in the virtual environment, participants in the TruckSim programme were required to complete two drives on the same route. Each simulator drive lasted between 15 and 20 minutes and encompassed rural, urban and high-speed driving sections. Between the two drives, participants were given training in fuel efficiency by a qualified truck driver trainer. For this programme 394 professional truck drivers participated and their performance in each of the key measures was recorded. On average, drivers made 12.5 per cent fewer gear changes, used 3.33 per cent less fuel and took 6.59 per cent less time to complete the route. These differences are comparable to those observed in on-road SAFED training. However, these improvements may have been an artefact of better *simulator* driving rather than better *truck* driving. Therefore, a study was undertaken to demonstrate that skills learned in the simulator would transfer to behind the wheel driving techniques.

Transfer of Training

This next phase of the work was again funded by the DfT. TRL recruited 60 truck drivers to experience fuel efficiency training in TruckSim on three separate occasions over a period of six months. The drivers came from eleven different companies that operated in a variety of industries, including automotive, food and hazardous chemicals. To investigate real-world driving, the fuel efficiency of the participating drivers was monitored for a working week before and after each of their three visits to TRL. This meant their on-road performance could be tracked in relation to their training in the simulator. For comparison and control the fuel efficiency of an additional 60 drivers with similar profiles to those undergoing simulator training and from the same company were monitored. Finally none of the drivers participating in the study received any other driver training for the duration of the project.

The simulator experience received by drivers coming to TRL consisted of a short period of familiarisation before a first attempt at driving a mixed rural and urban route, observed by a fully qualified driver trainer. Having completed the exercise, the trainer gave drivers instructions on how to improve their driving style in order to complete the route with greater fuel efficiency. Each driver then had a second attempt at the exercise giving an opportunity to demonstrate improved fuel efficiency. For all drives, drivers operated a simulated version of the Mercedes Actros 2544 articulated (6 × 2 axle configuration) lorry unit with 100 per cent load (estimated gross vehicle weight 44 tonnes).

In a further development of the training package, an analysis tool was developed to evaluate performance in the two exercises automatically. The simulator recorded data about various aspects of each drive at 60Hz. These data included information about the actions of the driver (e.g. steering angle, accelerator/brake/clutch depression); the behaviour of the vehicle (e.g. speed, lateral/longitudinal/rotational acceleration); the behaviour of the vehicle in the context of the external environment (e.g. distance to vehicle in front, lateral position) and the simulated engine and transmission characteristics (e.g. fuel used, revolutions per minute [RPM], engine torque). Using the data of drives completed in the previous projects and correlating these measures against fuel efficiency, it was possible to determine benchmark values for good and bad fuel-efficient driving practices. The analysis tool then scored drivers on a variety of criteria to give an indication as to how well they drove the vehicle in terms of fuel efficiency. A grade was given for each aspect: green for good, yellow for fair and red for poor fuel-efficient driving behaviour. The instructor could then provide tailored feedback to the driver about the aspects of their simulated and real-world driving on which they should concentrate to make the biggest improvements.

Results in the simulator showed that drivers made an 11 per cent improvement in their fuel efficiency over the three visits to the simulator, with the biggest gain being made during the first visit. It was also clear that drivers retained what they had learned from one visit to another as fuel efficiency did not deteriorate between visits. The simulator data revealed that drivers were handling the vehicle in a much more efficient manner. Average RPM observed during periods of acceleration dropped by 22 per cent resulting in the engine operating in a more efficient region and generating 45 per cent higher torque. There were also 29 per cent fewer gear changes over the course of the drives. It would be easy to assume that drivers simply slowed down to achieve these improvements but the data show that drivers were actually around 8 per cent faster overall.

The key question then was would this behaviour be transferred to the drivers' real-world driving back in their everyday work? It was found that relative to the control group, the simulator-trained drivers showed a progressive improvement in their fuel efficiency, returning a 16 per cent improvement in MPG after the third training session.

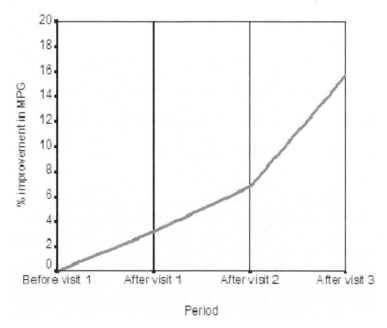

Figure 20.5   Mean percentage improvement in on-road fuel efficiency

Note that this improvement was greater than that achieved in the simulator. It was also higher than the improvement reported by Strayer and Drews (2003) who found that drivers trained in a simulator showed an average improvement of 2.8 per cent in fuel efficiency for the six-month period following training. It was concluded (see Parkes and Reed, 2005) that skills learned in TruckSim had indeed transferred to behind the wheel behaviour, providing the basis for a commercial driver fuel-efficiency training package.

Allied Bakeries Participation

The first company in the UK to use TruckSim to provide training for their truck drivers on a commercial basis was Allied Bakeries (AB). This commenced with a pilot study, supported by the Welsh Assembly Government, in which AB committed six drivers to a simulator training programme, similar to that described in the cohort study but in which drivers visited TRL for training on two occasions (rather than three). AB were able to keep detailed fuel consumption records for the six drivers involved in the study and to minimise other factors that may affect fuel efficiency. For example, AB ensured that each driver always drove the same vehicle and always drove on the same routes throughout the period of study. The drivers were chosen to represent a range of driving styles and to cover different normal driving environments.

Training was provided by an Approved Driving Instructor (ADI) for all forms of commercial vehicle. Participants visited TRL for two training sessions. In addition to this all participants visited TRL for a familiarisation visit in order to see the simulator facility and to have a brief drive of TruckSim to help reduce any feelings of anxiety that they might have about training on the simulator at TRL. This familiarisation visit was conducted on 10 April, 2008. The first training visit was conducted in July 2008; the second was conducted in September 2008.

Drivers were asked to operate a simulated Mercedes Actros 2541 rigid (6 × 4 axle configuration) lorry unit with 50 per cent load (estimated gross vehicle weight 17 tonnes). This vehicle type was selected to be most similar to the type of truck and load typically driven by the AB drivers. The training route was a mixture of rural and urban driving with a number of events designed to challenge the driver and provide opportunities to display fuel-efficient driving practices. Drivers received feedback on their performance using a revised version of automated assessment system, tailored to suit the type of simulated vehicle driven in this training programme.

A potential problem that may affect all training providers was the difficulty in scheduling drivers for training sessions. Training on the simulator required the co-ordination of trainees to attend the facility on specified dates. Issues such as illness, injury, leavers from the company and annual leave all affected driver bookings for the training programme. However, on completion of the training programme, results demonstrated that trainees produced a 25.6 per cent improvement in simulated fuel efficiency. Exploration of the changes in behaviour indicated that this accompanied by a reduction in RPM (reduced by 33.4 per cent) resulted in greater engine torque (increased by 50.5 per cent) when accelerating and a reduction in fuel wastage when slowing the vehicle. Drivers also made significantly fewer gear changes in the simulator after training (down by 26.4 per cent). This also contributed to greater torque values and better vehicle sympathy but may have been due partially to increased familiarity with the simulator gearbox.

Results in the real world showed that simulator training was associated with a mean fuel-efficiency improvement of 7.3 per cent. Biggest improvements were seen for drivers who completed mixed driving routes, whereas the smallest improvements were seen for drivers who usually drove in an urban environment. When the observed real-world fuel-efficiency improvements are applied to annual fuel usage the cost benefits are considerable and CO_2 emissions would be reduced by over 250 tonnes. Based on that assumption, the improved fuel efficiency of the trained drivers would provide a return on investment in a little over four months.

This study supported the results of the previous simulator fuel-efficiency training programme conducted at TRL, demonstrating that simulator training can achieve significant and pragmatic fuel efficiency savings on a commercial basis.

Further Training

Following the success of the pilot driver training programme, AB commissioned training of a further group of 32 drivers from their Cardiff depot, all of whom were to repeat the same training process as completed by the drivers in the pilot study. Convinced by the success of the pilot programme, AB did not apply the restrictive constraints whereby trained drivers would keep to the same route and vehicle. Consequently, it is difficult to assess the success of this programme at the level of the individual driver. However, partway through the training programme, AB reported a steady improvement in the overall fuel efficiency of the fleet. They have also observed a notable reduction in the number of accidents and their drivers report feeling less stressed at the end of a shift and having a greater sense of esteem following training. However, these are anecdotal findings and should therefore be treated with due care. It is now also a year since on-road fuel efficiency data were first collected for the AB drivers involved in the original pilot. Only one driver has remained on the same route and with the same vehicle type throughout this period. His mean fuel efficiency improvement following training was 15 per cent and over the year this has gradually reduced to a sustained improvement of 7 per cent. This would give AB a clear return on investment over the cost of the simulator training. A study in which the real-world fuel-efficiency of cohorts of drivers trained in the simulator at different frequencies (every three months; every six months; annually) was continuously monitored (with an associated cost–benefit analysis) would establish the optimal training regime.

Discussion

The TruckSim facility at TRL has had a steady transformation from an initial curio to the naissance of a genuine, cost-effective business opportunity to provide advanced training to professional truck drivers. The initial phases of the programme provided a strong foundation for the understanding of how TruckSim might best be used for training and provided the data that would go on to be used in creating the benchmarks for the automated driver assessment system. In the second phase of TruckSim, a shift in thinking led to significant advances in training methodology and simulator use. As a matter of course, a simulator generates vast quantities of data about the driving situation; pedal depression; steering angle; gear usage; vehicle speed; vehicle position; engine speed; engine torque and so on. Whilst the simulator can be used as a proxy for on-road training, these data provide a rich source of information that can be used to understand a driver's capabilities and training needs. Failing to exploit that information is to neglect a key area in which the simulator has a highly significant and functional benefit over the use of real trucks for training.

Furthermore, it was observed that the use of simulator data to diagnose driver faults appeared to produce a marked change in the dynamic between trainer and trainee. Instead of a trainer delivering drivers with subjective criticism of their performance (which may have been met with suspicion or even disdain), the simulator provided objective and impartial feedback on driver performance. The trainer is therefore able to focus on helping a driver to correct the areas of substandard performance. Removing this element of subjectivity seems to have improved engagement by trainees in the training process and more positive outcomes as a result.

The transfer of fuel-efficiency benefits trained in the simulator to real world driving has shown that simulator training can cause genuine improvements in real-world driving performance. The growing emphasis on corporate social responsibility (CSR) requires companies to demonstrate that they are taking reasonable steps to operate ethically and sustainably. The fuel efficiency improvements that have been observed following simulator training are associated with reductions in engine emissions thereby assisting companies in meeting CSR objectives.

The EU Driver Training Directive 2003/59/EC (EU Commission, 2003) makes it compulsory for drivers of lorries, buses, coaches and minibuses (when being used for hire and reward) to hold not only a driving licence but also a Certificate of Professional Competence (CPC) for drivers. The legislation requires existing drivers to undertake 35 hours periodic training over a period of five years (from 10 September 2008 for drivers of passenger carrying vehicles and from 10 September 2009 for drivers of large goods vehicles). Training sessions must be in periods of not less than seven hours. Fulfilling this requirement allows a driver to receive the CPC. The directive permits a proportion of the driver training for the CPC to be completed on a 'top-of-the-range simulator'. Subject to certification by the relevant authority (in the UK, this is the Joint Approvals Unit for Periodic Training),

training courses provided on TruckSim may form an appealing component of the CPC training process.

To conclude, the UK's first full-mission, high-fidelity truck simulator, TruckSim, has matured from an initial novelty into an effective training tool that can deliver cost-effective driver training. The simulator training process has been gradually refined and the development of automated driver assessment software to support the simulation has been fundamental in the success of the simulator training programme. TruckSim has carved itself a niche for professional driver training. The extent to which this niche will develop remains to be seen.

References

DfT (2003). *The Safe and Fuel-efficient Driving (SAFED) Standard. Good Practice Guide 2100*. London, Department for Transport.

DfT (2005). *Companies and Drivers Benefit from SAFED: A Selection of Case Studies*. TE213 Transport Energy Best Practice. London, Department for Transport.

EU Commission (2003). Directive 2003/59/EC of the European Parliament and of the Council of 15 July 2003 on the initial qualification and periodic training of drivers of certain road vehicles for the carriage of goods or passengers, amending Council Regulation (EEC) No 3820/85 and Council Directive 91/439/EEC and repealing Council Directive 76/914/EEC. *Official Journal of the European Union*, L 226, 10/09/2003 P. 0004–0017.

Parkes, A.M. and Reed, N. (2005). Transfer of fuel efficient driving technique from the simulator to the road. In D. de Waard, K.A. Brookhuis and A Toffetti, eds, *Developments in Human Factors in Transportation, Design and Evaluation*. Maastricht, The Netherlands, Shaker Publishing, pp. 163–176.

Strayer, D.L. and Drews. F.A. (2003). Simulator training improves driver efficiency: transfer from the simulator to the real world. In *Proceedings of the Second International Driving Symposium on Human Factors in Driver Assessment, Training, and Vehicle Design*. Park City, Utah.

Chapter 21

The Utility of Psychometric Testing for Predicting Bus Driver Behaviour

Wendy Lord and Joerg Prieler

Introduction

The prediction of driver behaviour and accident proneness in bus drivers is clearly important given the responsibility that bus drivers must take for public safety. A previous study (Prieler 2003) suggested that key predictors for success are intellectual reasoning plus a number of personality traits; namely, sociability, abstract thinking, dutifulness and deliberation. The criterion used in the Prieler study was the likelihood of candidates being recommended for employment as bus drivers. The present study which took place between 2005 and 2008 evaluates the relationship between psychometric test results and actual performance in the role. Thus by considering the results of these two studies there is the scope to compare what those involved in selection consider to be important and what actually is important.

The results support the previous findings but refine them such that broad intellectual reasoning was found to be less important than a more specific application of intellectual power – that of selective attention; the ability to selectively attend to relevant stimuli and to sustain that selective attention over time. In terms of behavioural style the present study supports the importance of the predictors identified by the previous study and adds additional insights.

Furthermore this study provides evidence for the utility of psychometric tests in improving selection of bus drivers and, perhaps most importantly, it demonstrates that the utility of psychometric tests in predicting job effectiveness depends not only on choosing the right tests but also on using the tests in accordance with good practice guidelines.

Background

The present study took place in Nottingham City Transport (NCT). This UK company runs a fleet of 386 buses throughout the city of Nottingham and outlying areas. NCT employs a team of more than 800 drivers supported by 160 technical and engineering staff. Between them, they carry 50 million passengers and cover over 12.5 million miles per year. NCT's mission is to provide those they serve with a safe and reliable public transport system. The two most important requirements in bus drivers are safe driving behaviour and effective customer management. NCT do not tolerate poor driving and they expect high standards of customer care from their drivers.

Applicants to the role of bus driver usually have no experience or qualifications in bus driving. After selection they attend a six-week training course. The initial three weeks is 'off the road' during which they learn how to drive a bus, how to operate ticket machines and customer service skills. This is followed by three weeks 'on the road' supervised bus driving.

It costs NCT £3000 to train each new recruit so mistakes at the selection stage are very costly. In 2005 NCT recognised that they were making a lot of expensive mistakes in selection: 1 in 3 new recruits were leaving or failing training. Of those who passed the training, 28 per cent left the company or were dismissed within six months. This six-month period had been identified as critical. Bus drivers still in the role six months after training tended to stay with the company.

An additional issue was the number of physical assaults on drivers (44 per annum) resulting in claims for compensation either from the driver or the customer. Not surprisingly, this indicated to the company a need to develop better customer management skills in drivers. NCT designed and rolled out a training package appropriately titled 'How Not to Get Hit'. In a review of the success of this programme NCT recognised that not all staff had the temperament to practice what had been learned. This observation and the high attrition rate led to this study being commissioned to identify better ways to select the raw material required for effective performance in bus drivers.

Understanding the Issues

In seeking to address the issues it was important to have a clear understanding of the factors underpinning them. We began by exploring the reasons for dismissal and the reasons for choosing to leave the company. The three main reasons for dismissal were:

- Failing the bus driving test at the third attempt
- Being abusive to other road users
- Poor driving record.

In terms of why bus drivers choose to leave, an analysis of exit interviews suggested that reasons for choosing to leave (as opposed to being dismissed) were less to do with the stress of driving or difficulties dealing with passengers in isolation but more to do with having to deal with these two job requirements at the same time. In other words, those who chose to leave left because of the complexity of the task of simultaneously attending to driving and customer management. The next step was to establish the extent to which the existing selection procedure predicted this dual capacity.

Review of the Selection Process

In 2005 the NCT selection procedure involved five steps. Failure to pass any one step precluded progression to the next. The steps were as follows:

1. Literacy test aimed to ensure a literacy level adequate to understanding written instructions related to changed routes etc.
2. Test of understanding of road signs
3. Numerical test aimed to assess ability to add up multiple fares and give correct change
4. Driving assessment: given that most applicants are not yet qualified to drive a bus, this takes place in a car – experienced bus drivers assess the potential to drive a bus based on driving behaviour in a car
5. Interview to assess customer service skills.

The review suggested that, in terms of measuring customer service skills, the process was very much geared to operating fare transactions rather than actually measuring effectiveness at dealing with customers. This latter capability was assessed only by interview. Furthermore the assessment of effective and safe driving was subjective and did not take place in a bus. Clearly what we needed to find was a more objective measure of safe driving behaviour and a more objective measure of the potential to manage customers effectively.

Empirical Study

Based on what had been learned about key competencies in this role and taking account of past studies (e.g. Prieler, 2003) we identified a number of metrics designed to measure what seemed to be relevant constructs; reasoning ability, selective attention, customer service skills and generic personality traits. These tests were administered to a sample of 60 bus drivers and correlated with performance criteria; managerial ratings and driving safety records.

Two of the psychometric tests that were trialled proved to be especially predictive of job success in this study:

1. The d2 Test of Attention. The d2 was developed over 40 years ago specifically as a test to differentiate between safe and unsafe drivers. It was developed at the Institute for Safety in Mining Industry and Transport in Essen in Germany. It is a measure of selective attention in the form of what is commonly referred to as a 'cancellation test'. The test taker must

scan lines of characters (consisting of a letter d or p marked with one, two, three, or four small dashes) and cross out all occurrences of the letter d with two dashes while ignoring the other characters as irrelevant. There are a range of different scores that can be obtained from the test. In this study there was a clear relationship between the score defined as 'number of items processed minus errors' and driving record in the sample. This score combines speed of processing with accuracy of processing. In driving safely it is important to process data both speedily and accurately so this result was not surprising. Based on the analysis between d2 scores and driving record in the sample from NCT a cut-off score was established at the 75th percentile for this parameter on the d2 test.

2. The NEO PI-R. The NEO PI-R is a personality test which measures 30 aspects of behavioural style. Seven of these aspects proved to be important predictors of job effectiveness in bus drivers; angry hostility; warmth; trust; compliance; achievement striving; self-discipline; deliberation. In some scales such as warmth and compliance the optimal score position was within the average range; too high a level of warmth may manifest as social distractibility (chatting too long with customers for example).

Revision of the NCT Selection Process

As a result of the empirical study, the d2 test of attention and the NEO PI-R were incorporated into the NCT selection process as follows:

1. Literacy and numeracy tests
2. The d2 test of attention
3. The NEO PI-R
4. Practical driving assessment
5. Interview.

Applicants had to obtain scores in the predictive range for all scales in order to progress to the next stage.

Outcome

Since the start up of the new selection procedure up until January 2007, 195 new drivers were recruited onto the training programme. Eleven out of twelve passed the training course (compared to previously when one out of three failed): 91 per cent of those who passed training were still employed after six months (compared to only 72 per cent previously). Assaults were down from 44 per annum to 12 per annum. The financial saving of the new selection procedure was calculated by NCT as being £154,000.

Then Something Seemed to Go Wrong ...

During the summer of 2007 attrition rates started to rise again. An investigation was undertaken. In the intervening period, the original HR manager had left the company and a new HR manager was in place. The new HR manager was using the tests but ignoring the selection guidelines. When the profiles of 45 leavers were analysed, it was found that 39 of them would not have been recruited in the first place had the empirically validated guidelines for selection on the basis of the tests been followed.

The current situation

NCT are now back on track with overall staff turnover at 11.5 per cent (including leaving due to retirement or relocation) compared to an industry average of 25–30 per cent. In addition, as a result of the reduction of attrition and assaults, respect for the HR function has increased.

A lesson learned

Applying good practice with psychometric tests is of paramount importance to ensure that the right tests are chosen, that they are used in accordance with good practice and that their use is monitored in relation to scores on the test and performance in the role.

Discussion

The predictive factors found here support to some extent the findings in a previous study (Prieler 2003) but there are divergences. It is interesting to consider why the divergence arises. There are two key differences in the methodology of the two studies. Firstly the criteria used to predict performance were different. In the Prieler study the criterion was the likelihood of being recommended for employment as a bus driver whereas in the present study the criterion was actual performance on the job. The likelihood of being recommended for a position is clearly less strong as a criterion than actually knowing how people perform in the role so it is interesting that, nevertheless, despite the different criteria there was considerable concordance between findings. For the same reason it is not surprising that the present study, related as it was to actual work performance, was able to clarify and refine predictive factors over and above what was found in the Prieler study.

Of course the metrics used also play a role in output from studies such as these. The d2 for example measures a more specific aspect of intellectual performance than the general intelligence measure used in the Prieler study. Also the metrics used to measure personality (16PF in the first study and NEO in this one) are slightly different in how they structure personality traits. Regardless of the metric used, common to both studies are factors relating to extraversion and conscientiousness. Both studies identified as important a concern to do a good job (conscientiousness factors). In terms of extraversion, this study refined the nature of the predictive relationship between personality facets related to interpersonal style and job effectiveness. This is likely to be partly due to the different metrics used. The Prieler study used the 16PF which compared to NEO PI-R is less focused on separating out the specific behaviours arising from extraversion. However, the criteria used are also relevant to different nuances in the results. Sociability (identified in the previous study as important in observer recommendations) is more easily observable than for example trust or underlying emotional attitude towards others (angry hostility). This raises the issue that factors that are observable to those responsible for recommending individuals for employment do not tell the whole story about what actually drives success in the role.

References

Brickenkamp, R. and Zilmer, E. (2003). *The d2 Test of Attention Manual*, ninth edn. Goettingen, Hogrefe and Huber.
Costa, P.T. and McCrae, R.R (1992). *Revised NEO PI-R and NEO Five Factor Inventory: Professional Manual*. Tampa, FL, PAR Inc.
Lord, W. (2006). *The NEO PI-R Guide to Interpretation and Feedback in a Work Context*. Oxford, Hogrefe UK
Prieler, J.A. (2003). *Predicting Driver Behaviour in Uses and Abuses of Intelligence*. New York, Royal Fireworks Press

Chapter 22

Identification of Barriers to and Facilitators for the Implementation of Occupational Road Safety Initiatives

Tamara Banks and Jeremy Davey

Introduction

Identifying the restraining forces for opposing a safety initiative and the driving forces for accepting a safety initiative allows strategies to be implemented to maximise the effects of the facilitators, while minimising the effects of the barriers. Previous research conducted in the area of organisational change has identified that organisational, financial and professional factors may act as barriers to, or facilitators for, implementing change programmes (Blake et al., 2006; Weiner et al., 2008). Several studies have been conducted to explore the barriers to and facilitators for implementing safety initiatives.

For example, questionnaire research conducted on a sample of 115 Turkish food businesses identified seven barriers to implementing food safety programmes. These barriers included: a lack of understanding of the system; the safety system being too complicated; lack of time; high staff turnover; lack of employee motivation; complicated terminology; and a lack of personnel training (Bas et al., 2007).

Research conducted in the United Kingdom to explore barriers to implementing initiatives aimed at tackling musculoskeletal disorders identified seven additional barriers. These included: employee resistance to changing their behaviour; difficulties in obtaining senior managerial authorisation for changes; managers' lack of appreciation for the importance of health and safety initiatives; insufficient resources; difficulty finding appropriate space and equipment; industrial relations issues; and prioritisation of production over safety (Whysall et al., 2006).

A prioritisation of production over safety has also been identified as a barrier in the automobile manufacturing (Clarke, 2006) and agricultural industries (Australian Safety and Compensation Council, 2006). Other barriers to adopting safe behaviours and protective equipment in the Australian agricultural industry include farmers' current attitudes to safety. More specifically there was a general view that injuries were a normal and accepted part of farming operations. There was also a reluctance to accept safety standards imposed by health and safety personnel, as farmers believed this undermined autonomy and the farmers perceived that they had sufficient experience and common sense to manage their own practices (Australian Safety and Compensation Council, 2006).

Resistance to change, particularly by more experienced staff, has also been identified as a major barrier to the implementation of safe practices in health care (Blake et al., 2006). Other barriers to adopting safe behaviours in the United States Health Care industry include: mistrust and fear of punitive outcomes; poor communication between departments; time constraints; and use of contract staff (Blake et al., 2006).

In relation to facilitators, research conducted in the United Kingdom to explore factors that enhanced the implementation of initiatives aimed at tackling musculoskeletal disorders identified four facilitators. These included: supportive managers; change in management which had prompted a review of practices and prompted action; good communication between management and workers; and localised control over health and safety budget spending (Whysall et al., 2006).

The importance of management commitment has also been identified as a facilitator in two additional studies. Firstly, management commitment has been recognised as a facilitator for the implementation of safe practices in health care (Blake et al., 2006). Other facilitators for adopting safe behaviours in the United States Health Care industry include: regular audits and feedback to reinforce compliance; education aimed at raising awareness; confronting resistance and discussing expectations; presence of a change champion; staff involvement in implementation; external pressure to enhance safety; and presence of safety reminders such as posters (Blake et al., 2006).

Secondly, management commitment has been recognised as a facilitator for the implementation of incentive programmes (Wilde, 1994). Other conditions believed to maximise the effectiveness of incentive programme efficacy included: simple rules; attractiveness of the rewards; attainability; short incubation period; staff involvement; and rewarding all levels of the organisation (Wilde, 1994).

While these studies provide some guidance, little is known about whether the facilitators and barriers observed in industries such as hospitality and agriculture can be generalised to the to implementing of occupational road safety initiatives. Previous researchers have proposed the following barriers to managing occupational road safety in Australia and the United Kingdom: limited interaction between fleet managers and occupational health and safety personnel; perceived lack of resources; limited status/authority of the person primarily responsible for managing fleet safety; operational procedures and structures; lack of senior management commitment; reactive focus on injury prevention; and claims-led rather than safety-led procedures (Haworth and Senserrick, 2003; Haworth et al., 2008; Murray et al., 2001). To assist practitioners in understanding conditions that may influence the effectiveness of occupational road safety initiatives, this chapter will explore both the barriers to, and the facilitators for, accepting and implementing occupational road safety initiatives.

Method

Interviews were conducted with 24 participants sourced from four Australian organisations. Participants from within each organisation comprised four front line employees and two managers. Given the real-world context of this qualitative study, the selection of participants was a convenience sample with care taken to ensure that the participants selected were representative of each organisation's driving workforce. Participants ranged in age from 24 to 58 years. As a majority of the drivers within the researched organisations were male, 87 per cent of the employees selected to participate in this study were male. All participants reported regularly driving a vehicle for occupational purposes.

Interview questions were developed based on adaptations to questions used in previous research pertaining to implementing initiatives aimed at managing musculoskeletal disorders (Whysall et al., 2006). The questions asked in this study were:

- Has your company already taken any action that you know of to reduce work-related road safety risk? What and when?
- Is your company currently taking any actions that you know of to maintain work-related road safety within your company? What?
- What do you think are the main barriers or difficulties experienced when making or attempting to make safety changes in your organisation?
- If applicable, how have (or how could) these barriers be overcome?
- What do you think are the main facilitators or things that have helped in implementing safety changes in your organisation?
- What have been the outcomes (actual and perceived) of the safety changes that have been made?
- What do you think were the main reasons for this outcome?

Interview data were transcribed verbatim and then analysed using a three-phase approach as described by Miles and Huberman (1994). Firstly, data were organised via cutting and pasting material into meaningful collections that corresponded with the interview questions. Secondly, emerging themes were identified and patterns within and between themes were explored. This phase involved summarising the data under each theme and selecting verbatim quotes to illustrate the themes. Thirdly, conclusions were drawn after interpretations of the data were verified against the interview transcripts and existing literature.

Results

Themes are presented in order of their strength. The strength of a theme was determined by the number of interview participants that presented information pertaining to that theme.

Barriers

Key themes that emerged as perceived barriers to implementing occupational road safety initiatives included: prioritisation of production over safety; complacency towards occupational road risks; diversity; insufficient resources; limited employee input in safety decisions; and a perception that road safety initiatives were an unnecessary burden.

Nine participants cited conflicts in priorities between production and safety as a barrier to managing occupational road risks. Two sub-themes emerged in this theme. These included mixed messages from management and self-imposed production pressure. In regards to mixed messages from management, participants from all four organisations described how their managers conveyed that safety was the highest priority. However, the high workloads set by managers encouraged drivers to focus more on production targets. Several employees described their frustration with attempting to manage the juggling act of meeting high production and safety goals. For example one front line employee stated that:

> safety changes normally impact with a negative. Normally, changes will be to not drive as long or far, but increased work loads always conflict. All layers of management are aware of the situation about staff shortages and extra distances to travel. I believe the company does have a commitment to driver safety but is willing to overlook its own policy when it comes to a situation of resources and money.

In regards to self-imposed production pressure, some staff reported being highly committed to their work and at times believed that their work commitment motivated them to engage in risky driving practices. For example, one manager reported how she chose to attend a late work appointment even though it meant that she would have to engage in night driving which she acknowledged was becoming challenging for her due to her degenerating vision.

Nine participants cited complacency towards occupational road risks as a barrier to the intervention process. Although participants acknowledged that there were risks involved with driving, they believed that most people accepted these risks. One manager hypothesised that the high risk tolerance observed in his area was probably due to the risk being considered as '*more of a chronic risk rather than a catastrophic one*'. An indifference towards road risks was reported by both front line employees and managers, for example '*All they did was reverse into a post, you know. They didn't hurt anybody.*' Several participants commented on how occupational road safety was not treated as seriously as other safety issues in their organisation. Employees cited complacency as a barrier when describing reasons for non-compliance with an occupational road risk management process. For example one employee commented '*You are supposed to write down your own hazards... and no-one classes driving as a significant hazard.*' Managers also believed that complacency was a hurdle that needed to be overcome when implementing road safety initiatives. In relation to rolling out a planned initiative, the manager anticipated that due to employee complacency '*I think people won't take notice of it.*'

Seven participants cited diversity across their organisation as a barrier to managing occupational road risks. Two sub-themes emerged in this theme. These included diversity within the organisation and diversity in vehicle ownership. In regards to diversity within the organisation, participants commented that differences between geographical regions and services provided created difficulty in producing risk-management strategies that would be applicable organisation-wide. For example '*it's probably not as applicable to other divisions because we transport our kids; whereas no-one else transports their clients.*' In regards to diversity in vehicle ownership, two participants perceived that the use of private vehicles for occupational purposes presented challenges in applying consistent risk management practices as the organisation did not have as much authority over the use of these vehicles as they did fleet vehicles.

Six participants cited the cost of managing occupational road risks as a barrier. Participants reported that there was only a very limited budget, or in some organisations no budget, for managing occupational road risks. Managers believed that limited resources had delayed the implementation of initiatives. Several managers also reported pressure to justify initiatives in terms of the cost versus potential outcomes. One manager described his experience in trying to gain approval from senior management to implement a $10,000 driver training programme as '*I need to strongly convince my manager. He's a good manager, but he's an accountant.*'

Five participants cited limited involvement in decision-making as a barrier to accepting risk management initiatives. Front line employees reported a feeling of 'us' versus 'them'. Management and administrative personnel were perceived to make knee-jerk decisions that were not appropriate for operational staff. Employees from all of the organisations reported that many decisions were made without consulting them and that this had resulted in ineffective safety initiatives. For example, one front line employee reported how he perceived that a management decision to reduce vehicle loading weight had made his job harder and less safe. He stated that '*since the weight has been off them, it's slipping all over the shop*'.

Five participants cited perceptions of unnecessary burdens as a barrier. Participants reported that there was a mentality of '*it was ok last time*' and that this made it difficult to change people's habits and work routines. Senior managers described how middle managers often perceived occupational road safety initiatives as '*another burden, another thing that they've got to try and fit in their roster, fit in their budget*'. Some front line employees expressed concerns relating to additional work associated with engaging in safe practices. For example one participant stated that it '*ends up taking longer to accomplish tasks*'.

Seven additional factors were also perceived to hamper the implementation of risk management initiatives. As each of these factors was only cited once, only a brief description is provided for each barrier. The factors were: limited data systems; quality vehicles; lack of knowledge; reduced client rapport; perceptions of initiative effectiveness; change fatigue; and hierarchical organisational structure.

One manager perceived that the limited incident data system in use in his organisation restricted his ability to identify and monitor high risk employees and vehicles. He noted that the incident data that was currently collected was collected for insurance purposes. He believed that this was a problem because it was not detailed and it did not include incidents that cost less than the insurance premium to repair or were unlikely to receive an insurance payout for example '*if it's the baseball bat on the car*'.

One employee perceived that the high quality of vehicles supplied by his organisation tempted him to drive at '*160km on a straight, flat road in the middle of nowhere because you can*'. He commented that historically it felt unsafe driving an old Ute on a dirt road, but with the improvements in road surface and vehicle designs over the years, he now felt safer driving and this encouraged him to operate vehicles at higher speeds. He reported that '*it's not because you want to go faster. It's just because it feels so slow. It feels so agonisingly slow to sit on 130 on a highway in an SV6.*'

One manager recognised knowledge deficiencies as a barrier. This manager believed that the health and safety officers were interested in managing occupational road risks but believed that they did not '*have any knowledge of road safety or fleet safety*'.

Another employee described how his team had resisted the implementation of cargo barriers between the front and rear seats due to a fear of reduced client rapport. He explained that while his team recognise that management was fitting the barriers to protect them from violent clients, the staff resisted the implementation because they perceived that '*it would be an impediment to either their conversation or their relationship with clients*'.

One manager believed that people's perceptions of initiative effectiveness could be a barrier. She commented that '*people think if you're going to do something about road safety it's driver training ... It is what makes sense to the lay person*'. She viewed this as a barrier to implementing other road risk management strategies such as '*policies and induction*' because staff only wanted driver training and resisted other initiatives because they did not recognise their value.

Another manager believed that employees in his organisation may resist change because they are tired of constantly having to change their practices and processes. He believed that people

> just want things to stay the same for three days in a row ... they just need to get their job done and they've got to stop being told how they can do it better and just have some time to actually do the work.

Finally, an employee believed that the hierarchical structure in his organisation slowed down the implementation of risk management initiatives. He commented that there's so many chiefs that if you say something to this fellow and he agrees – he's got to go through ten chiefs before he gets to the next – to the big chiefs and by that time it's all changed and it's not worth shit any way. The whole place is just too top heavy.

When asked how the perceived barriers could be overcome, participants provided several suggestions. These included: having a well-designed overall change in management approach; increasing employee involvement in decision-making; reducing work demands; and increasing awareness and knowledge of occupational road risks.

Facilitators

Three key themes emerged as perceived facilitators for implementing occupational road safety initiatives. These included: management commitment; the presence of existing systems that could support the implementation of initiatives; and supportive relationships.

Seven participants cited management commitment as a facilitator in the intervention process. Participants believed managers were committed to driver safety and described how this had assisted in the implementation of occupational road safety initiatives. For example one manager commented '*if you have your executive team on board, it just happens*'.

Four participants recognised the presence of existing systems as a facilitator. Systems perceived to be beneficial in implementing road safety strategies included communication systems and risk-management systems. Participants believed that the presence of good communication structures allowed safety messages to be easily conveyed to staff. Other participants discussed how the implementation of initiatives could be enhanced by drawing upon the risk-management framework already operating within their organisation. For example a manager noted that '*the risk-management paradigm is part of our core business*'. Participants believed that these existing systems could facilitate the implementation of road safety initiatives.

Three participants cited supportive relationships as a facilitator in the intervention process. For example one front line employee described how the supportive culture within his organisation provided a positive environment that could be tapped into when implementing initiatives. He described the presence of '*a community kind of atmosphere in work. So we all look after each other.*' He believed that this '*kind of mutual responsibility feeling*' would facilitate employees embracing road safety initiatives that involved employees looking out for each other.

Four additional factors were also perceived to assist in the implementation of risk-management initiatives. As each of these factors was only cited once, only a brief description is provided for each facilitator. These included: autonomy;

community road safety campaigns; openness to change; and a culture of accountability and governance.

One manager perceived that the smaller organisation he now worked in allowed him the autonomy to implement initiatives rapidly. Unlike in a previous organisation that he worked for, '*that required going to the media and marketing people and doing this and that and it would take a week*', in his current organisation he is now able to write driving safety messages and publish them the next day. He commented on how '*things happen a lot quicker and I have much more control, more responsibility and I see that as a definite plus*'.

One employee believed that community road safety campaigns provided a good foundation for her organisation to build upon when implementing safety initiatives. She commented that '*the road signs have helped out there ... They put big signs up saying 'passenger, is your driver alert?' So they're good. They actually make you think 'oh, wow, keep alert*'.

Openness to change was cited as a facilitator. One employee believed that many employees in his area embraced change. He commented that '*I think people just look forward to the opportunities that change brings. So they pretty much grab onto anything.*' He believed that this open attitude towards change would definitely assist with the implementation of risk-management initiatives.

Finally, one manager perceived that the culture of accountability and governance within his organization was beneficial. He perceived that '*if somebody at a very high level says it has to happen, then it happens. Or if there's a policy released with regards to something, then it generally happens.*' He believed that this culture would facilitate the implementation of road safety initiatives that were endorsed by senior management.

Discussion

All of the key themes that emerged as perceived barriers to implementing occupational road safety initiatives in this study have been identified as barriers in previous studies, except for diversity. More specifically, prioritisation of production over safety has been identified as a barrier when implementing risk management strategies targeting musculoskeletal disorders (Whysall et al., 2006), safe automobile manufacturing (Clarke, 2006) and safe agricultural practices (Australian Safety and Compensation Council, 2006). Complacency has been identified as a barrier in the agricultural industry, where farmers view injuries to be a normal and accepted part of farming operations (Australian Safety and Compensation Council, 2006). Insufficient resources have been identified as a barrier to implementing a range of health and safety initiatives (Bas et al., 2007; Blake et al., 2006; Haworth et al., 2008; Whysall et al. 2006). Limited employee input in safety decisions has been identified as a barrier in the agricultural industry, where farmers reported a reluctance to accept safety standards imposed by safety personnel as they perceived that they had sufficient experience and common sense to manage their own practices (Australian Safety and Compensation Council, 2006). Finally, perceptions that safety initiatives are an unnecessary burden has been identified as a barrier when implementing risk-management strategies targeting musculoskeletal disorders (Whysall et al., 2006).

Three of the less cited themes that emerged as perceived barriers to implementing occupational road safety initiatives in this study have been identified as barriers in previous studies. More specifically, limited data systems have been identified as a barrier to managing occupational road risks. Murrray and colleagues (2001) identified barriers to managing risks when incident data are collected for insurance claims purposes rather than risk-management purposes. Lack of knowledge has been identified as a barrier in the hospitality industry. Bas and colleagues (2007) identified employees' lack of understanding of a new food safety system as the main barrier to implementing a new international strategy to reduce food-borne disease. Finally, difficulties in obtaining senior managerial authorisation for changes has been identified as a barrier when implementing risk management strategies targeting musculoskeletal disorders (Whysall et al., 2006).

Of the seven themes that emerged as perceived facilitators to implementing road safety initiatives in this study, only three themes have been identified as facilitators in previous studies. These included: management commitment; the presence of existing systems that could support the implementation of initiatives; and autonomy. More specifically, management commitment has been identified as a facilitator to implementing risk-management strategies targeting: musculoskeletal disorders (Whysall et al., 2006); safe practices in health care (Blake et al., 2006); and a range of safety behaviours through the use of incentives (Wilde, 1994). The presence of existing systems that could support the implementation of initiatives, including good communication systems between management and workers, has been identified as a facilitator to implementing risk-management strategies (Whysall et al. 2006). Finally, autonomy in regards to localised control over health and safety budget spending has also been identified as a facilitator to implementing risk-management strategies in previous research (Whysall et al. 2006).

This research has important implications for academics and practitioners. For academics, the authors suggest that the findings from the current study may be applied to assist in the understanding of research results pertaining to the effectiveness of initiatives. It is hypothesised that the organisational environment including the presence of any change barriers or facilitators may influence initiative outcomes. To assist in the interpretation of study findings, future researchers may briefly acknowledge the presence of any change barriers or facilitators that may impact upon the results they obtain. The authors also suggest that future studies expand upon this exploratory research by applying the same methodology with a more diverse sample to determine if the barriers and facilitators identified in this study generalise to organisations operating in different industries and in different countries.

For practitioners, when implementing occupational road safety initiatives, the authors suggest that considerations be given to managing the key barriers identified in this study. These included: prioritisation of production over safety; complacency towards occupational road risks; insufficient resources; diversity; limited employee input in safety decisions; and a perception that road safety initiatives were an unnecessary burden. The authors also suggest that practitioners consider proactively developing and utilising the key facilitators identified in this study. These included: management commitment; the presence of existing systems that could support the implementation of initiatives; and supportive relationships.

References

Australian Safety and Compensation Council (2006). *Beyond Common Sense: Report on the Barriers to Adoption of Safety in the Agriculture Industry.* Canberra, Commonwealth of Australia.

Bas, M. et al. (2007). Difficulties and barriers for the implementation of HACCP and food safety systems in food businesses in Turkey. *Food Control*, 18, 124–130.

Blake, S. et al. (2006). Facilitators and barriers to 10 national quality forum safe practices. *American Journal of Medical Quality*, 21, 323–334.

Clarke, S. (2006). Safety climate in an automobile manufacturing plant: The effects of work environment, job communication and safety attitudes on accidents and unsafe behaviour. *Personnel Review*, 35(4), 413–430.

Haworth, N. et al. (2008). *Improving Fleet Safety – Current Approaches and Best Practice Guidelines.* Sydney, Australia, Austroads.

Haworth, N. and Senserrick, T. (2003). *Review of Fleet Safety and Driver Training: Current Practices and Recommendations.* Unpublished report prepared for the Department of Treasury and Finance, Monash University Accident Research Centre.

Miles, M.N. and Huberman, A.M. (1994). *An Expanded Sourcebook. Qualitative Data Analysis.* London, Sage.

Murray, W. et al. (2001) Overcoming the barriers to fleet safety in Australia. In *2001 Road Safety Research, Policing and Education Conference, 18–20 November 2001, Melbourne, Victoria.*

Weiner, B. et al. (2008). Conceptualization and measurement of organizational research and other fields readiness for change: A review of the literature in health services. *Medical Care Research Review*, 65, 379–436.

Whysall, Z. et al. (2006). Implementing health and safety interventions in the workplace: An exploratory study. *International Journal of Industrial Ergonomics*, 36(9), 809–818.

Wilde, G.J. (1994). *Target Risk: Dealing with the Danger of Death, Disease and Damage in Everyday Decisions.* Toronto, PDE Publications.

PART 5
Human Factors and Driver Attention

Chapter 23

An Observational Survey of Driving Distractions in England

Mark J.M. Sullman

Introduction

The issue of driver distraction is an important one, with the National Highway Traffic Safety Administration (NHTSA) estimating that driver inattention contributes to 25–50 per cent of all crashes (NHTSA, 1997). Although there is some debate about the definition of driving distraction, one common-sense definition provided by Ranney (1994) is any secondary activity that draws the motorist's attention away from the main task of driving. Whilst driving, there is a very large number of potential distractions, with one of the most obvious being a mobile phone. There has been an extensive amount of research on the impact of using a mobile phone while driving, which is thought to increase the risk of crash involvement by at least fourfold (Redelmeier and Tibshirani, 1997; Violanti and Marshall, 1996).

Although there is evidence to suggest that drivers are aware that using a mobile phone while driving is dangerous (Gras et al. 2007; Sullman and Baas, 2004), and is illegal in many countries, many drivers continue to use a hand-held mobile phone while driving. For example, research in New Zealand using survey methodology (where using a hand-held mobile phone while driving is not illegal) found that over 57 per cent of the drivers surveyed used a hand-held mobile phone at least 'occasionally' while driving (Sullman and Baas, 2004). Similarly, a study conducted in Spain, where this behaviour is illegal, found that 60 per cent of the drivers surveyed reported using a hand-held mobile phone while driving (Gras et al., 2007). Further commonality between these two studies can be shown in that mobile phone use was higher amongst younger drivers. Although Sullman and Baas found that males engage in this behaviour more often than females, this finding was not replicated in the Spanish research. One criticism of these studies is that they were based on self-report, rather than observations of actual behaviour.

Unfortunately there have been very few naturalistic studies of driving distraction and those which do exist have mostly been focused on use of a hand-held mobile phone while driving. For example, Horberry et al. (2001) investigated the use of a hand-held mobile phone by drivers in Perth (western Australia). They found that overall 1.5 per cent of the observed road users in Western Australia were using a hand-held mobile phone while driving. They also found that those observed using a mobile phone were more likely to be male and younger. Interestingly they did not observe any differences according to the time of day. A number of these findings were also replicated by another Australian study, which was conducted in Melbourne (McEvoy et al. 2007). McEvoy and colleagues found that 1.6 per cent of the observed drivers were using a hand-held mobile phone and that the rate was higher amongst males and younger drivers. However, in contrast to Horberry and colleagues research, they found that mobile phone use was higher in the morning than in the afternoon.

The observational studies which have looked broadly at driving distraction have both used in-vehicle cameras in a relatively small number of vehicles to collect their data. For example, in one of the only peer reviewed studies to look more broadly at the issue of driving distraction in the general population, Stutts et al. (2005) installed cameras into 70 cars in America. They found that driving distraction was a common occurrence in everyday driving and that although using a mobile phone while driving was one of the many distracters; it was not even in the top five in terms of frequency of engagement. They found that the most frequently observed distractions were eating and drinking, followed by distractions in the vehicle (reaching for an object or manipulating controls). However, although this research gives exceptionally good in-depth data on driving distractions, it is difficult to generalise these results to the driving population, particularly due to the small sample size.

A second study which investigated driver distraction in a naturalistic setting was conducted amongst long-haul truck drivers (Hanowski et al., 2005). This again used in-vehicle cameras and was used to collect data from 41 long-haul truck drivers using two instrumented trucks. However, they were only interested in looking at critical incidents and the distractions which lead up to the critical incidents. Therefore an analysis of the prevalence of driving distractions amongst these professional drivers was not an aim of the research.

As these two studies are the only published research which measures driving distractions in a naturalistic manner, and only one of these included private vehicle drivers, much more research is needed. Therefore, the present research set out to investigate the proportion of UK drivers who engage in a secondary (distracting) activity whilst driving. The research also investigated the type of secondary task and whether there were any differences according to age and gender, along with time of the day and day of the week.

Method

Procedure

The data were collected via roadside observation using a clipboard, form and pen. The observer noted every vehicle that drove past and whether they were engaged in a secondary activity in addition to driving. Each of the observational sessions lasted for 60 minutes and a total of 20 hour-long observations took place in St

Albans (Hertfordshire). The approximate age of the driver, gender, time of day and day of week were also measured.

The observer was positioned so that the cars and the drivers were clearly visible, while at the same time being as unobtrusive as possible. In most cases the observer was not visible, in advance, to the motorist. On all occasions the traffic on the same side of the street (as the observer) coming towards the observer was observed. The observed secondary tasks had to fall into the operational definitions to qualify as a distracter. These definitions were established in advance using the distractions identified by previous research and were refined following pilot testing.

Definitions

> Primary task only – the motorist was engaged in driving only.
> Mobile phone use – the driver was holding a phone to the ear in a clearly visible manner.
> Texting/keying numbers – the driver was clearly holding the mobile phone and pushing the keys in a manner to send a text message or dial numbers.
> Drinking – the driver was holding or drinking some form of beverage in a clearly visible manner.
> Eating – the driver was holding or eating some form of food in a clearly visible manner.
> Other – all other secondary activities which did not fit into one of the previously mentioned categories. This included such things as: talking to a passenger, map reading, using a telecommunication device (other than a mobile phone) or reaching for something or adjusting controls (e.g. heater, radio).

Locations

The observational sites were selected with the aid of an online random number generator. In the first step of this procedure a map of St Albans was obtained and every street within the city limits was given a number. Following this ten random numbers were generated using an online random number generator and these were matched to the corresponding road on the map.

Timing

All ten sites were observed twice, firstly from 7a.m. to 8a.m. and secondly from 2p.m. to 3p.m. All observations were undertaken on a Monday or Friday. As the observations took place in the British spring/summer, the daylight and weather conditions allowed a clear view of the drivers on all occasions.

Results

In total 12,214 drivers were observed, with 5.5 per cent of those observed to be undertaking a secondary activity whilst driving. The most frequently observed secondary task was using a mobile phone, which was observed in 2.6 per cent of the cases (2.2 per cent talking and 0.4 per cent keying numbers or texting). Smoking was the next most commonly observed secondary activity (0.9 per cent), followed by eating (0.8 per cent) and drinking (0.6 per cent). The 'other' category contained the remaining 0.5 per cent.

Table 23.1 shows that the majority of both genders were engaged in the primary task (driving) only, but also that there were fewer males (4.8 per cent) than females (6.2 per cent) who were engaged in a secondary activity. A chi-squared test revealed that these differences were statistically significant X^2 (1, 12214) = 11.1, $P < 0.001$.

Table 23.1   Driving only vs involvement in a secondary task

Gender		Driving	Secondary	Total
	Count	6183	314	6497
Male	Expected	6141	356	6497
	% within gender	95.2	4.8	100
Female	count	5362	355	5717
	Expected	5404	313	5717
	% within gender	93.8	6.2	100
Total		**11545**	**669**	**12214**

A more detailed breakdown revealed that the most frequently occurring secondary task was use of a mobile phone, for both genders (Table 23.2). The second most frequently observed secondary activity among males was smoking, while for females it was eating. However, these differences were not significantly different (X^2 (5, 669) = 8.44, $P = 0.134$).

Table 23.3 shows the number of drivers who were observed driving only, versus those observed engaging in a secondary activity for each of the three age groups. This shows that overall there were more younger drivers engaged in a secondary activity than expected (264 vs 212). A similar difference was also observed in the middle age group (354 vs 293), while the older age group was much less frequently observed engaged in a secondary activity than expected (51 vs 164). These differences were statistically significant, X^2 (2, 12213) = 109.31, $P < 0.001$.

Table 23.2   Secondary activity while driving, by gender

		\multicolumn{6}{c}{Secondary task}	Total
		Phone	**Text**	**Eat**	**Drink**	**Smoke**	**Others**	
Male	Observed	132	18	40	28	60	36	314
	Expected	127.7	21.1	48.3	32.9	52.6	31.4	314
Female	Observed	140	27	63	42	52	31	355
	Expected	144.3	23.9	54.7	37.1	59.4	35.6	355
Total		**272.0**	**45.0**	**103.0**	**70.0**	**112.0**	**67.0**	**669**

Table 23.3   Driving only vs involvement in a secondary task, by age group

Age group (years)		**Driving**	**Secondary**	**Total**
<30 years	Observed	3605	264	3869
	Expected	3657.1	211.9	3869
30–50 years	Observed	4996	354	5350
	Expected	5056.9	293.1	5350
>50 years	Observed	2943	51	2994
	Expected	2830	164	2994
Total		**11544**	**669**	**12213**

In order to reach the minimum required cell count the two mobile phone secondary tasks (phoning and texting) had to be combined. The observed secondary activities in the older drivers (> 50 years old) were considerably lower than expected in all cases, which was particularly notable in the use of a mobile phone (Table 23.4). Conversely, the observed engagement in secondary activities for the young drivers was considerably higher than expected in all cases except for the 'other' category. The differences between the age groups was statistically significant, X^2 (2, N = 12214) = 51.00, $P < 0.001$.

Table 23.5 shows that there appeared to be more engagement in secondary activities in the afternoon than would be expected and less during the morning. However, a chi-square test revealed that the difference between the morning and afternoon observation times was not statistically significant (X^2 (1, 12214) = 2.68, $P = 0.102$).

Table 23.4   Secondary activity while driving, by age group

Age group (years)		Secondary task					
		Mobile	Eat	Drink	Smoke	Other	Total
Young <30 years	Observed	140	43	26	40	15	264
	Expected	125.1	40.6	27.6	44.2	26.4	264.0
Middle 30–50 years	Observed	171	50	37	61	35	354
	Expected	167.7	54.5	37.0	59.3	35.5	354.0
Older >50 years	Observed	6	10	7	11	17	51
	Expected	24.2	7.9	5.3	8.5	5.1	51.0
Total		317	103	70	112	67	669

Table 23.5   Driving only vs involvement in a secondary task, by time of day

Time of day		Driving	Secondary	Total
Morning	Observed	6173	336	6509
	Expected	6152.5	356.5	6509
Afternoon	Observed	5372	333	5705
	Expected	5392.5	312.5	5705
Total	**Count**	**11545**	**669**	**12214**

Table 23.6 shows that regardless of the time of day, using a mobile phone to talk was the most common distraction, followed by smoking and eating. There were also several notable differences, such as texting and 'other', being higher in the afternoon, while phoning and eating were higher in the morning. Interestingly these differences were statistically significant (X^2 (5, 669) = 25.3, $P < 0.001$).

Discussion

The present research found that although the vast majority were engaged in the primary task (driving) only, a substantial proportion of drivers (5.5 per cent) were also engaged in a secondary activity when observed.

In terms of engagement in specific secondary tasks, the present study found that using a mobile phone was the most commonly observed distraction (2.6 per cent), with speaking on a hand-held phone being much more common (2.2 per

Table 23.6   Secondary activity while driving, by time of day

Time of day				Secondary task				
		Phone	Text	Eat	Drink	Smoke	Others	Total
Morning	Observed	142	11	59	41	61	22	336
	Expected	136.6	22.6	51.7	35.2	56.3	33.7	336
Afternoon	Observed	130	34	44	29	51	45	333
	Expected	135.4	22.4	51.3	34.8	55.7	33.3	333
Total	**Count**	**272**	**45**	**103**	**70**	**112**	**67**	**669**

cent) than texting or keying in numbers (0.4 per cent). This is a worrying finding, as it shows that not only are a substantial proportion of the drivers ignoring the law, but also that they are putting themselves and others at an increased risk of crash involvement. This needs to be addressed by some means, such as education and enforcement.

Unfortunately, although it is not possible to compare these findings with those of previous research using in-vehicle cameras, there were some similarities noted. For example the present study found that eating and drinking were amongst the most commonly observed secondary behaviours, as was the case in the research by Stutts et al. (2005). However, the finding that mobile phone use was the most common secondary activity is not in agreement with the findings of Stutts and colleagues. This could be due to differences in the methodology, such as the sample size (70 vs 12,214), data collection method (in-vehicle cameras vs roadside observation), research design (longitudinal vs cross-sectional) or even the country (USA vs UK).

The rate of mobile phone use found in the present study (2.6 per cent) is also slightly higher than the 1.5 per cent found in Perth (Horberry et al., 2001) and the 1.6 per cent found in Melbourne (Taylor et al., 2007). There could be a number of reasons for this, such as observer differences, differences in traffic culture, or even differences in the level of enforcement or education. Future research should be undertaken in other areas of the UK to confirm whether there is in fact a higher level of mobile phone use amongst UK drivers.

The present research also found that females were significantly more likely to be undertaking a secondary task than males. Although the sample size of Stutts et al. (2005) was too small to test for gender differences in most cases, in agreement with the present research they found that females were more likely to engage in two of the distractions they measured (grooming and attending to things outside the vehicle). However, these findings are in contrast to those from the Australian studies, which both found males were more likely to use a mobile phone while driving. However, the two Australian studies were both only comparing mobile phone use by gender, whereas the present study also measured several other types of secondary tasks. Nevertheless, if we look solely at the raw numbers, it seems that if there were any real differences it would be the reverse of what was found in Australia. Future research should investigate whether this was simply an aberration or whether females in the UK are more likely to use a mobile phone than males.

As was found in the two Australian studies (Horberry et al., 2001; Taylor et al., 2007) the present study found that younger drivers were more likely to be engaged in a secondary activity than older drivers. This finding also supports data obtained using self-report surveys (e.g. Pöysti et al., 2005; Sullman and Baas, 2004). The higher rate amongst young drivers is particularly concerning as they already have a higher risk of crash involvement than older drivers. Therefore targeting educational and enforcement resources specifically at younger drivers appears to be necessary.

The level of distracted drivers observed here is likely to be an underestimate of the true level of driver distraction, as only visible distractions could be observed. Therefore, any distractions which did not involve a visible action (e.g. talking on a mobile phone using a hands-free device) were not recorded.

Another limitation of this study is that in some cases the driver may have noticed the observer and changed their driving behaviour. However, as the observer was as unobtrusive as possible and in most cases was not likely to have been noticed by the driver, this is unlikely to have greatly affected the data. It should also be noted that the estimation of age was very difficult as it was very subjective and most prone to error. Therefore the age groups presented here should be regarded as approximate only.

References

Gras, N.E., Cunill, M., Sullman, M.J.M., Planes, M. and Font-Mayolas, S. (2007). Predictors of seat belt use amongst Spanish drivers. *Transportation Research Part F*, 10, 263–269.

Hanwoski, R.J. et al. (2005). Driver distraction in long-haul truck drivers. *Transportation Research Part F*, 8, 441–458.

Horberry, T. et al. (2001). Drivers' use of hand-held mobile phones in Western Australia. *Transportation Research Part F*, 4, 213–218.

McEvoy, S. et al. (2007). The contribution of passengers versus mobile phone use to motor vehicle crashes resulting in hospital attendance by the driver. *Accident Analysis and Prevention*, 39, 1170–1176.

NHTSA (National Highway Traffic Safety Administration) (1997). *An Investigation of the Safety Implications of Wireless Communications in Vehicles*. Report DOT HS 808-635. NHSTA, Washington, DC.

Pöysti, L. et al. (2005). Factors influencing the use of cellular (mobile) phone during driving and hazards while using it. *Accident Analysis and Prevention*, 37, 47–51.

Ranney, T.A. (1994). Models of driving behavior a review of their evolution. *Accident Analysis and Prevention, 26*, 733–750.

Redelmeier, D.A. and Tibshirani, R.J. (1997). Association between cellular telephone calls and motor vehicle collisions. *New England Journal of Medicine*, 336, 453–458.

Stutts, J.C. et al. (2005). Drivers' exposure to distractions in their natural driving environment. *Accident Analysis and Prevention*, 37, 1093–1101.

Sullman, M.J.M. and Baas, P. (2004). Mobile phone use amongst New Zealand drivers. *Transportation Research Part F*, 7, 95–105.

Taylor, D.M.D., MacBean, C.E., Das, A. and Rosli, R.M. (2007). Handheld mobile telephone use among Melbourne drivers. *The Medical Journal of Australia*, 187(8), 432–434.

Violanti, J.M. and Marshall, J.R. (1996). Cellular phones and traffic accidents: An epidemiological approach. *Accident Analysis and Prevention*, 28, 265–270.

Chapter 24

Calibration of an Eye-tracking System for Variable Message Signs Validation

M. Claudia Guattari, Maria Rosaria De Blasiis, Alessandro Calvi
and Andrea Benedetto

Introduction

It is well known that a driver's attention can be distracted by many environmental elements and this feature has to be taken into account in order to define road safety standards.

In the literature, this subject has been tackled from different points of view. In particular many authors have studied driver eye movements in relation to environmental stimuli. Nakayasu et al. (2007) evaluated a tool measuring the driver's visual skill and motor behaviour in various traffic situations. The relationship between eye movement and the response properties for trained, untrained and older drivers was examined. The results showed that in situations where there were few objects to be noted, for example on a straight road, the amount of eye movements was smaller, and fixation durations were longer than in situations where there were many objects, for example a crosswalk. The results also showed that eye movements depend on a driver's experience.

Anttila and colleagues (2000) used the analysis of eye movements to compare visual demand of variable message signs (VMS). Specifically, three VMS types were evaluated: a sign displaying the same messages alternately in Finnish and Swedish (2 seconds in each language), a sign displaying the same messages simultaneously and a sign displaying air and road surface temperatures in Finnish. Data were collected by recording the eye movement of 38 drivers during highway driving. Results showed no significant differences between percentages of subjects who fixed their attention on the sign. The highest percentage (100 per cent) was found for a bilingual sign displaying the messages simultaneously, followed by a bilingual sign displaying the messages alternately (95 per cent), by the temperature display (92 per cent), and by the second temperature display (89 per cent). Martens and Fox (2007) demonstrated that the repeated exposure to the same road environment changes eye movement behaviour. Participants drove a low-cost simulator while their eye movements were recorded. With repeated exposure participants' glances at traffic signs along the route were shorter through having a better recollection of the traffic signs along the route. At the last drive, the priority situation at an intersection was changed. The results of this study are clear:

when people drive along a road they develop expectations about traffic signs. The traffic sign recognition test resulted in better performance for those who drove the road several times compared with people who drove the road only once. Driving speeds for the yield situation were lower, and therefore more adequate, for people who drove the road for the first time compared with people who had driven the road several times.

Some methods have been studied to locate and track eye movements of drivers. For example, Eriksson and Papanikolopoulos colleagues (2000) described a system that tracks the eye movements. The purpose of such a system is to detect driver fatigue. By mounting a small camera inside the car, they can monitor the face of the driver and look for eye movements which indicate if the driver is no longer in a fit condition to drive due to sleepiness as measured by eye closing events. They described how to track eye movement. The system uses a combination of template-based matching and feature-based matching in order to localise the eyes. During tracking, the system is able to recover and resume the proper tracking. The system is also able to check if the eyes are open or closed, and this facilitates the detection of fatigue. To record the eye movements numerical and geometrical models have been implemented. Tock and Craw (1996) described a computer vision system for tracking the eyes of a car driver in order to measure the eyelid separation. These measures are significant to detect when a driver is becoming drowsy. The system consists of a colour CCD camera mounted behind the steering wheel of the car, connected to an advanced system of hardware and software. Although the system must adapt to different drivers in different driving positions, allowing movements of the seat and steering wheel, these remain constant for long periods. The outcome of this study showed that to obtain a correct result the system requires a dark environment. Villanueva et al. (2006) tried to accomplish the first step of a more extensive model and presented a simple expression for pupil orientation based on a framework of physical parameters. The proposed model requires alternative calibration strategies depending on the number of parameters employed to obtain an efficient measure of behaviour. It exhibits lower errors than other mathematical expressions, which normally need more calibration points to construct a competent model. The research models a video-oculographic system and once the model is derived, the work addresses different ways to obtain a simpler and more efficient form. Finally an experiment to validate the system is presented.

Many mathematical approaches to this subject have been proposed. Kawato and Tetsutani (2004), for example, proposed a new algorithm to extract and track the positions of eyes in a real-time video stream. For their application, they could take face images on a fairly large scale. Therefore, they could detect the eyes using blink detection based on the difference between successive images. To allow head movement during blink detection, they implemented a head movement cancellation algorithm, which works quite effectively. To obtain significant results by an eye tracking system the calibration of the software is very important. Morimoto and Mimica (2005) presented a review of eye gaze tracking technology and focused on recent advances that might facilitate its use in general computer applications.

The new solutions simplify or eliminate the calibration procedures and allow free head movement.

In addition, the method to design the simulated scenario is very important. Giannopulu and colleagues (2008) compared subjects' visuomotor strategies on similar video-projected environments in a fixed-base driving simulator. Experienced car drivers are exposed to two visual environments: a real traffic urban scenario prerecorded on video, and the 3D simulation of the same scene. The results indicate that eye movements differ between prerecorded and virtual environment.

The aim of this work is to set the calibration and validation of an eye-tracking system. This system is able to locate the gaze position in the scenario through x–y co-ordinates.

Method

Signs and equipment

Ten different variable message signs (VMS) have been tested (see Table 24.1).

Table 24.1   VMS tested

Variable Message Sign	
= 10 KM	1
FRANCAVILLA → ORTONA	2
ATTENZIONE	3
LAVORI IN CORSO ATTENZIONE	4
FRANCAVILLA ORTONA	5
KM 135 → KM 126	6
CODA PER INCIDENTE IN AUMENTO	7
= 8 KM FRANCAVILLA ORTONA	8
PRUDENZA	9
5 KM →	10

The messages consist of pictures and sentences. The signs were alternately displayed staying on the screen for 5 seconds. In between two messages the screen is black. The images have been implemented on the STI driving simulator (Benedetto et al., 2004) at the Virtual Reality laboratory of the Inter-universities Research Centre for Road Safety (CRISS). The simulator is put in a real vehicle and the images are projected in front of the car and sideways to cover a visual angle of 135°. Furthermore, sound is fed into the engine to reproduce the best acoustic environment.

The eye movements and the images on the screen have been recorded with two high-resolution cameras.

Participants

Twenty-one participants, approximately half male and half female, with a mean age of 25 years old, were recruited as volunteers from the Department of Sciences of Civil Engineering at the University Roma Tre via direct contact. Driving experience was about 5,000 km per year in urban road and 10,000 km per year in rural road. The authors asked drivers to observe carefully one sign at a time, minimising the head movements to maximise the accuracy of the experimental results.

Eye-tracking system

During the simulation the eye movements were tracked by a high-definition camera situated close to the steering wheel of the car and the resultant video is produced via software; the scenario is produced by a camera behind the car. The eye-tracking software extracts the images from the video with a frequency of 35 frames per second (see Figure 24.1).

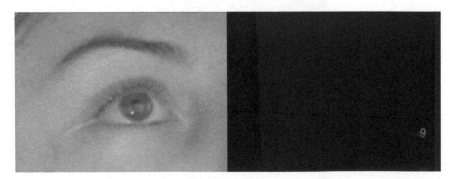

Figure 24.1   Frame extracted with eye tracking software

The software outcomes are the gaze position in the scenario (see Figure 24.2) and x–y co-ordinates of the centre of the eye.

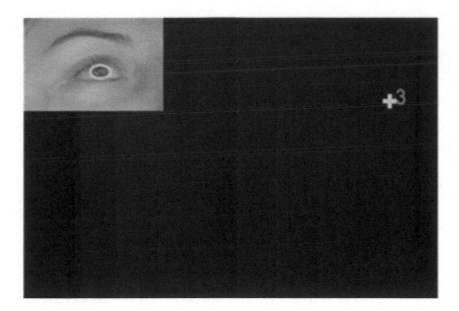

Figure 24.2   Output of eye tracking software (frame by frame)

Study design

The aim of this work is to verify which element in a VMS is not well or immediately understood during driving, once the best set of calibration has been established.

Drivers attended the experiment one at a time. To calibrate the eye-tracking system drivers were requested to look at nine points appearing consecutively on the screen for 2 seconds, in prefixed positions (see Figure 24.3).

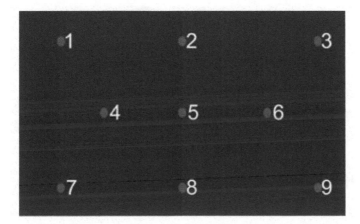

Figure 24.3   Points displayed during calibration

The distances between the centre of the gaze position and each of the nine points that should be focused on have been calculated. The average and the standard variation were calculated too. Comparing these statistics the authors were able to detect threshold values to determine whether or not the calibration is acceptable.

After calibration the ten VMS were shown to the driver one at a time for 5 seconds each, with a 2-second break; during the break the display is black.

Results

The distribution of distances between the gaze focus and the target points displayed showed that when the display's background was lit the calibration result was poor in all the cases. The distributions have been characterised by distances standard deviation > 60 pixels (see Figure 24.4).

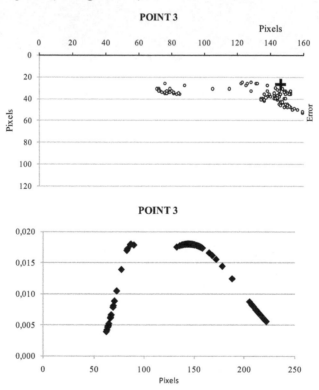

Figure 24.4   Example of not acceptable calibration

However, the distributions of such distances in the experiments with a dark background showed a smaller variance and a standard deviation of ≤ 60 pixels (see Figure 24.5).

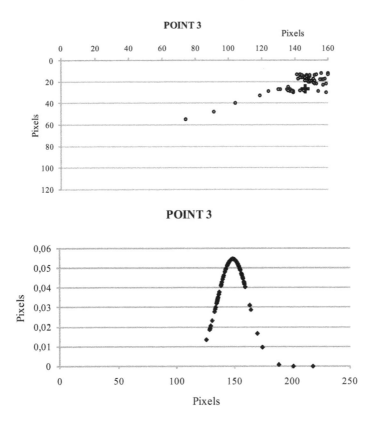

Figure 24.5   Example of acceptable calibration

Assuming such a threshold (60 pixels) for calibration, the percentage of time that the driver focused their attention on the picture or on the words is the same for all the VMS 3, 4, 6, 7, classified as understood; otherwise, the percentage of gaze time was quite different for the VMS1, 2, 5, 8, 9, 10, classified as not understood (see Table 24.1).

Comprehension of the VMS was established in a previous study based on the answers of a sample of drivers to a questionnaire submitted both while driving and at rest. The questionnaire investigated the driver's interpretation of the VMS. The percentage of time spent on the pictures or on the words confirms previous results in terms of VMS interpretation of complexity (see Table 24.2).

For example, in the case of VMS 9, the drivers fixated their gaze for a longer time (61 per cent) on the picture and for a shorter time (39 per cent) on the words. This result seems to confirm the response given by drivers to a questionnaire about the same signs. In this case the driver does not appear to understand the picture very easily. In the case of VMS 10, the comprehension of the words seems to be

Table 24.2   VMS comprehension

Variable message sign	Percentage of comprehension	Percentage of comprehension
	While driving	**At rest condition**
1	43	93
2	21	36
3	100	100
4	100	100
5	29	21
6	100	100
7	100	100
8	29	21
9	19	38
10	29	36

more difficult: drivers focused their gaze on the words for 72 per cent of time and for 28 per cent on the picture. In the case of VMS 3, 4, 6, 7, classified as understood, the drivers focus equally on pictures and text (from 48 to 50 per cent each) (see Table 24.3).

Discussion and Conclusion

Results of the calibration set demonstrated that the environment has to be dark to prevent corneal reflex and the drivers have to avoid free head movement to eliminate system errors.

Finally, it is evident that in order to obtain consistent standards for road safety, the influence of different elements of surrounding road on driver behaviour has to be taken into account in any assessment and virtual reality simulation equipped

Table 24.3   Percentages of time spent on VMS (picture and sentence)

Variable message sign	Time on the picture (%)	Time on the sentence (%)
1	56	44
2	26	74
3	50	50
4	49	51
5	51	49
6	41	59
7	48	52
8	41	59
9	61	39
10	28	72

with an eye-tracking system is an advanced tool able to analyse this particular human factor.

References

Anttila, V. et al. (2000). Visual demand of bilingual message signs displaying alternating text messages. *Transportation Research Part*, F 3, 65–74.

Benedetto, A. et al. (2004). Reliability of standards for safe overtaking: Advances using real time interactive simulation in virtual reality. Proc. 82 *TRB Annual Meeting*, Washington, DC.

Eriksson, M. and Papanikolopoulos, N.P. (2001). Driver fatigue: A vision-based approach to automatic diagnosis. *Transportation Research Part C*, 9, 399–413.

Giannopulu, I. et al. (2008). Visuomotor strategies using driving simulators in virtual and pre-recorded environment. *Advances in Transportation Studies XIV Issue*, Section B, 49–56.

Kawato, S. and Tetsutani, N. (2004). Detection and tracking of eyes for gaze-camera control. *Image and Vision Computing*, 22, 1031–1038.

Martens, M.H. and Fox, M.R.J. (2007). Do familiarity and expectations change perception? Drivers' glances and response to changes. *Transportation Research Part F*, 10, 476–492.

Morimoto, C.H. and Mimica, R.M. (2005). Eye gaze tracking techniques for interactive applications. *Computer Vision and Image Understanding*, 98, 4–24.

Nakayasu, H. et al. (2007). Visual perception and response behaviour by driving simulator and eye tracking system. Proceedings of the International Conference Road Safety and Simulation, RSS2007, Rome.

Tock, D. and Craw, I. (1996). Tracking and measuring drivers' eyes, image and vision. *Computing*, 14, 541–547.

Villanueva, A. et al. (2006). Eye tracking: Pupil orientation geometrical modeling. *Image and Vision Computing*, 24, 663–679.

Chapter 25

Visual Behaviour of Car Drivers in Road Traffic

Carmen Kettwich, Stefan Stockey and Uli Lemmer

Introduction

Nowadays the car driver has to deal with an increasingly complex environment. In spite of an information overflow the driver has to separate relevant from less relevant information in order to navigate the car safely on the road. Good orientation in road traffic is only possible if the attention of the vehicle driver is used for the perception of significant objects and events. Minor mistakes as well as inattention can have serious consequences.

Approximately 86.6 per cent of all accidents are caused by erratic behaviour of the car driver (Statistisches Bundesamt, 2008). According to several references the risk of crash involvement at night is two or three times higher in comparison to daytime (Sigthorsson, 1996). For this reason it is very important to study a driver's gaze behaviour so that optical elements of road traffic, like headlamps and road lighting, can be optimised.

The presented investigation of gaze behaviour during driving identifies objects and other occurrences which attract the car driver's attention during daytime and at night.

Method

Apparatus

A test vehicle from Mercedes Benz was used throughout the investigation. Driver's gaze direction and its duration were logged with a remote eye-tracking system during the whole test. The eye-tracking system consists of three cameras mounted at the dashboard plus a so-called scene camera. The latter records the traffic scene in front of the car with 60 frames per second. Each frame is marked with a definite timestamp. For the purpose of studying the gaze behaviour both films were overlaid and recorded with a computer in the boot of the car. Eye movements and fixation durations were determined from the logged data.

Participants

Fifteen subjects (five female and ten male) between 23 and 64 years of age took part in this study. The average age of all subjects was 32.5 years (SD 13.06). Five out of fifteen had visual aids to correct their eyesight. One-third of all participants had more than ten years of driving experience.

Procedure

Every subject had to perform two test runs – one during daytime and one at night. The test track was 32.6 km long and comprised urban roads, rural roads and motorways. The first kilometres of the test track were mainly for vehicle familiarisation. Table 25.1 shows the distribution of road signs and traffic lights along the test track.

Table 25.1   Distribution and frequency of traffic signs and traffic lights along the test track

	Road sign + traffic light	Lengths of test track (km)
Urban road	287	11.4
Rural road	180	8.5
Motorway	97	12.7
Total	**564**	**32.6**

Test runs were only conducted under dry weather conditions so that reflections could not distract the driver from the driving task and have an influence on the gaze behaviour of the driver.

Data analyses

Automatic analyses of the video film were not possible so each film was analysed frame by frame in order to extract the desired information. The evaluation of the viewing directions and objects was limited to three typical parts of urban and rural roads and motorways with a total length of 580 metres, because of the large amount of data.

The number of frames at a certain place or object determines the dwell time. The latter depends on several variables, like traffic situation, road complexity and mental workload of the driver. Dwell times on several objects and six viewing directions – front left, front middle, front right, rear left, rear middle and rear right – were calculated for all test runs. Figure 25.1 depicts these six viewing directions. The cut-off line of the car's headlamps determined the boundary between front

Figure 25.1 Viewing directions

and rear lines of vision. After dwell-time determination, gaze behaviour during the day was compared to gaze behaviour at night.

In order to compare the dwell times of all subjects, dwell times of each viewing direction and dwell time on defined objects were summed up for each driver, divided by the driving time of the corresponding section and averaged over all subjects afterwards. The actual result is an averaged dwell time of different viewing directions and objects in percentage terms.

To determine average dwell times of each viewing direction and several objects in seconds, all dwell times of the respective viewing directions and objects were averaged.

Results and Discussion

Gaze behaviour in an urban environment

Figure 25.2 shows the average dwell times of each viewing direction. Viewing direction front middle, front right, rear middle and rear right are fixated longer in comparison to the other viewing directions. On closer inspection of the daytime data the car driver fixates more front middle and front right. Subjects infrequently focus on the far field during the day. At night the percentage of dwell time in the far field increases (see Figure 25.3). Often it is not possible for the car driver to

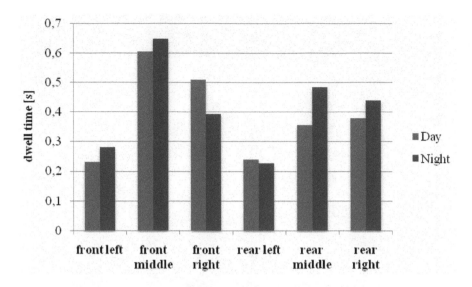

Figure 25.2   Average dwell time of each viewing direction in an urban environment

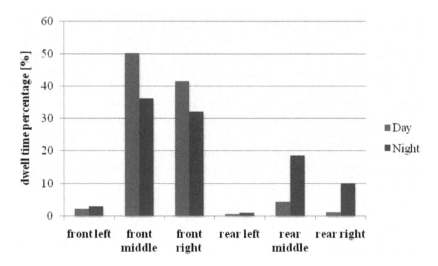

Figure 25.3   Dwell time percentage of the driving time for different viewing directions in an urban environment

look for objects in the far field in an urban environment, because of the traffic in the driver's lane as well as oncoming traffic. That could be one reason for the low percentage of dwell times in the far field. At night the traffic volume drops down to 17.5 per cent (Stock, 2005). For this reason the driver has more opportunities to look at the far field at night. Traffic becomes more noticeable in the dwell times of the far field by night.

The car driver focuses mainly on inline traffic, parking cars and traffic lights (see Table 25.2). In comparison to night-time the dwell-time percentage on inline traffic during the day is higher due to traffic volume.

Table 25.2 Dwell times and dwell-time percentage in an urban environment during the day and at night

	Inline traffic		Parking cars		Traffic lights	
	Day	**Night**	**Day**	**Night**	**Day**	**Night**
Dwell time (ms)	686	799	659	345	350	463
Dwell time percentage	43.9	16.4	13.8	9.3	6.4	10.7

Gaze behaviour in a rural environment

On average, car drivers fixate the far field longer at night compared to the day (see Figure 25.4). At night subjects focus in one-third of all cases on the viewing directions rear middle and rear right (see Figure 25.5). Furthermore objects are perceived later, because of a lower object-background contrast and ambient luminance (Bäumler, 2002). For this reason the driver tries to identify critical objects, like pedestrians and deer, as early as possible.

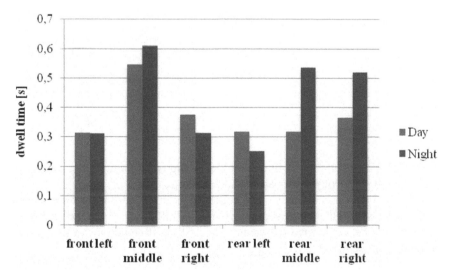

Figure 25.4 Average dwell time of each viewing direction in a rural environment

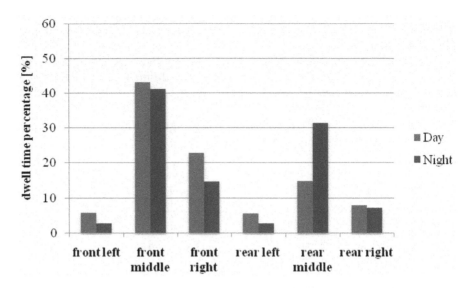

Figure 25.5 **Dwell-time percentage of the driving time for different viewing directions in a rural environment**

During the day and at night oncoming lane and nearside lanes are rarely fixed. The driver focuses mainly on inline traffic, traffic signs and traffic lights (see Table 25.3).

Table 25.3 **Dwell times and dwell-time percentage in a rural environment during the day and at night**

	Inline traffic		Traffic sign 274		Traffic lights	
	Day	Night	Day	Night	Day	Night
Dwell time (ms)	612	981	298	291	332	333
Dwell time percentage	28.2	14.3	6.3	4.7	7.8	4.8

On the one hand the dwell time on inline traffic is longer at night than during the day. Rear lights of the cars ahead, for example, attract the driver's attention more at night than during the day and could be the result for longer dwell times. Then again the dwell time percentage on inline traffic during the day is higher compared to the night. One reason for that is the difference in traffic volume during the course of the day.

Gaze behaviour while driving on a motorway

With the exception of the dwell time front left, the dwell times of the other viewing directions at night exceed the dwell times during the day (see Figure 25.6). Due to limited field of view at night it is more difficult for the driver to assess the vehicle speed of other road users. As a result of this, the car driver needs more time to gather relevant information. The latter results in a longer dwell time at night.

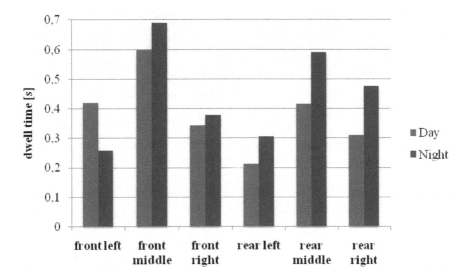

Figure 25.6   Average dwell time of each viewing direction while driving on a motorway

Subjects do not look at the overtaking lane very often (see Figure 25.7). Basically traffic signs and inline traffic are more likely to be considered during a motorway ride (see Table 25.4). Like the dwell times of different viewing directions, the dwell time on objects at night is longer than the dwell time on objects during the day. Lower ambient luminance and less contrast leads to longer dwell times, because it takes longer to perceive relevant driving information (Diem, 2005).

Despite the relative high dwell times on traffic signs at night, the dwell time percentage on traffic signs at night is lower compared with daytime. According to Diem 20–25 per cent of all traffic signs are fixed in the daytime compared with 10 per cent at night (Diem, 2005). In spite of the variances caused by the test design and other subjective factors, Diem's results could be validated.

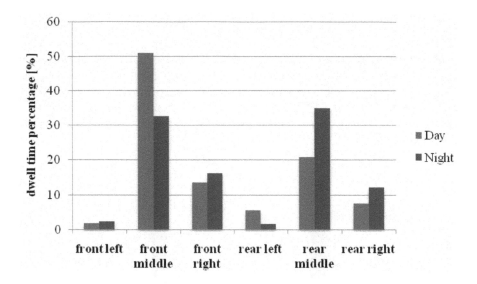

Figure 25.7   Dwell-time percentage of the driving time for different viewing directions while driving on a motorway

Table 25.4   Dwell times and dwell-time percentage on a motorway during the day and at night

	Inline traffic		Traffic sign 450		Traffic sign 448+449	
	Day	**Night**	**Day**	**Night**	**Day**	**Night**
Dwell time (ms)	619	974	323	333	399	570
Dwell time percentage	38.3	20.0	5.8	3.8	11.9	7.8

Conclusion

At night dwell times of the viewing directions rear middle and rear right are higher in comparison to the dwell times during the day. Furthermore the dwell time at night percentage of the driving time in the viewing directions rear middle and rear right exceeds the at night dwell-time percentage of the total driving time during the day. This is valid for nearly all traffic situations. Ambient luminance, contrast and adaptation level play an important role in object recognition (Bäumler, 2002). Due to low ambient luminance and contrast at night the reaction time increases (Hartmann, 1979) and objects are perceived later compared to daytime.

A driver should be able to see as much as possible at night, but should not glare other road users with the use of headlamps. For this reason headlamp technologies and road lighting should be improved. High beam is only used in 5 per cent of all traffic situations (Hella, 2000). A glare-free high beam would help to strengthen the high-beam operation. Thus objects could be recognised at an earlier stage and potential accidents may be avoided (Eichhorn, 2005).

References

Bäumler, H. (2002). Vergleichende Untersuchung von Fußgängerunfällen bei Tag und Nacht unter Berücksichtigung der Reaktionszeitproblematik bei Dunkelheit, 42–51, dissertation.

Diem, C. (2005). Blickverhalten von Kraftfahrern im dynamischen Straßenverlauf. Herbert Utz Verlag, 85–110, dissertation.

Hartmann, E. (1979). Reaktionszeit im Straßenverkehr. In J. Krochmann, (ed.), *Reaktionszeit von Kraftfahrern.* Berlin, Institut für Lichttechnik der Technischen Universität Berlin, pp. 7–17.

Eichhorn, K. (2005). Aktives Licht – Innovative Ansätze für die nächste Scheinwerfer-Generation. Automobiltechnische Zeitschrift, 11, 978–983.

Hella KG Hueck and Co (2000). *Hella Licht Research and Development Review, 68.* Lippstadt, Hella KG Hueck and Co.

Sigthorsson, H. (1996). Unfallgeschehen bei Helligkeit, Dämmerung und Dunkelheit. *Zeitung für Verkehrssicherheit,* 42(4), 149–155.

Statistisches Bundesamt (2008). *Statistisches Jahrbuch 2008.* Wiesbaden, Statistisches Bundesamt, p. 440.

Stock, W. (2005). Aufteilung von Eckwerten nach Pkw- Fahrtenmerkmalen gemäß MiD und KiD. DLR-Workshop, Neue Mobilitätsdaten für die Wissenschaft'.

Chapter 26

Icons for Actions in a Driving Simulator

Robert H Barbour

Introduction

The motor vehicle provides one of a number of globally accepted means of quickly moving people and products from place to place with very few constraints. There is a continued expansion of traffic across international borders with substantially different driving practices and conventions. With wide acceptance comes greater use, to the extent that driver behaviour is a focus of interest where use is such that congestion is common in particular localities. The major issue in driver training is the development of vehicle control skills required to keep to the carriageway and avoid damage to the vehicle from other objects (either stationary or moving). Almost without exception this fundamental position remains implicit in driver training in many countries. There are, of course, variations from country to country, reflected in a growing interest in the psychology of driving, but very little research into driving behaviour in the social context of other drivers exists. The study of the social aspects of everyday driver interactions is very much in the early stages. There has been some study of simulated driving environments in high-cost commercial and research-based driving simulators. This research is specific to particular vehicles and tracks. Data and findings are not accessible to the research community and international research in the area is not directly replicable. The major constraints on augmented reality driving simulation studies are the issues of how valid and reliable the data generated in the simulations are in relation to real-world data (Blana, 1996). Validity is difficult to establish where real and virtual worlds are compared because the essence of a simulation is that it is not the real world but generates approximately the same experiences for participants. There is little doubt that studies using the same environment in both software and hardware should facilitate reliable and comparable studies. However, developments in low-cost consumer-level computers, associated graphics processors, force feedback control devices and serious games software opens up new possibilities for comparable studies of group behaviour, as is described below.

The first section briefly outlines related literature sources and provides a justification for studies that employ simulations for investigating group driving behaviour. The section concludes with a specification for the capabilities of simulation systems required for studies of group driving contexts. The second section describes a set of low-cost driving simulation systems that report effects according to the specification outlined in the first section. The third section describes the data processing issues that emerge when simulation studies are undertaken. The fourth section describes an existing icon-based schema. The fifth section describes extension to the schema that accommodates for data attributes available in simulation environments. The chapter concludes by outlining future work in the area of social driving studies.

Relevant Literature

Instruction in early social uses of vehicles was considered with knowledge of the control and other mechanisms that are part of vehicles plus a few simple 'rules of the road' such as 'keep left/right' and the 'three-second rule'. Driver training was based on single persons driving single vehicles in environments that were considered essentially empty. Commercial driving simulators follow this pattern, placing individual drivers in a high-fidelity environment generating a strong sense of presence (Nunez and Blake, 2001). Bibliographic records of driving simulation research include generic documents (Kronfol and Grimm, 1990) and reports of dedicated research institutions (Anon 2006). Distinctions between different types of simulation games has been reported in the literature (Narayanasamy et al., 2006), while the potential contribution of simulators to the study of aspects of driving is illustrated and extended in Figure 26.1 (Ossen, 2008). Ways in which driving simulations could be improved were considered in the mid-1990s (Cremer et al., 1996) and more specifically detailed lists of attributes have been suggested (Green, 2005). Despite a focus on single persons in a single machine, driving is experienced in the real world in a social context. Driving is most often a group activity. An early two-vehicle linked simulation explored collision avoidance behaviour (Hancock and De Ridder, 2003). A group augmented reality driving simulation environment is considered in abstract in the next section.

Outline specification

A brief specification is given for the technical attributes required for a replicable group driving simulation that is practically suitable for researching driving behaviour across national and international boundaries. By far the most important attribute is that the system should provide a common underlying and agreed physics model. That model of vehicle–environment interaction should be linked to a common environment where environmental variables can be configured and the configurations reported for the events under study. The attributes of the vehicles should be configurable and configurations should be reportable as an attribute file. Any instance of a group driving interaction should be reportable as a verbatim record (a video clip) suitable for both manual and automated analysis. The simulation environment, including hardware, software, drivers and tasks should follow a common nomenclature so that researchers share a common descriptive and analytic vocabulary. The particular computer systems used for controlling

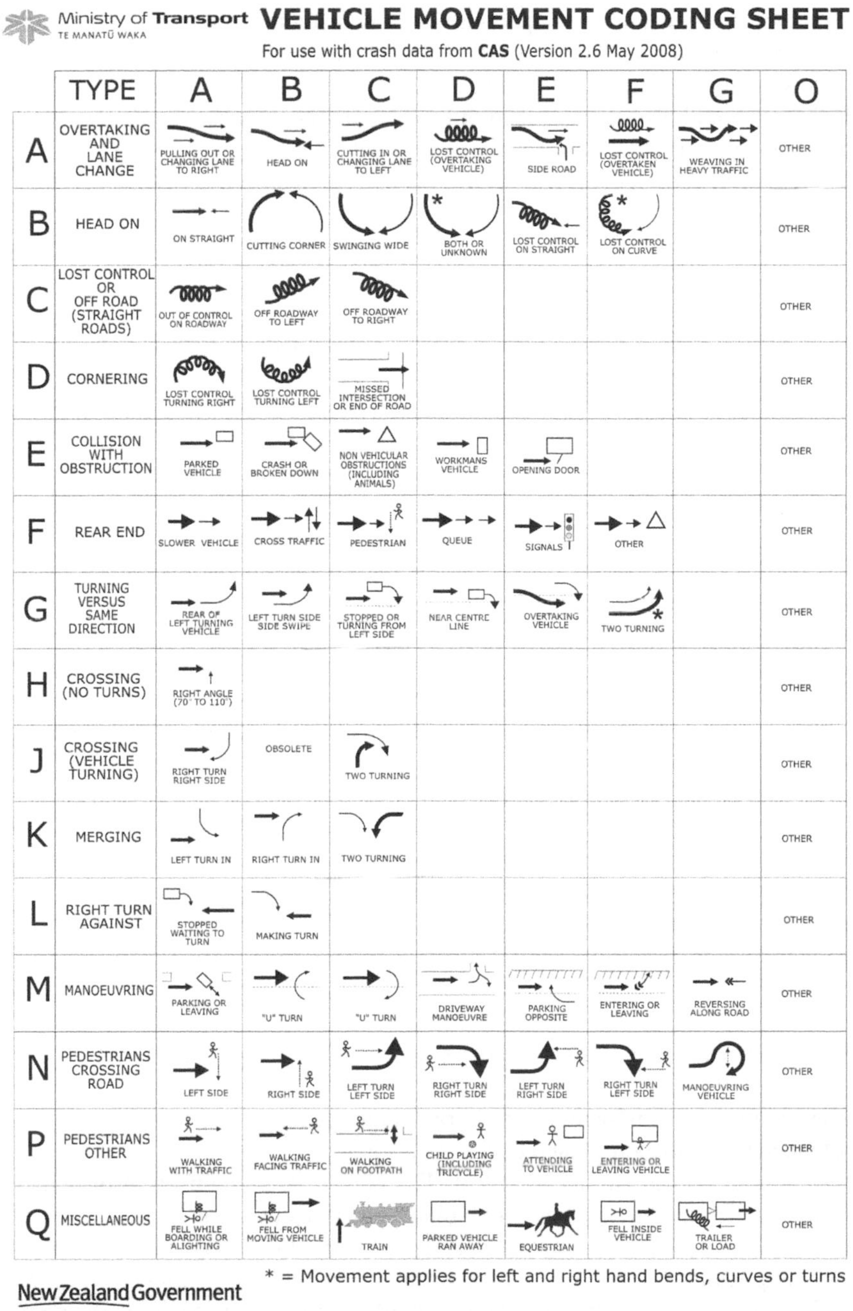

Ministry of Transport TE MANATŪ WAKA **VEHICLE MOVEMENT CODING SHEET**

For use with crash data from **CAS** (Version 2.6 May 2008)

TYPE	A	B	C	D	E	F	G	O
A	OVERTAKING AND LANE CHANGE	PULLING OUT OR CHANGING LANE TO RIGHT	HEAD ON	CUTTING IN OR CHANGING LANE TO LEFT	LOST CONTROL (OVERTAKING VEHICLE)	SIDE ROAD	LOST CONTROL (OVERTAKEN VEHICLE)	WEAVING IN HEAVY TRAFFIC	OTHER
B	HEAD ON	ON STRAIGHT	CUTTING CORNER	SWINGING WIDE	BOTH OR UNKNOWN	LOST CONTROL ON STRAIGHT	LOST CONTROL ON CURVE		OTHER
C	LOST CONTROL OR OFF ROAD (STRAIGHT ROADS)	OUT OF CONTROL ON ROADWAY	OFF ROADWAY TO LEFT	OFF ROADWAY TO RIGHT					OTHER
D	CORNERING	LOST CONTROL TURNING RIGHT	LOST CONTROL TURNING LEFT	MISSED INTERSECTION OR END OF ROAD					OTHER
E	COLLISION WITH OBSTRUCTION	PARKED VEHICLE	CRASH OR BROKEN DOWN	NON VEHICULAR OBSTRUCTIONS (INCLUDING ANIMALS)	WORKMANS VEHICLE	OPENING DOOR			OTHER
F	REAR END	SLOWER VEHICLE	CROSS TRAFFIC	PEDESTRIAN	QUEUE	SIGNALS	OTHER		OTHER
G	TURNING VERSUS SAME DIRECTION	REAR OF LEFT TURNING VEHICLE	LEFT TURN SIDE SIDE SWIPE	STOPPED OR TURNING FROM LEFT SIDE	NEAR CENTRE LINE	OVERTAKING VEHICLE	TWO TURNING		OTHER
H	CROSSING (NO TURNS)	RIGHT ANGLE (70° TO 110°)							OTHER
J	CROSSING (VEHICLE TURNING)	RIGHT TURN RIGHT SIDE	OBSOLETE	TWO TURNING					OTHER
K	MERGING	LEFT TURN IN	RIGHT TURN IN	TWO TURNING					OTHER
L	RIGHT TURN AGAINST	STOPPED WAITING TO TURN	MAKING TURN						OTHER
M	MANOEUVRING	PARKING OR LEAVING	"U" TURN	"U" TURN	DRIVEWAY MANOEUVRE	PARKING OPPOSITE	ENTERING OR LEAVING	REVERSING ALONG ROAD	OTHER
N	PEDESTRIANS CROSSING ROAD	LEFT SIDE	RIGHT SIDE	LEFT TURN LEFT SIDE	RIGHT TURN RIGHT SIDE	LEFT TURN RIGHT SIDE	RIGHT TURN LEFT SIDE	MANOEUVRING VEHICLE	OTHER
P	PEDESTRIANS OTHER	WALKING WITH TRAFFIC	WALKING FACING TRAFFIC	WALKING ON FOOTPATH	CHILD PLAYING (INCLUDING TRICYCLE)	ATTENDING TO VEHICLE	ENTERING OR LEAVING VEHICLE		OTHER
Q	MISCELLANEOUS	FELL WHILE BOARDING OR ALIGHTING	FELL FROM MOVING VEHICLE	TRAIN	PARKED VEHICLE RAN AWAY	EQUESTRIAN	FELL INSIDE VEHICLE	TRAILER OR LOAD	OTHER

* = Movement applies for left and right hand bends, curves or turns

NewZealand Government

Figure 26.1 CAS: Vehicle Movement Coding Sheet (reproduced with Permission of Ministry of Transport, New Zealand)

simulation systems are less important than the attributes of the user-independent reportable experience. That is, frame rates, screen size and resolution in pixels, shadow attributes and colour attributes. Within this broad specification software attributes should also be described in replicable ways. By far the easiest way to accommodate for these requirements is to select an augmented driving simulation system that already meets these requirements. The next section describes such a system.

Group driving simulation system

A six-person LAN-based VR environment is described in Barbour (2009). The paper describes hardware (PC P4 or better CPU, 256 Mb graphics processors, 1280 × 1024 17 inch LCD monitors, Logitech G25 Wheel™, pedals and gear lever) and setup aspects of the system. Software (Simbin™ GTL™, GTR 2™, Race-07™ and GTR- Evo™) and the advantages of the system are also described. What distinguishes the current research underway from other investigations is the focus on social behaviour as seen in vehicle movement in a common virtual environment. The 11 kilometres of rural dual-carriageway modelled in the system is part of a serious driving simulation game that uses the Pacejka model (Pacejka, 2005) in common with the physics models of commercial simulators. The environment provides high-fidelity graphics and sound, engaging and responsive vehicles, selectable and rapid changes in light conditions, variable weather and a replay feature. Multiple vehicles are available and interactions can be with groups of vehicles controlled by avatars or other drivers. It is the capability to replay interactions from multiple points of view including from an aerial position, ground level front and rear as well as cockpit and linked TV cameras that enables the externalisation and close study of the interactions of drivers. These views are user selectable and provide the essential elements for turning entertainment into a research tool. All six participants can see and respond to each other and the shared driving world in a similar way to that which occurs in the real world.

Why Networked Simulation Studies Can Inform Driving Behaviour

Other dimensions influencing both real and augmented driving experiences include vehicle attributes that reflect a continuing development and understanding of vehicle dynamics and their control. The physics and engineering of vehicles are well understood to the extent that many functions (such as fuel management, braking, traction and stability) can be assigned to digital management and control. These applications of control systems form part of an ongoing global experiment in a human–machine interaction. The study of the consequences of the interaction of technology aids in everyday group driving situations is also very much under development. Almost nothing is reflected in drivers' behaviour on public roads of the consequences for groups of drivers in mixed technology streams of traffic.

To put this claim more exactly, if a stream of traffic in a multi-lane motorway is moving at a constant velocity of about 60 mph/100 kph then the travelling distances between individual vehicles in the streams of vehicles will, all other things being equal, need to vary from one vehicle to the next. That distance will reflect the attributes of the environment, vehicle and the driver. If the combination of driver/vehicle/road conditions/weather is taken into account some vehicles will be at risk of being rear-ended. The stopping distance of some vehicles is such that a vehicle behind cannot stop in the available space in most congested motorways where normal travelling behaviour is observed. Other vehicles will be at risk of hitting the vehicle in front, again because the space left between vehicles is too limited to allow for reaction time, braking distance, and the behaviours of other vehicles in front. The problem is exacerbated by the additional loading on the stream of traffic added by large trucks and passenger service vehicles. The issue identified here is that driving behaviour needs to be studied at the social level implicit in the stream-of-traffic concept rather than the simplistic distance between any two vehicles.

Changing ambient environmental conditions of light, temperature and moisture on the carriageway alter the way vehicles respond to driver control. While these influences are named, few people can describe the consequences of such changes for vehicle control in a group driving context. Even fewer drivers respond to such changes in thoughtful ways in real-world driving contexts.

Of the many parameters that could have been chosen to introduce in this chapter, the above brief outline points to a conundrum of space and time. The consequence of each of the complex items suggested, moves the time for response available to drivers into the range where humans are at the upper limit of their response times and well beyond their real-time cognitive capacities. The driving context becomes too much, too soon, too fast, for human response times to manage effectively in the group driving context. The specific motorway contexts described above become problems because the changes themselves and the behaviour required to accommodate for them need first to be studied, then understood, predicted in context, responses initiated and expedited under circumstances where clues that changes in vehicle control are required are almost entirely absent. The evidence required to put investigations on a firm foundation cannot easily be gained in real traffic situations. The cost to property and person would not be acceptable. Serious driving games, as indicated, provide high-fidelity playback of driving interactions. In their raw form, these records have their own inherent problems, as discussed in the next section.

Vehicle Interaction, Description and Analysis

Byte level system records of single vehicle movements in a single simulator generate very large data-files running to many megabytes in size for interaction sequences of realistic tasks. For example, a follow the leader interaction lasting 20 minutes generates a file 23 Mb in size. Multiple individual interactions generated in group contexts generate much larger files. How to describe the interactions presents problems of data abstraction that seem, at first, intractable. An investigation was carried out to determine how other researchers had addressed the issue. Not surprisingly, not much was found that was directly relevant to describing the behaviour of multiple users in a group driving simulation. Other disciplines are also concerned with vehicle movement. Transport engineers, road safety engineers and traffic crash analysts all have tools and techniques for abstracting key elements from the large amounts of data available. The New Zealand Crash Analysis System (CAS) (Hewitt et al., 1960) was developed for reporting the attributes of vehicle movements in vehicle crashes in an iconic and coded form. CAS, Figure 26.2, has been developed and refined over a period of 40 years (Hewitt, personal communication, 2009) It has been updated recently and is available with an accompanying coding sheet that is now linked to a computer based system used for recording the results of crash analyses. The main use of the tool is in studies designed to reduce road trauma and for reporting in court proceedings. A glance at the sheet indicates that it covers, with icons and text, many of the attributes of traffic-related events that analysts looking into driver behaviour in simulators are interested in. It is particularly suited to event analysis. Extensions are required to Ossen's (2008) criteria in order to cover the events observed in a networked multi-participant system. These extensions to existing tools are needed to communicate richer descriptions in the international driving context than is shown in the CAS system, to promote understanding and to educate road users about vehicle-related social interaction outcomes. The following section sketches the extensions required.

Further Work: Extensions to CAS

Instructions for using CAS are found at the New Zealand Ministry of Transport website (http://www.landtransport.govt.nz/research/cas/). These instructions were used as a starting point for interpreting augmented reality interactions in networked environments. The first suggested improvement for a generic augmented reality system is the addition of an element indicating the front of the vehicle. A simple 'V' on the box representing the vehicle indicates the normal direction of travel. The 'Plot box and whisker' model (Desai et al., 2007; Massart et al., 2006) is used to represent vehicle movement attributes and could be extended by adding a tail the length of which indicates speed at a point in a journey (see Ossen, 2008). Vehicle types are poorly represented by simplistic icons in the CAS system. The vehicles are represented by common elements summarised in the form of a set of icons for the major vehicle types shown in Figure 26.1. No indication is given as to passengers' or drivers' attributes in the vehicles. Age, gender and ethnicity are obvious additions that could be annotated on or under the vehicle icon. Fuel type could be shown with a triple letter acronym. While extensions beyond PET were

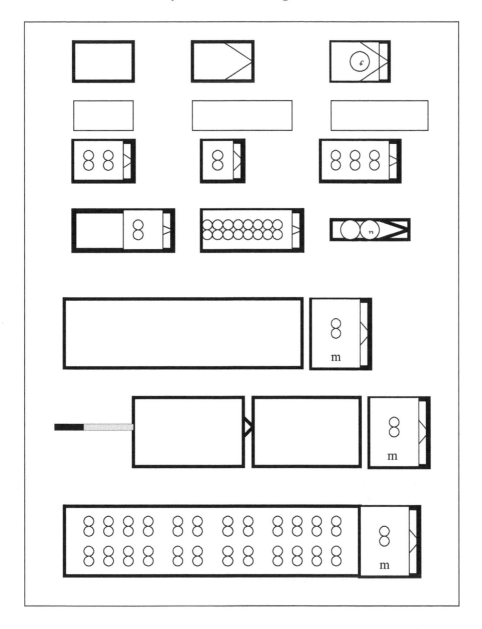

Figure 26.2 Vehicle icons: the icon elements, a saloon car, a two-seater, a sports utility vehicle, a van, a mini bus, a motorcycle, a truck, a truck and trailer (with a 60/100 kph tail, a passenger bus)

not required in earlier times, DES for diesel, LPG for liquid petroleum gas, CNG for compressed natural gas, ALC for alcohol, HYB for hybrid, ELE for electric, HFC for hydrogen fuel cell, HYD for hydrogen and AT1(fission) and AT2 (fusion) for atomic power seem possible starting points. Green alternatives could be GPT, GDE, GNG, for petrol, diesel and gas. Icons are required for weather (light levels, temperature, humidity, visibility, wind and cloud cover) and environmental attributes such as road surface.

For research to be internationally comparable, standards and conventions are required for reporting data on maps of carriageways in general and specific roadways in particular. These standards could be achieved either by adopting existing augmented reality environments such as those widely used in serious games or by developing open-source definitions with internationally accepted configurations and use-by dates for particular system configurations. There is clearly much work required here to shift the study of driving behaviour from one-off high-cost idiosyncratic systems to widely available low-cost but comparable systems such as is described here. There is also a need for defining common scenarios so that the driving performances being reported are as free as possible from confounding variables. These suggested changes should bring about a body of research that, because it is low cost and shares all but the drivers in common, allows for generalisation beyond the reported sample groups. The suggested changes also make the study of group driving behaviour accessible to wider groups of researchers, requiring a much smaller skills set than the multidisciplinary teams assembled for driving research centres.

Conclusion

The need for the international study of group driving behaviour studies using networked systems has been described. An existing augmented reality driving system has been outlined. A national Crash Analysis System has been illustrated that could form the basis of a common abstraction system for reducing the complexity of simulation data sets. 'Iconifying' vehicle attributes reduces the possibility of meaning being 'lost in translation'. The extensions required for augmented reality studies prerequisite for internationally comparable study of the details of group driving behaviours have been outlined.

References

Anon (2006). *Publications on STISIM and the STI Vehicle/Tire Dynamics Model.* Hawthorn, California, Systems Technology Inc: 9. Retreived 28 June 2010 from http://www.systemstech.com/.

Barbour, R.H. (2009). *Practical Low-Cost Augmented Reality Research into Group Driving Behaviour.* CITA-09. Kuching, Malaysia, UNIMAS.

Blana, E. (1996). *Driving Simulator Validation Studies: A Literature Review.* Leeds, Institute of Transport Studies, University of Leeds, Working Paper 480.

Cremer, J.K. et al. (1996). Driving simulators: Challenges for VR technology. *IEEE Computer Graphics and Applications*, 16(5), 16–20.

Desai, A.V. et al. (2007). The utility of the AusEd driving simulator in the clinical assessment of driver fatigue. *Behavior Research Methods*, 39(3), 673–681.

Green, P. (2005). *How Driving Simulator Data Quality Can be Improved.* Orlando, FL, University of Transportation Research Institute.

Hancock, P.A. and De Ridder S.N. (2003). Behavioural accident avoidance science: Understanding response to collision incipient conditions. *Ergonomics*, 46(12), 1111–1135.

Hewitt, C., Badger, S. et al. (1960–). *Vehicle Movement Codes System.* Retrieved 28 June 2009 from http://www.nzta.govt.nz/resources/guide-to-treatment-of-crash-location/docs/appendix-b1.pdf28.

Kronfol, R. and Grimm A. (1990). *Driving Simulators: A Classified Bibliography.* Ann Arbor, MI, University of Michigan, Transport Research Institute 35.

Massart, D.L., Smeyers-Verbeke, J. et al. (2006). *Visual Presentation of Data by Means of Box-Plots.* Chester, LC-GC Europe.

Narayanasamy, V., Wong, K.W., Fung, C.C. and Rai, S. (2006). Distinguishing games and simulation games from simulators. *Computers in Entertainment*, 4(2), 141–144.

Nunez, D. and Blake E. (2001). Cognitive presence as a unified concept of virtual reality effectiveness. *ACM*, 115–118.

Ossen, S.J.L. (2008). Longitudinal Driving Behavior: Theory and Empirics. Faculty of Civil Engineering and Geosciences, Transport & Planning Department, The Netherlands TRAIL Research School Delft University of Technology. Unpublished PhD.

Pacejka, H.B. (2005). *Tire and Vehicle Dynamics.* Warrednale, AE International and Elsevier.

Chapter 27
Contributory Factors for Incidents Involving Local and Non-local Road Users

Linda Walker and Paul S. Broughton

Introduction

Road traffic crashes have a high economic cost, including cost of clear up, medical costs and loss of output from those killed or injured. Much greater are the human costs to friends, family and society. Where the victims are from outside the local area, it may also have an impact on the image and tourism potential of the area. Tourism is estimated to be worth £4.2 billion to the Scottish economy, supporting over 9 per cent of total employment (www.visitscotland.org). Apart from this additional potential cost to an area in terms of reduced tourism income, there may also be ethical considerations and perhaps even a duty of care, to ensure, as far as is realistically possible, the safety of visitors while in the area. An estimated 16 million overnight tourist trips are made in Scotland every year with 70 per cent choosing the car as their mode of transport (VisitScotland, 2009). Previous research has indicated that visitors to an area may not have a greater propensity for having crashes but that these crashes tend to be more serious and, more importantly in terms of this research, they appear to have different characteristics (Walker and Page, 2004; Walker, 2007). If the nature and type of crashes is different to those of local drivers, there may be different underlying causes. As Walker and Page's research concentrated on a relatively small area of Scotland (the Forth Valley area), there is a need to further explore this issue to identify if the same patterns are evident in the wider statistics for Scotland. This chapter reports on an exploratory analysis of road traffic crash data for the whole of Scotland and seeks to examine this data in relation to Ray Fuller's Risk Allostasis Theory (RAT). The resulting analysis should allow a clearer insight into the direction of future policy planning in relation to reducing visitor crashes.

Targeting a Message

Road traffic crashes are generally preventable (WHO, 2004). Social marketing literature suggests that prevention measures have a greater chance of being successful if they are targeted (Kotler and Lee, 2008). In order to be targeted though, there must be an appreciation of the specific groups and the factors leading to, in this case, road traffic crashes. Recent years have seen an appreciation of the need to target road safety measures and design interventions at specific groups; campaigns that have emerged from this approach have included ones aimed specifically at young drivers and ones aimed at motorcyclists.

The policy behind such approaches to road safety measures has allowed for a more focused approach that tailors messages to the behaviours of those groups. This requires an understanding of these groups. This approach and its premise is the philosophy behind this research.

Previous Research

Despite the continued growth in tourism and its importance to the global economy, there is very little research on safety issues relating to visitors (Walker, 2007). This research tends to focus on disease, activities, crime, or terrorism. There is a very limited literature base relating to road safety issues despite it being identified as a major cause of mortality amongst visitors and tourists globally (Heggie et al., 2008). This dearth of research makes exploratory research such as this vital to opening up debate amongst academics, practitioners and policy-makers on ways of preventing unnecessary death and injury amongst this group.

This chapter is the result of analysis of data on road traffic crashes comparing data for visitors (defined as more than 40km from home at the time of the incident) and local residents. This information will then be analysed in light of Fuller's Risk Allostasis Theory (RAT). RAT is a recent adaptation of Fuller's widely accepted model of Task-Capability Interface Model (2000) and will be used to assess the data relating to visitor crashes to seek a clearer understanding of this group and the issues that relate to them.

Data

The analysis presented here is from STATS19 data for all of Scotland from 1999 to 2007; the STATS19 data collection forms are completed by police in the event of any road traffic crash resulting in injury within the UK. The dataset analysed for this chapter, comprising a total of 153,110 vehicle users/pedestrians with information on collision contributing factors, is available for crashes from 2005 onwards.

Vehicle Types

Within the data there are 16 categories of vehicle, ranging from cars to ridden horses. The analysis reported here is only for crashes involving cars; this will prevent the analysis of the data being distorted by other vehicle types that have different profiles, for example HGVs, which will often be greater than 40km away from the driver domicile due to a work trip rather than a leisure journey. The focus on cars is suitable also as it is the main form of transport used by visitors (VisitScotland, 2007).

Distance from Domicile

Data on the distance that the crash occurred from the driver's domicile were available from the dataset; the mean distance was 18.2 km (95 per cent CI 17.8–18.6; SD = 74.2). These data allow each person to be classified as either a local or visitor by virtue of how far they were from their domicile. Previous explorations of these effects have used 40km in the UK as the threshold above which a person is defined as being a visitor to the area (Fletcher and Sharples, 2001; Walker and Page 2004; Walker, 2007); this is based on data from the National Transport Survey. For this chapter the visitor grouping was constructed with all drivers over 40km being assigned as visitors along with those elsewhere in the UK (N = 13673; 10.6 per cent) and those from outside the UK. Within this group 692 of these car drivers were overseas visitors. Those drivers that were below the 40km threshold were assigned as local, with all other data set as missing. A total of 115,443 people were classified as local (89.4 per cent).

When Do Crashes Occur?

The peak day for crashes is on Fridays when 16 per cent of all crashes occur (Table 27.1). From this the number of crashes that occur at the weekend and on weekdays can be calculated, with 27 per cent of all car-involved crashes happening at the weekend (the weekend is defined as Saturday and Sunday). The main differences between visitor and local crashes occur on Saturdays and Sundays (33 per cent and 26 per cent respectively: χ^2 (1 df, N = 126217) = 326.526, $P < 0.001$; Cramer's V = 0.051). On the weekend (Saturday/Sunday) the mean distance between domicile and crash site is 20.2km, compared to 17.1 for weekdays (t(52355) = 6.386, $P < 0.001$; $r = 0.028$).

Around 10 per cent of crashes occur between midnight and 8a.m.; between 8a.m. and 4p.m. 47 per cent of crashes occur with 43 per cent between 4p.m. and midnight. Figure 27.1 also shows that there is an increase in crashes for locals on weekdays at around 8a.m.; this is not seen on the weekend or for visitors.

Who are Having Crashes?

Male drivers are involved in the majority of crashed cars (97,534 compared to 55494 females), a 64 per cent male and 36 per cent female split. Visitor males

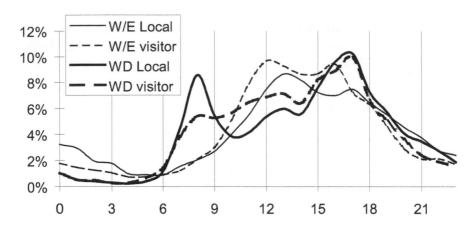

Figure 27.1 **Crashes for locals and visitors by weekend and weekday**

Table 27.1 **Days of week crashes occur for car drivers**

	Local	Visitor	Total	Local %	Visitor %	Total %
Sunday	12906	2099	15005	11	15	12
Monday	15718	1782	17500	14	13	14
Tuesday	16071	1683	17754	14	12	14
Wednesday	16300	1637	17937	15	12	14
Thursday	16766	1761	18527	15	13	15
Friday	18752	2432	21184	17	18	17
Saturday	15875	2435	18310	14	18	15
Total	**112388**	**13829**	**126217**	**100**	**100**	**100**

(χ^2 (6 df, n=126217) = 402.345; P < 0.001; Cramer's V = 0.056).

are more likely to be crash involved than local males, 73 per cent to 62 per cent respectively (χ^2 (1 df, N = 91028) = 433.463, P < 0.001; Cramer's V = 0.069).

There is a difference between male and female driver crash involvement on weekdays and the weekend, with 69 per cent of the cars involved in crashes on the weekend being driven by males compared to 62 per cent for weekdays (χ^2 (1 df, N = 149101) = 539.268, P < 0.001; Cramer's V = 0.060). There is no significant difference for visitors crash involvement when comparing gender and weekend/ weekday. However, local female drivers are less likely to be involved in a crash during the week compared to the weekend (33 per cent and 40 per cent respectively; χ^2 (1 df, N = 92573) = 53.486, P < 0.001; Cramer's V = 0.040).

The mean age for car drivers involved in crashes is 38.3 years (95 per cent CI 38.21–38.36; SD = 15.5); the mean age for car drivers involved in a crash is 38.1 years for locals and 40.0 for visitors ($t(12920) = -11.944$, $P < 0.001$; $r = 0.105$) and the age difference of weekend to weekday crashes is 37.2 years and 38.7 years respectively ($t(151010) = 16.301$, $P < 0.001$; $r = 0.042$).

Where are Crashes Occurring?

The majority of crashes involving visitors occur on A roads (65 per cent compared to 44 per cent for locals) while locals crash more on unclassified roads (34 per cent compared to 14 per cent for visitors: χ^2 (6 df, N = 91367) = 3476.958, $P < 0.001$; Cramer's V = 0.195).

Locals are more likely than visitors to have crashes within an area that has a 30mph speed limit (58 and 26 per cent respectively); within the 60mph speed limit locals have 27 per cent of their crashes with visitors having 50 per cent (χ^2 (9 df, N = 87776) = 4187.801, $P < 0.001$; Cramer's V = 0.218).

Casualties and Severity

The majority of crashes are only slight (82 per cent), with 16 per cent being serious and under 2 per cent fatal (N = 159822). There is a difference in severity of crashes between locals and visitors with visitors suffering slightly more KSI crashes (22 and 17 per cent respectively – χ^2 (1 df, N = 126127) = 243.487, $P < 0.001$; Cramer's V = 0.044). KSI crashes are more likely for both visitors and locals on the weekends, increasing from 16 per cent during the week to 19 per cent on the weekend for locals (χ^2 (1 df, N = 112388) = 128.991, $P < 0.001$; Cramer's V = 0.034) and from 21 to 25 per cent (χ^2 (1 df, N = 13829) = 19.971, $P < 0.001$; Cramer's V = 0.038) for locals.

The mean number of casualties in a crash is 1.35 (95 per cent CI 1.34–1.35; SD 0.87), however the mean is higher for visitors than locals (1.61 and 1.41 respectively – t(16028) = 18.100; $P < 0.001$; r = 0.142).

Contributory Factors

In the assessment of STATS19 crash contributory factors odds ratios (OR) were calculated. An odds ratio of greater than one here indicates that visitors have a greater chance of being affected by the contributory factor. A value of less than one would indicate that locals have the higher probability of having the contributory code assigned to them.

Fatigue as a contributory factor is much more of an issue for visitor drivers than local drivers (OR = 3.24; 95 per cent CI 2.59–4.05 $P < 0.001$), with crashes that relate to inexperience with the vehicle being driven also being far more likely to occur to visitors than locals (OR = 1.74; 95 per cent CI 1.27–2.40 P < 0.001). The driver having a crash because they are distracted by something outside the vehicle has an OR = 1.32 (95 per cent CI 1.03–1.72 P = 0.032). Other contributory factors that are more likely to be reported in crashes involving visitors are: an animal or object in the road (OR = 1.54; 95 per cent CI 1.22–1.95 P < 0.001); nervousness, uncertainty or panic (OR = 1.49; 95 per cent CI 1.19–1.87 P < 0.001); swerved (OR = 1.52; 95 per cent CI 1.30–1.76 P = 0.038); sudden braking (OR = 1.26; 95 per cent CI 1.12–1.40 P < 0.001); travelling too fast for the conditions (OR = 1.17; 95 per cent CI 1.06–1.28 P = 0.001); travelling too close (OR = 1.32; 95 per cent CI 1.20–1.46 P < 0.001); poor turn or manoeuvre (OR = 1.15; 95 per cent CI 1.05–1.26 P = 0.002); slippery road (OR = 1.32; 95 per cent CI 1.23–1.46 P < 0.001) and loss of control (OR = 1.40; 95 per cent CI 1.30–1.51 P < 0.001).

Locals have a higher propensity for aggressive driving to be a contributory factor (OR = 0.77; 95 per cent CI 0.60–0.98 p = 0.038) as well as exceeding the speed limit (OR = 0.75; 95 per cent CI 0.63–0.89 P = 0.001); being impaired by alcohol (OR = 0.67; 95 per cent CI 0.57–0.80 P < 0.001); being an inexperienced or a learner driver (OR = 0.76; 95 per cent CI 0.64–0.90 P = 0.001); disobeying stop or give way signs (OR = 0.70; 95 per cent CI 0.60–0.83 P < 0.001) and failing to look properly (OR = 0.83; 95 per cent CI 0.78–0.89 P < 0.001).

The contributory factors were categorised into four groups: misjudgement errors (for example, poor turn or manoeuvre); illegal activity (for example, exceeding the speed limit); personal factors (for example, fatigue) and environmental factors (for example, an animal or object in the road). Using these factors, visitors are more likely to make a misjudgement error (OR= 1.35; 95 per cent CI 1.29–1.42; P < 0.001) but are less likely to carry out an illegal activity (OR= 0.81; 95 per cent CI 0.77–0.87; P < 0.001). The other two factors were not significant.

Weather

Average weather data were obtained for each month covering all the years of the dataset (average temperature in degrees centigrade, average rainfall in millimetres and average hours of sunshine).

For this analysis it is considered hot when the average temperature is above 7.7° centigrade (median value 7.70° C). When it is hot, aggression is a contributing factor for visitors 85 per cent of the time, yet only 56 per cent for locals (χ^2 (1 df, N = 789) = 20.879, P < 0.001; Cramer's V = 0.163). Inexperience with vehicle is also more likely to be a contributory factor for visitor drivers when it is hot (70 per cent compared to 50 per cent χ^2 (1 df, N = 249) = 4.962, P = 0.026; Cramer's V = 0.141); similarly junction restarts are more likely to be a contributory factor when it is hot for visitor drivers (70 per cent for visitors and 52 per cent for locals – χ^2 (1 df, N = 504) = 4.657, P = 0.014; Cramer's V = 0.109).

For this analysis it is considered wet when the monthly average rainfall exceeds 129mm (median value of rainfall is 128.94mm); when it is wet spray is a significant crash contributory factor for all drivers but more so for locals for whom 88 per cent had this as a factor compared to 50 per cent for visitor drivers (χ^2 (1 df, N = 54) = 8.483, P = 0.004; Cramer's V=0.395). Inadequate/masked signs or road markings is also a significant factor when wet, 72 per cent for locals compared to 46 per cent for visitors (χ^2 (1 df, N = 243) = 6.777, P = 0.009; Cramer's V = 0.167).

For this analysis it is considered sunny if the monthly average sunshine is above 3.30 hours (median value 3.295 hours) and when it is sunny visitors are more likely to have the contributing factor of failed/misleading signal (68 per cent verses 50 per cent, χ^2 (1 df, N = 409) = 5.617, P = 0.018; Cramer's V = 0.117) and aggressive driving (70 per cent verses 51 per cent, χ^2 (1 df, N = 789) = 8.202, P = 0.004; Cramer's V = 0.102).

Discussion

Exposure issues

In examining the data in relation to RAT, there is no reason to assume any differences in terms of driver capability for visitors and locals as a group. Fuller (2008) describes driver capability as arising '*from more distal to more proximal influences, from the driver's basic physiological characteristics, education, training and experience*'. In assessing the information gained from analysing the data, it can be assumed that the capability of visitors and local residents is likely to be similar, particularly as the majority of the visitors in the data are from within the UK and therefore will have similar backgrounds due to aspects such as training requirements to pass their test and basic culture aspects. Arguably it could be said that they are likely to differ in experience insomuch as visitors may have less familiarity with the specific roads in the area but in terms of general experience, this is likely to be similar with an exposure in their driving experience of a range of similar road types, weather conditions, etc. Fuller expresses that '*This capability arms the driver with strategies for information acquisition and the capability of pre-adaptation to anticipated changes in task demand.*' If the assumption is made that both groups have similar variability in terms of their capabilities across the group, then the assumption can be made that differences are caused by differences in task demand between the two groups.

If this assumption is made, then we can assess the differences in terms of the differences in task demand. What are these likely to be? As with young drivers, crashes involving visitors are more likely to have a higher number of people in the car.

Some of the differences between the crashes that locals and visitors have can be explained by exposure. For example it is more likely that people will travel further on the weekend so that they can take part in leisure activities. The peak that occurs for locals at 8a.m. on weekdays (see Figure 27.1) may just reflect the extra local traffic as people commute to work. If the 8a.m. data are discounted then there is only a small difference between locals and visitors in terms of 'time of crash' patterns.

Male drivers have more crashes, but they do more of the driving (Mason and Prior, 2005; Dapartment for Transport, 2005); therefore this is also caused by the extra exposure. From this it may be concluded that on longer journeys, those above 40km, males are even more likely to be the driver which supports earlier work by Walker (2007).

Exposure may also be a factor in where accidents occur (Walker, 2007). Locals are more likely to have a crash in a town compared to visitors, as this is a reflection that to be more than 40km away from their domicile, major roads were probably used. Where crashes occur with reference to the speed limit, data reflects this with locals being more likely to crash in a 30mph zone and visitors in a 60mph limit area.

Another visitor/local difference that may be an exposure issue is that visitors are more likely to be crash involved with a contributory factor of inexperience in hot weather compared with locals under the same condition. The use of hire cars for leisure activities could be the cause of this; however this aspect will be explored in further research.

Risk Allostasis Theory

When investigating behavioural differences then an understanding of why people behave in the way that they do within the environment being considered is needed. There are many theories surrounding road user behaviour, and for this paper Risk Allostasis Theory (RAT) will be used (Fuller 2008).

One of the aspects of Risk Allostasis Theory is that drivers are in control provided that their capability is not outstripped by task demands placed upon them (Fuller 2000). When driving there can be many tasks placed upon the driver (Panou et al., 2005; Stradling and Anable 2007) including route choice, travel time, navigation tasks, road positioning, making progress while avoiding collisions and managing mood (avoiding boredom or anxiety). Within RAT 'drivers experience task difficulty in the same way as they experience feelings of risk'; thus the feeling of risk can be a surrogate for task difficulty (Fuller et al., 2008).

Drivers have a level of task difficulty that they would like to maintain. However, this level is not a constant value and temporary factors, such as anger, aggression, social influences and the pressure of being late can alter this target task difficulty. The goals of the journey and the motivation to put effort into the driving task affect this task difficulty/risk threshold (Fuller 2005). Vehicle characteristics, route chosen and time of day when the road is being used also have an effect on the perceived task demand. Perceived, and real, capability can be affected by human factor variables such as tiredness and impairment by alcohol or drugs. When a driver is a visitor to a location then some of the above factors may be different from when they are in their own locality, and with that the risk and nature of crashes that they suffer may also change.

Knowledge of the road is a factor that can adjust task difficulty, with lack of knowledge increasing task demand. Some of the contributory factors may reflect this with locals being more likely to be involved in a crash because they disobeyed a stop sign or not looked at a junction as familiarity with the driving environment convincing them that breaking these rules does not increase their risk as they may do it regularly. On the other hand visitors are more likely to have being nervous or uncertain as a contributory factor because lack of knowledge of the road can increase perceived task demand and this may be demonstrated with slower, and more nervous driving. Related to this is sudden braking; an act that often demonstrates that the driver is out of control (Stradling et al., 2008; Broughton and Walker 2009) and that task demand is outstripping capability. Drivers who are nervous and unsure can easily find themselves in this type of situation where their perceived capability cannot cope and the quickest way to reduce the task demand is to slow down. Fatigue, the contributory factor with the highest odds ratio (OR), affects driver capability and therefore puts drivers in a position where it is easier for demand to outstrip capability and put a driver into an out of control state that may lead to a crash. Inexperience with the vehicle being driven is another contributory factor that increases task demand, and therefore a driver can enter an out of control situation easier. Visitors are more likely to have unfamiliarity with the vehicle as a contributing factor; this may be due to visitors hiring cars for longer trips.

It was recently reported that certain landmarks might distract drivers and put them at a higher risk of having a crash (BBC, 2009). Certainly any distraction of the driver increases the task demand by adding at least one extra task to the driving activity. It might be expected that those who regularly drive past such landmarks would be less likely to be distracted and therefore external distractions would be a major difference in causes of crashes between local and visitor drivers. There is a difference between the two groups, but when considering the 95 per cent confidence intervals of the odds ratio, the effect size is fairly small.

Conclusion and Further Research

This preliminary analysis of the STATS19 data with regards to whether the car driver was a local or visitor indicates that there are two main reasons that explain why there is a difference in the type and number of crashes between these groups. Exposure is the first of these factors; for example visitor drivers are less likely to be driving on unclassified roads and this is reflected in the crash statistics. The second factor is behavioural; that is a driver's behaviour changes when they are a visitor compared to how they drive when on roads close to their domicile, for example driving in a nervous manner due to being unfamiliar with the roads. These behavioural factors can be discussed using theories relating to task difficulty where factors affect the difficulty level for visitors differently to locals.

Some of the gender differences for local and visitor drivers in collision contributing factors may also be explained using RAT; for example the differences that relate to following too close may be explained by the difference in how the genders react in certain situations (Fessler et al., 2004). Therefore, with the higher ratio of male to female drivers amongst visitors, this may in part account for different crash patterns.

Visitors are significantly more likely to be involved in a KSI crash than locals. That fact alone should focus the attention of anyone who is looking to reduce the KSI figures as this group is more at risk. Not only are they more likely to be involved in a KSI, but the number of casualties per crash is also likely to be higher for visitors, probably due to exposure, as they are more likely to have a higher number of passengers per vehicle (Walker, 2007). This suggests that targeting this group with effective measures could be more effective in reducing the actual number of casualties for the same reduction in crashes.

In terms of cost to the economy, as visitor crashes are predominantly associated with rural roads and such incidents are more expensive to deal with, together with the cost to tourism-related businesses if the area is considered unsafe, adds to any moral or ethical impetus that there may be. The policy implications of this research are that the specific needs of visitors and tourists need to be addressed as a distinct issue. This is not a problem specific to Scotland; road safety is a global issue (WHO, 2004) and therefore the implications of this research are not confined to Scotland.

References

BBC (2009) Wallace tower poses accident risk. Available at http://news.bbc. co.uk/1/hi/scotland/tayside_and_central/7952534.stm, accessed 1 May 2009.

Broughton, P.S. and Linda Walker (2009). *Motorcycling and Leisure: Understanding the Recreational Ptw Rider*. Aldershot, Ashgate Publishing.

Department for Transport (2005). *Focus on Personal Travel 2005 Edition*. London, Department for Transport

Fessler, D.M.T. et al. (2004). Angry men and disgusted women: An evolutionary approach to the influence of emotions on risk taking. *Organizational Behavior and Human Decision Process*, 95, 107–123.

Fuller, R. (2008). What drives the driver? Surface tensions and hidden consensus. International Conference on Traffic and Transport Psychology. Washington, DC, 31 August–4 September 2008.

Fuller, R. et al. (2008). Task difficulty and risk in the determination of driver behaviour. *Revue européenne de psychologie appliquée*, 58, 13–21.

Fuller, R. (2005). Towards a general theory of driver behaviour, *Accident Analysis and Prevention*, 37(3), 461–472

Heggie, T.W. et al. (2008). Recreational travel fatalities in US National Parks. *International Society of Travel Medicine*, 15(6), 404–411.

Kotler, P. and Lee, N.R. (2008). *Social Marketing: Influencing Behaviors for Good*, 3rd edn. Thousand Oaks, CA: Sage

Mason, M. and Prior, M. (2005). *Road Users' Satisfaction Survey 2004–2005*. UK: Highways Agency.

Panou, M. et al. *Modeling driver behaviour in EU and international projects*. Ispra, Luxembourg, Office for Official Publication of the European Communities

Stradling, S. et al. (2008) *Understanding Inappropriate High Speed: A Quantitative Analysis. Road Safety Research Report 93*. London, Department for Transport

Stradling, S.G. and Anable, J. (2007). Individual travel patterns. In R.D. Knowles et al., (eds), *Transport Geographies: An Introduction,* pp. 179–195. Oxford: Blackwell Publishers.

VisitScotland (2009). *Tourism in Scotland 2007*. www.visitscotland.org, accessed 20 April 2009.

Walker, L. and Page, S.J. (2004). The contribution of tourists and visitors to road traffic accidents: A preliminary analysis of trends and issues for central Scotland. *Current Issues in Tourism*, 7(3), 217–241.

Walker, L. (2007). Scoping the dimensions of visitor well-being: A case study of Scotland's Forth Valley. Unpublished PhD Thesis.

WHO (World Health Organisation) (2004). *World Report on Road Traffic Injury Prevention*, edited by Margie Peden et al. Geneva, WHO.

Chapter 28

Severity of Injury Outcomes for Older Drivers Involved in Intersection Crashes

Peter Hillard

Introduction

With ageing populations, the number of older drivers in most OECD countries will increase significantly over the next few decades. Crash data show that older drivers are more likely to be involved in intersection crashes than other age groups. Therefore, as the driver population ages it seems likely that there will be a commensurate increase in the incidence of these types of crash. Intersection crashes typically result in corner and side impacts. Current systems of occupant protection are considerably less effective for side impact and corner impacts than other types of crash. Intersection collisions frequently involve rapid rotation or spinning of the struck vehicle after the crash. This spinning motion can induce undesirable loading of the chest by the safety belt. Older adults tend to be less able to sustain such loads than younger adults due to calcification of cartilage in the chest that tends to occur with ageing (Kent and Patrie, 2005). Moreover, serious chest injuries are three times more likely to be fatal to elderly occupants than to younger occupants (Bergeron et al., 2003; Bulger et al., 2000; Taylor et al., 2002). The aim of this study is to investigate whether intersection crashes are particularly dangerous for older drivers in comparison to other crash types.

Background

Demographic changes

Over the next four to five decades, there will be a substantial increase in both the number and proportion of older people in most industrialised countries. The Organisation for Economic Cooperation and Development (OECD) has estimated that by 2030 one in four persons will be aged 65 years or over in most OECD countries (OECD, 2001). In Australia, the proportion of persons aged 65 years and older is predicted to increase from 11.1 per cent in 2001 to 24.2 per cent in 2051, with the greatest growth for those aged 80 years and over (Australian Bureau of Statistics, 1999). It is also anticipated that there will be an increase in older drivers' licensing rates (Hakamies-Blomqvist, 1993) and that they will be undertaking longer and more frequent journeys (OECD, 2001). These changes will combine to produce a marked increase in the number of older drivers on the road. With increased numbers and increased driving exposure for older drivers, a commensurate increase in future crash levels has been predicted. In both Australia (Fildes et al., 2001) and the United States (Hu et al., 2000), older driver fatalities have been predicted to triple by 2025, relative to 1995 levels.

Functional decline in older adults

The ageing process encompasses many individual differences. However, as a generalisation, even relatively healthy older adults are likely to experience some level of decline in sensory, cognitive and physical areas (Langford and Koppel, 2006). The most important sensory declines with 'normal' ageing that impact on driving ability are visual including: (1) narrowing of the sensory visual field, (2) impaired detection of angular motion (which affects the judgement of the velocity and distance of approaching vehicles) and (3) declining contrast sensitivity which impairs the ability to drive at night and in bad weather (Scialfa et al., 1987). The most important common forms of cognitive decline are narrowing of the attentional visual field (i.e. when the attention is focused on a task at the fovea, events in the peripheral field may not be perceived) and increasing slowness of information processing. In the physical area, decline of oculomotor functioning (which impairs the ability to scan the traffic scene) and conditions such as arthritis that reduce joint mobility (which may reduce the individual's capacity to execute vehicle control actions) are the most important (Janke, 1994; Oxley et al., 2006).

With some types of functional decline, older drivers can compensate for the decline by consciously, or unconsciously, adjusting their driving pattern. For example, older drivers with decreased contrast sensitivity may avoid driving at night and in bad weather (Langford and Koppel, 2006). With other types of functional decline this strategy may not be possible. In particular, it is not possible to compensate for impaired detection of angular motion. This affects the older driver's ability to judge the velocity and distance of approaching vehicles and so can lead to difficulty in identifying appropriate gap selection at intersections (Scialfa et al., 1987). Similarly, it is difficult to compensate for narrowing of the attentional visual field, especially if this is accompanied by reduced oculomotor function and/or reduced neck mobility. These types of decline reduce the older driver's ability to perceive other vehicles approaching from the side and so, again, may tend to increase their crash risk at intersections.

Method

The aim of this study is to investigate whether intersection crashes are particularly dangerous for older drivers in comparison to other crash types. Police-reported injury crashes that occurred in the state of Victoria, Australia, in the 15 years to the end of 2005 (N = 255,432) are analysed to investigate differences in the severity of injury outcome by crash type for three cohorts of crash-involved drivers: (i) those aged 60 years and over, (ii) those aged 75 years and over, and (iii) those aged between 30 and 49 years (the reference cohort).

Results

Older drivers' over-representation in intersection crash data

The 255,432 injury crashes in the dataset involved 395,542 drivers. Of these, 147,308 were aged between 30 and 49 years, 39,599 were aged 60 years and over and 10,225 were aged 75 years and over. Table 28.1 shows a breakdown of the major crash types for each age cohort expressed as a percentage of all the crashes that the cohort was involved in. From this it can be seen that the older drivers in the dataset were slightly less likely to have hit a pedestrian than the drivers aged 30 to 49 years. They were also less likely to have been involved in a head on (not overtaking), rear end or overtaking crash. For the cohort aged 75 years and over the percentage of crashes for these three crash types was approximately half that of the cohort aged between 30 and 49 years. The proportion of off-path crashes was approximately the same for all three age cohorts.

Table 28.1   Crash type by crash-involved driver age group (% of all injury crashes for cohort)

Crash type	Driver age group		
	30–49	60+	75+
Pedestrian	5.7	4.9	4.8
Head on (not overtaking)	5.3	4.1	2.7
Rear end	31.1	21.2	15.0
Overtaking	1.3	1.0	0.7
Off path	11.0	10.9	11.7
Cross traffic	10.2	15.5	17.4
Intersection with turning movement	18.6	23.3	25.7
Other	16.8	19.1	22.0
Total	**100.0**	**100.0**	**100.0**

Of particular interest to the present study are differences observed between the three cohorts in relation to cross traffic crashes and intersection crashes involving a turning movement (e.g. right-through crashes). Cross traffic crashes accounted for 15.5 per cent of all crashes involving drivers aged 60 years and over, rising to 17.4 per cent of all crashes involving drivers aged 75 years and over, compared to only 10.2 per cent for drivers aged between 30 and 49 years. Intersection crashes involving turning movements accounted for 23.3 per cent of all crashes involving drivers aged 60 years and over, rising to 25.7 per cent for crashes involving drivers aged 75 years and over, compared to 18.6 per cent of drivers aged between 30 and 49 years. Taken together these figures indicate that in Victoria in the period 1991 to 2005, crash-involved drivers aged 60 years and over were 35 per cent more likely to have been involved in an intersection crash than crash-involved drivers aged between 30 and 49 years. Crash-involved drivers aged 75 years and older were 50 per cent more likely to have been involved in an intersection crash. These results are in agreement with previous analyses of datasets from other Australian states and the United States (Lyman et al., 2002, and Meuleners et al., 2006).

Increased fragility of older drivers

Table 28.2 shows the differential fatality rates associated with the crash types discussed above for the two older cohorts relative to the cohort aged between 30 and 49 years (pedestrian crashes have been excluded because the number of associated driver fatalities was, understandably, very low). For all crash types fatality rates were higher for older drivers. Overall, drivers aged 65 years and over were 2.4 times more likely to have been killed than drivers in the reference cohort; this figure rose to 3.8 for drivers aged 75 years and over. However, there was significant variation in the differential rates for different crash types. Off-path crash types were associated with the lowest differential rates. Drivers aged 65 years and over and 75 years and over were 1.2 and 1.4 times, respectively, more likely to have been killed than the reference cohort. For head-on (not overtaking) crashes the differential rates rose to 2.5 and 3.9 times that of the reference cohort, and for overtaking crashes to 2.1 and 5.1. For the cross traffic crashes in the dataset, the differential fatality rates were 4.0 for the 60 years and over cohort and 7.3 for the 75 years and over cohort. For rear-end crash types these figures rose to 5.7 and 11.3. The highest differential rates were observed for intersection crash types involving turning movements. In these, drivers aged 60 years and over were 10.5 times more likely to have been killed and drivers aged 75 years and over were 20.6 times more likely to have been killed.

Table 28.3 shows the differential severe injury rates associated with the crash types discussed above for the two older cohorts relative to the cohort aged between 30 and 49 years (again pedestrian crashes have been excluded). For clarification, in the dataset used for the analysis a 'severe injury' is defined as one for which the victim required hospitalisation for one night or more. As with the fatality data, for all crash types severe injury rates were higher for the older drivers. However, the

Table 28.2 **Differential fatality rates by crash type and crash-involved driver age group (relative to cohort aged 30–49 years)**

Crash type	Driver age group		
	30–49	60+	75+
Head on (not overtaking)	1.0	2.6	3.9
Rear end	1.0	5.7	11.3
Overtaking	1.0	2.1	5.1
Off path	1.0	1.2	1.4
Cross traffic	1.0	4.0	7.3
Intersection with turning movement	1.0	10.5	20.6
Other	1.0	3.5	5.5
All crash types	**1.0**	**2.4**	**3.8**

Table 28.3 **Differential severe injury rates by crash type and crash-involved driver age group (relative to cohort aged 30–49 years)**

Crash type	Driver age group		
	30–49	60+	75+
Head on (not overtaking)	1.0	1.3	1.5
Rear end	1.0	1.6	2.3
Overtaking	1.0	1.3	2.0
Off path	1.0	1.2	1.4
Cross traffic	1.0	1.5	1.8
Intersection with turning movement	1.0	1.7	2.3
Other	1.0	1.4	2.6
All crash types	**1.0**	**1.5**	**2.0**

differences between the rates for the three cohorts were far less marked. Overall, drivers aged 65 years and over were 1.5 times more likely to have been seriously injured than drivers in the reference cohort; this figure rose to 2.0 for drivers aged 75 years and over. In relation to the severity of injury outcome associated with the various crash types, similar trends were observed to those seen in the fatality data with cross traffic, rear-end and intersection crashes involving turning movements being the most injurious, when the two older driver cohorts were considered together.

Police reported injury crashes in Victoria can also be assigned a third level of injury severity, 'other injury'. These are defined as injuries for which the victim required treatment at the scene, or as an outpatient, but were not as severe as to require their admission to hospital. When the other injury crash data was analysed in the same manner as the fatal and serious injury data presented above, no age-related patterns were observed.

Discussion

For all main crash types except off-path crashes, markedly higher fatality rates were observed for older drivers in comparison to the rates for younger mature adult drivers. Serious injury rates for older drivers were also generally elevated but not to the same extent as the fatality rates. No age-related patterns were observed in relation to 'other injury' rates. This apparent decrease in fragility with decreasing severity of injury outcome seems likely to be a function of crash severity. In general, older adults will certainly have lower biomechanical tolerances than younger mature adults but these tolerances may only be approached in crashes of higher severity. This could explain the observed absence of significantly higher fatality rates amongst older drivers involved in off-path crashes, i.e. the older drivers lost control of their vehicles at lower speeds than the younger mature adults and so were involved in crashes of lower severity. Crash severity was not corrected for in the present analysis because of the limitations of the dataset analysed. The only available variable which could have been used for correcting for crash severity would have been the speed limit of the road on which the crash occurred. In general, this is an imperfect method of accounting for crash severity. When making comparisons between older drivers and younger mature adult drivers it would probably be even more unsatisfactory due to age-related speed choice biases. Repeat of the present analysis using a dataset containing more objective measures of crash severity would allow for the testing of the above hypothesis.

Cross traffic, rear-end and intersection crash types involving turning movements were associated with particularly high differential fatality and serious injury rates for the older drivers. That cross traffic and intersection crashes should be so injurious to older drivers is not particularly surprising. As mentioned in the introduction, these types of crash tend to result in side and corner impacts which result in rapid rotation or spinning of the struck vehicle. This can induce significant loading of the occupant's chest by the seat belt which older adults are less able to sustain than younger adults. Additionally, the 'fatal' category in the dataset analysed includes drivers who died from their injuries up to 30 days after the crash and, as previously mentioned, serious chest injuries are three times more likely to be fatal to older adults due to an increased tendency to develop post-trauma complications such as pneumonia.

The observation that rear-end crashes should be so injurious to older drivers is more surprising. Rear-end crashes tend to be of relatively low severity because the major component of the vehicles' velocities are in the same direction and so the change of velocity experienced by the vehicles tends to be relatively low. The body area most typically injured in a rear-end crash is the neck and middle back. A possible hypothesis for the elevated fatality rates observed in the older drivers in the dataset would be reduced biomechanical tolerance in these regions in older adults. Osteoporosis affects most postmenopausal women to some degree and about 10 per cent of men in the same age group. Along with the bones of the wrist and hip, the vertebrae display the most marked reduction in mechanical performance through the structural changes which occur with osteoporosis. An alternative, or possibly complimentary, hypothesis would be that the mean travel speed of older drivers is lower than that of younger adult drivers and so if they are involved in a rear-end crash it tends to be a more severe crash than it would be if they were travelling at the mean speed of the surrounding traffic.

As in many other industrialised countries, in Australia the proportion of persons aged 60 years and over is predicted to substantially increase over the next few decades, with the greatest increase in the group aged 75 years and over. The results of the analysis presented in this paper indicate that cross traffic, rear-end and intersection crash types involving turning movements are particularly injurious to older drivers. However, involvement in rear-end crashes was observed to significantly decrease with increasing age whereas involvement in cross traffic and intersection crash types involving turning movements was observed to significantly increase with increasing age. To avoid a commensurate increase in the incidence of intersection crashes, with their associated high mortality and serious injury rates for older drivers, it would seem timely to target resources and develop effective countermeasures now.

References

Australian Bureau of Statistics (1999). *Older people: A Social Report*. Canberra, ABS.

Bergeron, E. et al. (2003). Elderly trauma patients with rib fractures are at greater risk of death and pneumonia. *The Journal of Trauma*, 54, 478–485.

Bulger, E.M. et al. (2000). Rib fractures in the elderly. *The Journal of Trauma*, 48, 1040–1046.

Fildes, B.N. et al. (2001). Older driver safety – a challenge for Sweden's Vision Zero. Australian Transport Federation Conference, Hobart.

Hakamies-Blomqvist, L.E. (1993). Fatal accidents of older drivers. *Accident Analysis and Prevention*, 25, 19–27.

Hu, P.S. et al. (2000). *Projecting Fatalities in Crashes Involving Older Drivers*. Oak Ridge, TN, National Highway Traffic Safety Administration.

Janke, M. (1994). *Age-related Disabilities that may Impair Driving and their Assessment: Literature Review*. Sacramento, CA, California Department of Motor Vehicles.

Kent, R. and Patrie, J. (2005). Chest deflection tolerance to blunt anterior loading is sensitive to age but not load distribution. *Forensic Science International*, 149, 121–128.

Langford, J. and Koppel, S. (2006). Epidemiology of older driver crashes – identifying older driver risk factors and exposure patterns. *Transportation Research Part F*, 9, 309–321.

Lyman, S. et al. (2002). Older driver involvements in police reported crashes and fatal crashes: Trends and predictions. *Injury Prevention*, 8, 116–120.

Meuleners, L.B. et al. (2006). Fragility and crash over-representation among older drivers in Western Australia. *Accident Analysis and Prevention*, 38, 1006–1010.

OECD (Organisation for Economic Cooperation and Development) (2001). *Ageing and Transport: Mobility Needs and Safety Issues*. Paris, OECD Scientific Expert Group.

Oxley, J. et al. (2006). Intersection design for older drivers. *Transportation Research Part F*, 9, 335–346.

Scialfa, C. et al. (1987). Age differences in judgements of vehicle velocity and distance. 31st Human Factors Society Conference, New York.

Taylor, M.D. et al. (2002). Trauma in the elderly: Intensive care unit resource use and outcome. *The Journal of Trauma*, 53, 407–414.

Index

Note: **Bold** page numbers indicate illustrations; *italic* page numbers indicate tables and graphs.

16PF personality test 273, 274

AAP (Australian Applied Psychology Ltd.) 132, 134, 136, 137
accident criteria 6
accident proneness 3–4
accident records:
  and exposure 4–5
  stability over time of 3–4
accident risk:
  and driver personality 89–90
  and motorcyclists 149, 161, 172–3, 175, 176, 179, 189, 193, 203–4, 205–7
  and young drivers 7–8, 14–17, 75
accidents *see* road traffic accidents
accompanied driving 16
achievement-striving (NEO PI-R attribute) *97*, 272
ACPO (Association of Chief Police Officers) 162
adolescence:
  and accident risk 7–8, 14–17
  neuropsychology 8–9, 76–7
  problem behaviour 16–17
  and risky behaviour 10–14, 76–7, 149–54
  see also young drivers
ADTA (Australian Drivers Training Association) 65
advocacy campaigns, on rear seatbelt wearing 122–3, *123*, 126, *126*, 127–8
af Wåhlberg, A.E. 3–6, 92, 248, 249
affect, and behaviour 24, 62
affection 91

age:
  and driver distraction 290–1, *291*, *292*
  and driving behaviour 14–15
  see also middle-aged drivers, older drivers, young drivers
aggression *68*, *69*, *70*, 79
  and driving style 63, 90, 91, *93*, 94, 96, *97*, 98, 99, **101**, 332
aggressive violations:
  DBQ factor 214, 244
  MDBQ factor *218*, 223
agreeableness (FFM factor), and driving style 89, 90, 91, *95*, 96, *97*, *98*, 99, 100, 101, **101**
agricultural industry, safety initiatives 28, 275–6, 282–3
Ajzen, I. 139
Akintola, Ladoke 193–208
alcolock devices 23
Allan, Sandy 179–91
Allied Bakeries (AB), and TruckSim training 264–5
altruism, and driving style 62, 63, 90, *97*
angry hostility (NEO PI-R attribute) 272, 274
  and driving style 94, 96, *97*, *98*, 99
antisocial activities 11
António, Patricia 149–57, 247
Anttila, V. 297
anxiety, and driving style 90, *97*, *98*
'Around the Corner' scheme 179–91, *182*, *185*, *186*, *187*, *188*, *189*
assaults on bus drivers 270, 272, 273
assertiveness *97*
assessment of driving *see* DPA (driver performance assessment)
asset management approach to safety data collection 242–3
attitude change 139

of motorcyclists 171, *171*, *172*, 187–9, *188*, *189*
Austin, J. 233, 234
Australia:
  agricultural industry safety initiatives 275–6
  demographic changes 339–40
  driver distraction studies 287–8, 293, 294
  occupational road safety initiatives study 277–84
  older driver study 340–5, *341*, *343*
  professional drivers, safety 215–24, *218*, *219*, *221*, 231, 235, 245
  road traffic accidents 213, 241–2
Australian Applied Psychology Ltd. (AAP) 132, 134, 136, 137
Australian Drivers Training Association (ADTA) 65
Australian Safety and Compensation Council 241
Austria, driver rehabilitation programmes 131–7, *133*, *135*
autonomy, and safety initiatives 282, 283
aversion to risk-taking 66, *68*, *69*, *70*, 71
avoidance coping 62, 65, *68*, *69*, *70*, *94*, *98*
awareness campaigns:
  motorcyclists *see* 'Around the Corner' scheme
  professional drivers 233

Baas, P 287, 294
backseat passengers *see* seatbelt wearing
Ballard, M.E. 79
Banks, Tamara 229–37, 275–84
Barbour, Robert 317–24
Bardodej, Julia 131–7
barriers to implementing safety initiatives 275–6
  occupational road safety study 277, 278–84
Bas, M. 283
base rate neglect 5
Bechara, A. 25
behaviour, relationship to intention 12
behavioural beliefs (motorcyclists) 165, 166, *166*

Benedetto, Andrea 297–305
BFI (Big Five Inventory) 89
bias in research data 4
Big Five Inventory *see* BFI
Biggs, Herbert 229–37
Bjerre, B. 235
booklets, in road safety campaigns 183
borderline personality disorder 154
Bos, J.M. 236
brain development in adolescents 8–9, 15, 76–7, 152
braking hard (motorcyclists) 186–7, *186*, *187*
Broughton, Paul S. 152, 155, 161–76, 179–91, 327–36
burden, of occupational safety initiatives 279, 283, 284
Burgess, Cris 161–76, 244
bus drivers 269–71
  psychometric testing study 271–4
  traffic enforcement in Malaysia 106–7

Calvi, Alessandro 297–305
cameras *see* in-vehicle cameras
Canada, professional driver safety 235
cancellation test 271–2
carbon dioxide *see* emissions
cargo barriers 280
CAS (Crash Analysis System) **319**, 322, 324
caudate nucleus 77
CBR (Dutch National Driving Examination Institute) 41
CBT (computer-based training) 260
Certificate of Professional Competence (CPC) 257, 266–7
change:
  change fatigue 280
  resistance to 276
Chattington, M. 89–101
chest injuries in road traffic accidents 339, 344
Chiang, W. 236
Chinese New Year celebrations, and road traffic accidents *see* Ops Sikap XVI
close following 76, 79, 213
  DAQ factor 244

coach drivers, and safety 246
cognitive control system 10, 11, *11*
cognitive decline in older drivers 340
cognitive dissonance theory 234
common method variance 4
compensation, workers' claims for 241
compensatory devices, in-vehicle 235–6
competence *97*
complacency, and occupational road risk
278, 282, 284
compliance (NEO PI-R attribute) *97, 98,*
272
computer-based training (CBT) 260
confrontive coping 65, *68, 69, 70*
and driving style 62, 63, 71, 94, *94,*
96, *98,* 99, 100, **101**
conscience 91
conscientiousness 274
and driving style 89, 91, *95, 97, 98,*
99–100, 101, **101**
consideration for other road users 39–40
consultation, lack of and occupational road
risk 279, 282, 284
contemporary risks (MDBQ factor)
217–18, *219, 221,* 222, 224
control beliefs (motorcyclists) 165, 166,
166, 167
control errors (MRBQ factor) 194, 195
control/safety (MRBQ factor) 199, *200–1,*
201, *203, 204,* 205
controlled driving 40
corporate social responsibility (CSR)
266
cortex 8–9, 77, 152
cost, and occupational road risk 279
covert traffic enforcement 106, 107–8
CPC (Certificate of Professional
Competence) 257, 266–7
Cranfield University 248
crashes *see* cross traffic crashes, head
on crashes, intersection crashes,
KSI crashes, off path crashes,
overtaking crashes, pedestrian
crashes, rear end crashes, road
traffic accidents
Craw, I. 298
CRISS (Inter-universities Research Centre
for Road Safety) 300

cross traffic crashes *341*, 342, 343, *343,*
344, 345
Crundall, D. 25, 32, 33–4
CSR (corporate social responsibility) 266
Cunill, Mònica 139–44
customer service, by bus drivers 270, 271
Cynk, Stephanie 257–67

d2 test of attention 271–2, 273
Dahlen, E.R. 77
Damasio, A.R. 24–5, 33, 34
DAQ (Driver Attitude Questionnaire) 243,
244–5
data recorders, in-car 236
Davey, Jeremy 213–24, 229–37, 241–51,
275–84
daytime driving, gaze behaviour study
309–14, *310, 311, 312, 313, 314*
DBI (Driver Behaviour Inventory) 90–1,
98, 99, 100
DBQ (Driver Behaviour Questionnaire)
76, 89, 90, 91, 92, 193, 194, 214,
243–4, 245, 247, 249
and driving style *95,* 96, 98, 100
and professional drivers 214–24, *218,*
219, 221
DCQ (Driver Coping Questionnaire) 65,
91, 92, 247
and driving style *94,* 96, *98,* 99, 100
DDDI (Dula Dangerous Driving Index)
79, 80, 81, 84
De Blasiis, Maria Rosario 297–305
deaths in road traffic accidents 7, 90, 119,
328
Malaysia *106,* 107, 116–17, 120, *121,*
124
motorcyclists 161, 172, 175, 179, 181,
182, 189, 191, 193
older people 340, 342, 343, *343,*
344–5
work-related driving 213, 241
young people 14, 75, 139, 151
decision-making, in adolescents 8, 11–14,
152
Deffenbacher, J.L. 243, 247
deliberation (NEO PI-R attribute) *97, 98,*
272
delivery drivers *see* professional drivers

demerit points 131, 220, 224
    see also penalty points
demographic changes 339–40
DePasquale, J.P. 247
depression, and driving style *97, 98, 99*
deviant beliefs (motorcyclists) 166, *166,*
    167, 171, *171, 172,* 173
diaries, driving 247
DiClemente, C.C. 165, 174
dislike of driving *68, 69, 70*
    and driving style 91, *93, 97,* 98, 99
dismissal, of bus drivers 270
distraction:
    and driving 216, 219, *219,* 222, 223,
        224, 247, 287–8, 297, 332, 335
        England study 288–94, *290, 291,*
            292, 293
        see also eye-movements of drivers
diversity, and occupational road risk 279,
    282, 284
Dorn, L. 90–1, 99, 100, 214, 247, 248
DPA (driver performance assessment) 39
    study 39–49, *43, 44, 45, 46, 47*
drink-driving 12, 13, 79, 140, *142,* 143
    and accident risk 161
    alcolock devices 23
    DAQ factor 244, 245
    driver rehabilitation courses 131–7,
        133, 135
    and motorcyclists 169–70, 173, *204,*
        207
drinking while driving 288, 289, 290, 293
Driver Anger Scale 243, 247
Driver Attitude Questionnaire *see* DAQ
Driver Behaviour Inventory *see* DBI
driver behaviour modelling 23–4, 61
Driver Behaviour Questionnaire *see* DBQ
driver coping strategies 62–3
    Driver Coping Questionnaire *see* DCQ
driver distraction *see* distraction
driver performance assessment *see* DPA
driver performance criteria 39–40
driver personality:
    and driving style 63, 89–91
        study 91–6, *93, 94, 95, 97,*
            98–101, *98, 99,* **101**
driver rehabilitation programmes, Austria
    131–7, *133, 135*

driver selection 231
    bus drivers 270, 271, 272–3
Driver Stress Inventory *see* DSI
driver training 15–16, 37–9
    and accident reduction 75
    Driver Training Stepwise (DTS) 42,
        45, 48
    mopeds 149–57
    study 39–49
Driving Attitudes Survey 64–5
driving behaviour, online assessment of
    247–8
driving competence, model of 38, **38**
driving diaries 247
driving frequency 63
driving instructors:
    and DPA study 40, 41, 42
    education of in Norway 51–6
driving lessons 55–6, *55*
driving skill 75
Driving Skill Inventory *see* DSI
driving style 76
    and driver personality 63, 89–91
        study 91–6, *93, 94, 95, 97,*
            98–101, *98, 99,* **101**
driving tests, prediction of results 44–8,
    44, 45, 46, 47
driving under the influence (DUI) *see*
    drink-driving
drug-driving 247
DSI (Driver Stress Inventory) 65–6, 90,
    92, 247, 248
    and driving style 63, *93,* 96, *97,* 98, 99
DSI (Driving Skill Inventory) 243, 247
DTS (Driver Training Stepwise) 42, 45, 48
dual process theories 9–10, 15–16
    and adolescent risk-taking 10–14
DUI (driving under the influence) *see*
    drink-driving
Dula, C.S., Dula Dangerous Driving Index
    79, 80, 81, 84
Dunne, Simon 75–84
Dutch National Driving Examination
    Institute (CBR) 41
dutifulness *97, 98*

EADS 257
    see also TruckSim

eating while driving 213, 216, 288, 289, 290, 292, 293
Efficacy scale 66, *68, 69, 70*
Elliott, M.A. 193–4, 207
EMG (electromyographic) responses 25
Emile (Rousseau) 8, 16
emissions 38, 40, 89, 265, 266
emotion-focused coping 62, 63, 65, *68, 69, 70, 98*
emotional and motivational systems of the brain, development of 8, 9
emotional responses during driving 25–34
empathy 54
environmentally responsible driving 40
EPQ (Eysenck Personality Questionnaire) 90, 91
Epstein, S. 24
Eriksson, M. 298
errors 8
    DBQ factor *95*, 214, 222, 244
    MDBQ factor 217–18, *218, 219, 221*, 222, 223–4
    MRBQ factor 199, *200–1*, 201, *202, 204*, 205
    omission 82–3, *83*
European Commission Directive on Training for Professional Drivers 257, 266
Evans, J. 9, 10, 15
excitement seeking, and driving style 62, 63, 96, *97, 98, 99*, 100, 101, **101**
experience induced changes 8
experienced drivers, hazard perception 26, 29–34, *29, 30, 31*
experiential development 15
explicit knowledge 52
exposure, and accident risk 4–5, 6, 15, 17, 75, 179, 214, 216, 220–1, 223, 241, 243, 245, 297, 333–4
extraversion 274
    and driving style 89, 90, 96, 99, 100
eye-movements of drivers 297–9
    gaze behaviour study 307–15, *308*, **309**, *310, 311, 312, 313, 314*
    variable message signs study 299–305, *299*, **300, 301**, *302, 303, 304, 305*
Eysenck Personality Questionnaire (EPQ) 90, 91

facilitators:
    of implementing safety initiatives 275, 276
    occupational road safety study 277, 281–2, 283, 294
failure to yield 76
Farley, F. 12, 13
fatalities *see* deaths in road traffic accidents
fatigue, and driving 24, 213–14, 216, 219, *219*, 222, 223, 224, 241, 244, 298, 331–2, 335
    fatigue management technologies 235
    fatigue proneness *68, 69, 70*, 96, *97*
FDRI (Fleet Driver Risk Index) 248
feelings, and driving behaviour 23–4
female drivers:
    accident risk 17, 75, 76, 149, 151, 329–30, 336
    and driver distraction 290, *290, 291*, 293–4
    driving behaviour 140–2, *141, 142*, 143
Festiger, L. 234
FFM (Five Factor Model) 89, 90, 91, 100
FJC programme (moped riders, Portugal) 149–50, 155–6
FKK (Krampen's Questionnaire for Locus of Control and Competence Beliefs) 134
Fleet Driver Risk Index (FDRI) 248
fleet drivers *see* professional drivers
fleet safety *see* occupational road safety
FMS (Full Mission Simulator) 257
following too close 76, 79
Font-Mayolas, Sílvia 139–44
food safety 283
Fox, M.R.J. 297–8
Freeman, James 213–24, 241–51
fuel efficiency:
    and driving style 40, 89, 101
    and simulator training 261–4, *263*, 266
Full Mission Simulator (FMS) 257
Fuller, R. 23–4, 79, 327, 328, 333, 334
functional decline, in older drivers 340
fuzzy trace theory 12, 13
Fylan, Fiona 161–76

Gandolfi, J. 247
gasoline vapour recovery devices 236
gaze behaviour *see* eye-movements of
    drivers
GDE (Goals for Driver Education) 37, 56
Geller, E. 231, 232, 234
gender:
    and accident risk 75, 76, 149, 151,
        329–30, 336
    and driver distraction 290, *290, 291,*
        293–4
    and driving behaviour 17, 140–2, *141,*
        142, 143, 144, 181
    and hazard perception 30, 32
    impulsivity study 81–3, *82, 83,* 84
Gerrard, M. 12–13
Giannopulu, I. 299
gists 13
Glendon, A.I. 246
Go/NoGo task study 78, 79–84, *81, 82, 83*
Goals for Driver Education (GDE) 37, 56
Gormley, Michael 75–84
Gottschalk, P. 52
Gras, Maria Eugènia 139–44, 214, 243,
    244, 287
Gregersen, N. 232–3, 234, 237
Griffin, M. 246
group discussions, and professional driver
    safety 232–3, 236, 237
group driving simulations 317–18, **319,**
    320–2, **323,** 324
group goal-setting, and professional driver
    safety 233
Guattari, M. Claudia 297–305

Harrison, A.H. 89–101
hazard monitoring *68, 69, 70, 97*
hazard perception 23–5
    study 26–34, **27, 28,** *29, 30, 31*
    see also risk perception
head on crashes 341, *341,* 342, *343*
health care, safety practices 283
'hearts and minds' campaigns, and
    motorcyclists 175, 180, 190
Helander, M. 25
helmet use (motorcyclists) *112, 113, 114,*
    121, 140, *142,* 143
Hewitt, C. 322

hierarchical structures, and safety
    initiatives 280–1
high-mileage drivers 4–5
Highway Code violations (MDBQ factor)
    217–18, *218, 219, 221,* 222–3
Hillard, Peter 339–45
Horberry, T. 287–8, 293
hormones, and behaviour 9
hospitality industry 283

icons, in driving simulations 318, **319,**
    322, **323,** 324
IGT (Iowa Gambling Task) 25
impulse control 152
impulsiveness 77–8, *97, 98*
    and driving behaviour 78–84
in-vehicle cameras 288
    eye-movement tracking studies 298,
        300, 307
in-vehicle compensatory devices 235–6
in-vehicle data recorders 236
inattention *see* distraction
incentive programmes, and safety
    initiatives 276, 283
incident data systems 280
individual differences driver research 5–6
inexperienced drivers:
    and accident risk 75–6
    hazard perception 26, 29–34, *29, 30,*
        31
information 52
information processing 9
Institute for Safety in Mining Industry and
    Transport 271
Insurance Institute for Highway Safety 75
intention, relationship with behaviour 12
Inter-universities Research Centre for Road
    Safety (CRISS) 300
Internet:
    'Around the Corner' safety scheme
        website 180, 184–6, *185,* 190
    web-based assessment of driving
        behaviour 247–8
    web-based risk management tools 231
intersection crashes 339, 340–2, *341,* 343,
    343, 344, 345
Iowa Gambling Task 25
Iversen, H. 61–2

J. Stork Suicide Risk Scale 156
Japan, professional driver training 232
Jonah, B.A. 77, 90

Kawato, S. 298
Kee, L.S. 105–17
Kelly, Steve W. 23–34
Kettwich, Carmen 307–15
King, Mark 229–37
Kinnear, Neale 23–34
Kirkpatrick, D. 132, *132*
Kjelsrud, Hilde 51–6
knowledge 52, 283, 335
    motorcyclists 165, *166*, 188–9, *189*
Kolb's model of learning 52–3, **53**
Kostela, J. 235
Krampen's Questionnaire for Locus of
    Control and Competence Beliefs
    (FKK) 134
KSI (killed and seriously injured) crashes,
    and motorcyclists 161, 175, 179,
    190–1

Lajunen, T. 90, 243, 246–7
landmarks, and accident risk 335
lapses (DBQ factor) *95*, 214, 244
learning process 52–4
learning to drive *see* driver training
Lemmer, Uli 307–15
LGVs (large goods vehicles) 257
licensing:
    graduated 7, 16
    of young drivers 7–8, 14
lie scale 249
likelihood of accident 66, 67, *68*, *69*, *70*,
    71
Litherland, D.K. 246
Llaneras, R 232, 235
local drivers, and road accidents 329, *330*,
    332, 335, 336
Lord, Wendy 269–74
Lothian and Borders Police, 'Around the
    Corner' scheme 179–91, *182*, *185*,
    186, *187*, *188*, *189*
Ludwig, T. 231, 232, 233, 234–5
Luke, T. 89–101

McEvoy, S. 288

Machin, M. 61–72, 246
Malaysia:
    Malaysian Institute of Road Safety
        Research (MIROS) 122
    Malaysian Road Safety Plan (2006–
        2010) 119–20
    new vehicle registrations 119, *120*
    road traffic accidents 119–20
    traffic enforcement 105–7, *106*
        study 108–9, *110*, 111–17, *111*,
        112, *113*, *114*, *115*, *116*
male drivers:
    accident risk 7, 17, 75–6, 149, 151,
        329–30, 334, 336
    and driver distraction 290, *290*, *291*,
        293–4
    driving behaviour 63, 70–1, 140–2,
        141, *142*, 143, 144
management commitment, and safety
    initiatives 276, 281, 283, 284
Manchester Driver Behaviour
    Questionnaire *see* DBQ
Martens, M.H. 297–8
Maslina, M. 105–17
Matos, Manuel 149–57
Matthews, G. 61, 62–3, 65–6, 90, 91, 98,
    99, 100, 247
MDBQ (Modified Driver Behaviour
    Questionnaire) 216–21, *218*, *219*,
    221
media campaigns, and road safety 107,
    122–3
Mejza, M.C. 231, 232, 235
middle-aged drivers, and driver distraction
    291, *292*
Mimica, R.M. 298
MIROS (Malaysian Institute of Road
    Safety Research) 122
mobile phone use whilst driving 140, 141,
    142, 143, 213, 216, 223, 241, 287,
    289, 290, 291, 292–3
Modified Driver Behaviour Questionnaire
    see MDBQ
Mohamed, Norlen 119–28
mopeds, rider training programmes 149–57
Moromoto, C.H. 298
motivation to ride safely (motorcyclists)
    165, *166*

motivation-volition (motorcyclists) 166, *166*, 171, *171*
Motorcycle Rider Behaviour Questionnaire *see* MRBQ
motorcycles, Power Index (PI) 181
motorcyclists:
    accident rates 149, 161, 172–3, 176, 179, 189, 193, 206–7
    commercial, Nigeria study 194–5, *195–7*, 198–9, *199*, *200*, 201, *201*, *202*, 203–8, *203*, *204*
    deaths in road traffic accidents 161, 172, 175, 179, 181, *182*, 189, 191, 193
    police 180, 182–3, 184, 185, 190
    rider education programmes:
        'Around the Corner' scheme 179–91, *182*, *185*, *186*, *187*, *188*, *189*
        National RIDE Scheme (UK) 161–76, *164*, *166*, *167*, *168*, *169*, **170**, *171*, **172**
        Portugal 149–50, 149–57, 154–7
    stunts (MRBQ factor) 194, 195, 199, *200–1*, *202*, *203*, *204*, 205
motorways 321
    gaze behaviour study *308*, 313–14, *313*, *314*
MRBQ (Motorcycle Rider Behaviour Questionnaire) 193–4
    Nigeria study 194–5, *195–7*, 198–9, *199*, *200*, 201, *201*, *202*, 203–8, *203*, *204*
multi-tasking, and driving 216, 222, 244
Murray, W. 231, 248, 283
musculoskeletal disorders, safety initiatives 275, 276, 277, 282, 283
myths in driver research 3–6

Nakayasu, H. 297
National Highway Traffic Safety Administration (NHTSA) 278
National RIDE Scheme (UK) 161–3
    evaluation study 163–76, *164*, *166*, *167*, *168*, *169*, **170**, *171*, **172**
Neal, A. 246
Nechtelberger, Franz 131–7
Nechtelberger, Martin 131–7

negative emotions, while driving 79
NEO PI-R personality inventory 91, 92, 272, 273, 274
    and driving style *95*, *97*, *98*, *99*
neurodevelopment *see* brain development
neuroticism, and driving style 89, 90, 91, 94, *95*, 96, *97*, *98*, 99, 100, 101, **101**
New Zealand:
    CAS (Crash Analysis System) **319**, 322, 324
    mobile phone use while driving 287
Newnam, S. 235
Nigeria, motorcycle use 193–4
    study 194–5, *195–7*, 198–9, *199*, *200*, 201, *201*, *202*, 203–8, *203*, *204*
night driving, gaze behaviour study 309–15, *310*, *311*, *312*, *313*, *314*
noise levels 40
non-problematic drivers 80, 81–5, *81*, *82*, *83*
Noradrenalina, Isah 105–17, 119–28
Nord-Trondelag University College, Norway 51, 54–6
norms (motorcyclists) 165, *166*, 167, 171, *171*
Norway, education of driving teachers 51–6
Nottingham City Transport (NCT), bus driver study 269–74

O'Brien, Fearghal 75–84
obsessive-compulsive disorder 6
occupational road safety:
    behavioural influences study 229–37
    implementation study 277–84
    risk assessment tools study 241–51
OECD (Organisation for Economic Cooperation and Development) 339
off path crashes *341*, 342, *343*, 344
older drivers:
    and driver distraction 291, *291*, *292*
    road traffic injuries study 339–45, *341*, *343*
Olson, R. 233, 234
omission errors 82–3, *83*
online assessment of driving behaviour 247–8

Onna, Marieke van 37–49
openness, and driving style 89, *95, 97, 98, 99*, 100
Ops Sikap XVI traffic enforcement programme 105–7, *106*
    study 108–9, *110*, 111–17, *111, 112, 113, 114, 115, 116*
Organisational Safety Climate Questionnaire 246
organisational safety culture assessment 246–7
Organisational Safety Culture Scale 246
osteoporosis 345
overt traffic enforcement 106, 107–8
overtaking, dangerous 79, 213
    crashes 341, *341, 343*
    DAQ factor 244, 245
Oz, B. 246–7

Pacejka, H.B. 320
Papanikolopoulos, N.P. 298
Parker, D. 9, 23, 90, 98, 174, 191, 193, 194, 214, 222, 243, 244, 246, 248
Parkes, Andrew M. 257–67
participation, lack of and occupational road risk 279, 282, 284
passengers, safe loading and unloading of 233, 234
pay rises, and driver safety 231–2, 236
PDQ (Problem Driver Questionnaire) 79, 80, 81, 84
pedestrian crashes 341, *341*
peers, influence on adolescent behaviour 11, 14, 76, 77, 151, 154
penalty points:
    perceptions of in Spain 139–44
    UK 161–2
    see also demerit points
perception of risk of being caught (traffic offences) 109, 111–17, *111, 112, 113, 114, 115*
performance feedback, and professional driver safety 232–3
personality disorder 154
Peters, F. 24
PFC (prefrontal cortex) 9, 152
Planes, Monserrat 139–44
pledge cards 234

Plint, Janna E. 61–72
'Plot box and whisker' model 322
Police Driver Risk Index 247
police, 'hearts and minds' campaigns 175, 180
police motorcyclists, and 'Around the Corner' scheme 180, 182–3, 184, 185, 190
Portugal, moped rider training programme 149–50, 154–7
positive emotions, and driving style 96, 99, *99*, 100, **101**
Power Index (PI), motorcycles 181
predictors of accidents 4
prefrontal cortex (PFC) 9, 152
Prieler, Joerg 269–74
problem behaviour 154
Problem Driver Questionnaire (PDQ) 79, 80, 81, 84
problematic drivers 80, 81–5, *81, 82, 83*
Prochaska, J.O. 165, 174
production, prioritisation over safety 278, 282, 284
professional drivers:
    and safety 213–14
        Australia study 215–24, *218, 219, 221*
        safety initiatives study 229–37
    selection 231
    training 232, 236–7
Propensity for Angry Driving Scale 247
protective equipment, and professional driver safety 235–6
prototype willingness model 12–13
proxy safety variables 5
psychologists, role in moped rider training programme 149–50, 154–7
psychometric tests, and bus drivers 271–4
psychopathology 153, 154, 156
psychoticism 90
PTWs (powered two wheelers) 161
    see also motorcyclists
public awareness campaigns, rear seatbelt wearing 122–3, *123*, 124, 125–6, *125, 126*

racing 79
Ranney, T.A. 287

RAT (Risk Allostasis Theory) 23, 327,
          328, 333, 334–5
reaction time 81–2, *82*
reappraisal coping 62–3, 65, *68, 69, 70,
          94, 98*
rear end crashes 341, *341*, 342, 343, *343*,
          344, 345
rear seatbelts *see* seatbelt wearing
Reason, J. 89, 90, 193, 198, 214, 243
Reed, Nick 89–101, 257–67
reflecting teams, role in driving instructor
          training 51, 54–6
reflection, and learning 53–4
reminder signs 234, 236
reward-seeking, in adolescents 10
rewards, and professional drivers 234,
          236, 237
Reyna, V.F. 12, 13
RHMF (Road Haulage Modernisation
          Fund) 257
Rider Risk Reduction scheme 162
Risk Allostasis Theory (RAT) 23, 327,
          328, 333, 334–5
risk management tools, web-based 231
risk perception 24, 61
          assessing 61–2, 66
risk prototypes 13
risk-taking 8, 67, 79
          in adolescents 8, 149–54
          aversion to and speeding 64–72
Road Behaviour Questionnaire 247
Road Haulage Modernisation Fund
          (RHMF) 257
road safety campaigns:
          and motorcyclists 207–8
          'Around the Corner' scheme
                    179–91, *182, 185, 186, 187,
                    188, 189*
          and occupational safety initiatives
                    282
          and professional drivers 234, 236
          targeting 327–8
road traffic accidents 327
          deaths 7, 90, 119, 328
                    Malaysia *106*, 107, 116–17, 120,
                              121, 124
                    motorcyclists 161, 172, 175, 179,
                              181, *182*, 189, 191, 193

older people 340, 342, 343, *343,
          344–5*
work-related driving 213, 241
young people 14, 75, 139, 151
MRDB as predictor of 220, *221*, 223
Scotland study 327, 328–36, *330*
Road Traffic Act (1988) 161
Road Transport Department, Malaysia
          (JPJ) 106
Robbins, R. 89–101
Robbins, S. 246
Rodriguez, D. 231–2
Roelofs, Erik 37–49
Rousseau, J-J. 8, 16
Rowland, Bevan 213–24, 241–51
Rundmo, T. 61–2, 63, 67, 90
rural roads, gaze behaviour study *308,*
          311–12, *311, 312*

safe driving 39
SAFED (safe and fuel efficient driving)
          standard 261, 262
safety belts *see* seatbelts
Safety Climate Questionnaire 246
safety criteria 5
safety culture assessment 246–7
safety equipment use (MRBQ factor) 194,
          195
safety policy development 231
Salminen, S. 232
Sankey, K.S. 62, 63, 66, 67, 71
Scheltema, K. 233, 234
Scotland, road traffic accidents study 327,
          328–36, *330*
SCR (skin conductance response) 25
          in hazard perception study 26, 27–8,
                    30–2, *31*, 33–4
seatbelt wearing 141, *142*, 143
          professional drivers 232, 233, 234
          rear seat passengers 140, *142*
                    promotion of in Malaysia 119–20,
                              122–8, *125, 126, 127*
          reasons for not wearing 127, *127*
secondary activity *see* distraction
secondary sources 6
Section 3 offences (Road Traffic Act 1988)
          161
selective attention 271–2

bus drivers 269
self-consciousness *97, 98*
self-discipline (NEO PI-R attribute) *97,
    98, 99,* 272
self-efficacy beliefs 134, 136
self-monitoring forms, and professional
    drivers 234
self-reported driving behaviour 4, 67, 90,
    101, 243–4, 247–9
    DBI (Driver Behaviour Inventory)
        90–1, 98, 99, 100
    DBQ (Driver Behaviour Questionnaire)
        76, 89, 90, 91, 92, *95,* 96, 98, 100,
        193, 194, 214–24, *218, 219, 221,*
        243–4, 245, 247, 249
    limitations of 224, 249–50
    MDBQ (Modified Driver Behaviour
        Questionnaire) 216–21, *218, 219,
        221*
    motorcyclists 165, 168–70, *168, 169,*
        173
    and personality 96
sensation-seeking 91
    and accident risk 77, 90
    in adolescents 10, 76, 77, 151
sensory decline, in older drivers 340
side impacts 339
simulators 247
    and driving styles 91–4, *95,* 96, *97,*
        98–101, *98, 99,* **101**
    and eye movement studies 297–8, *299*
    group driving simulations 317–18,
        319, 320–2, **323,** 324
    simulator sickness 260, 261
    TruckSim 257–60, **258,** *259,* **259, 260**
        evaluation study 260–7, *263*
    variable message signs study 299–305,
        299, **300, 301,** *302, 303, 304, 305*
skin conductance response *see* SCR
Skippon, S.M. 89–101
Slovik, P. 24, 25, 32–3
SMH (somatic marker hypothesis) 24–5,
    33
smoking while driving 290, 292
sociability 274
socio-emotional system 10, 11, *11,* 76,
    77
somatic markers 24–5, 33

Spain:
    mobile phone use while driving 287
    perceptions of penalty point law
        139–44
speeding 5, 76, 79, 140, *142,* 143
    and accident risk 161
    and aversion to risk-taking 61–8, *68,
        69,* 70–2, *70*
    DAQ factor 244, 245
    and driver personality 63
    factors influencing 63–4
    MDBQ factor 216, 217, *219*
    motorcyclists 168–9, *168,* 173, 194
    MRBQ factor 194, 195
    perception of risk of being caught 112,
        112, 113, *113, 114*
    Speeding scale 67, *68, 69, 70*
    speeding/impatience (MRBQ factor)
        199, *200–1, 202,* 203, *203, 204,*
        205
Stages of Change model 165, *169,* 170,
    170, 174
'stairs of learning' 53–4
STATS19 data 328
Steinberg, L 10, 11, *11,* 76–7, 84
STI driving simulator 300
Stockey, Stefan 307–15
stopping distances 321
Stradling, Steve 23–34, 161–76, 186, 246,
    334, 335
Strategic Road Safety Interventions and
    Potential Fatality Reduction
    (2007–2010),
    Malaysia 120, *121*
stunts (MRBQ factor) 194, 195, 199,
    200–1, 202, 203, 204, 205
Stutts, J.C. 288, 293
Sullman, Mark J.M. 139–44, 214, 219,
    223, 244, 246, 248, 287–94
Sumer, N. 91, 99–100
Summala, H. 243, 247
Sunday, Oluwadiya Kehinde 193–208
supervised driving 7
supportive relationships, and safety
    initiatives 281, 284
susceptibility to collision involvement
    (motorcyclists) 165, 166, *166,*
    167

susceptibility-control (motorcyclists) *166*,
     167, 171, *171*
Sweden, professional driver safety 232–3,
     234, 235
synaptic pruning 9
System 1 and System 2 processes 9, 10,
     11, 12, 13, 15–16
systems, and safety initiatives 281, 283,
     284

tacit knowledge 52
Taiwan, gasoline vapour recovery devices
     236
task difficulty homeostasis 23
Task-Capability Interface Model 328
task-focused coping 62, 65, *68, 69, 70,
     98*
taxi drivers, and safety 245, 246
Taylor, D.H. 25
Taylor, D.M.D. 293
Tetsutani, N. 298
texting while driving 289, 290, 292
Theory of Planned Behaviour *see* TPB
Theory of Reasoned Action *see* TRA
Thomson, James 23–34
thrill seeking *68, 69, 70*
     and driving style 63, 71, *93*, 94, *97*,
        98, 99, **101**
     and motorcyclists 165, *166*, 167
thrill-culture, motorcyclists *166*, 167, 171,
     171
Tiller, T. 53–4
time halo effect, in traffic enforcement 108
time pressures, and driving 213, 216, 219,
     219, 222, 223, 224
tiredness *see* fatigue
Tock, D. 298
tourism, and road traffic accidents 327,
     328, 329, *330*, 335, 336
TPB (Theory of Planned Behaviour) 10,
     11–12, 139–40
TRA (Theory of Reasoned Action) 11–12,
     13
traffic enforcement:
     Malaysia 105–7, *106*, 108–9, *110*,
        111–17, *111, 112, 113, 114, 115,
        116*

overt and covert methods 106, 107–8,
     114–16, *116*
     rationale for 107
     time halo effect 108
traffic errors (MRBQ factor) 194, 195
traffic flow, facilitating 40
traffic offences *112*, 113–14, *113, 114*
     MRDB as predictor of 220–1, *221*,
        223, 224
traffic sign recognition test 297–8
training:
     transfer of 262–4
     see also driver training
Transtheoretical Model of Behaviour Change
     see Stages of Change model
Trinity College, Dublin 78
TRL 257, 261, 262, 264, 266
truck drivers:
     and distraction 288
     safety, US studies 231–2, 233, 234–5
     TruckSim training 257–67
     see also professional drivers
TruckSim 257–60, **258**, *259*, **259, 260**
     evaluation study 260–7, *263*
trust:
     NEO PI-R attribute *97, 98, 99*, 272
     role of in learning 54
Turkish drivers, and accident risk 89
Twisk, Davera 7–17, 143, 247

UK, and MRBQ 193–4, *195–7*, 198,
     205–7
Ulleberg, P 63, 67, 90
University of Girona 140
University Roma Tre 300
University of Southern Queensland (USQ)
     64, 65, 67
University of Vienna 132
unsafe driving behaviours 76
urban roads, gaze behaviour study *308*,
     309–11, *310, 311*
US, truck drivers and safety 231–2, 233,
     234–5

validity of research 39
variable message signs (VMS) 287
eye-movements of drivers study 299–305,
     299, **300, 301**, *302, 303, 304, 305*

Vehicle Movement Coding Sheet **319**
Villanueva, A. 298
violations (DBQ factor) 214, 244
and driving style 91, 94, *95*, 96, 98, *99*,
101
visitors, and road accidents 327, 328, 329,
330, 335, 336
Visser, Jan 37–49
visual decline, in older drivers 340
Vlakveld, Willem 7–17

Walker, Linda 179–91, 327–36
warmth (NEO PI-R attribute) 272
and driving style 96, *97*, *99*
Watson, B.C. 237
weather, and road safety 174, 332–3
web-based assessment of driving behaviour
247–8
web-based risk management tools 231
websites, and 'Around the Corner' safety
scheme 180, 184–6, *185*, 190
White, J. 231
Wills, A. 245, 246
Wishart, Darren 213–24, 241–51

work schedules, and driving 213
work-related road safety *see* occupational
road safety
Worry and Concern scale 66, *68*, *69*, *70*,
71
Wouters, P.I. 236

yield, failure to 76
young drivers:
and accident risk 7–8, 14–17, 75
deaths in road traffic accidents 14, 75,
139, 151
and driver distraction 290, *291*, 294
driving behaviour 63, 70–1, 75–8,
139–40, 247
Spanish penalty point law study
140–4, *141*, *142*
impulsivity study 78–84
see also adolescence
Yusoff, Muhammad Fadhli Mohd 119–28

Zajonc, R.B. 24
Zulliger Projective Test 156
Zung Anxiety Self-Evaluate Scale 156

Milton Keynes UK
Ingram Content Group UK Ltd.
UKHW031126141024
449569UK00006B/407